The 35mm Film Source Book

The 35mm Film Source Book

Marc Levey

Focal Press
Boston London

Focal Press is an imprint of Butterworth–Heinemann.

∞ Recognizing the importance of preserving what has been written, it is the policy of Butterworth–Heinemann to have the books it publishes printed on acid-free paper, and we exert our best efforts to that end.

Library of Congress Cataloging-in-Publication Data
Levey, Marc.
The 35mm film source book/Marc Levey.
p. cm.
Includes index.
ISBN 0-240-80125-3
1. Photography—Films. 2. 35mm cameras—Equipment and supplies.
I. Title.
TR283.L48 1992
771'.5324—dc20 92-28873
CIP

British Library Cataloguing-in-Publication Data
A catalogue record for this book is available from the British Library.

Butterworth–Heinemann
80 Montvale Avenue
Stoneham, MA 02180

10 9 8 7 6 5 4 3 2 1

Printed in the United States of America

To Jeanne without whom I would be unable to appreciate the joy of the photograph discovered and through whose eyes I see anew.

Contents

Acknowledgments

Any book as ambitious as this would be impossible to complete without the cooperation, support, and encouragement of many individuals both inside and outside the photo industry. I would like to take this opportunity to publicly thank a number of people whose steadfastness through the duration of this process is greatly appreciated.

Thanks to Agfa's Paul Rinaldi and Steffan Hoffman, and their technical expert, Bill Freitag. Thanks also to Kodak's Mike Sullivan, fellow Penn Stater, Joe Runde, Cindy Traino, and a special thanks to the superlative Sue Nolan Green. As they have in the past, Doug Fachnie, Steve Heerstatt, and Paul Hudak from Fuji Film provided invaluable assistance. Thanks also to Edelman's Mindy Kramer and Cathy McGee; to Jim Appleby and Lynne Tuttle at 3-M Scotch; to Nadine Reichner at Ilford; to Paul Gordon at Konica; to James Murphy, Sam Leggerio, and Betsy Donovan at Polaroid; and to Bob Geist at GMI who introduced me to Forte film.

My thanks also to Rich Parker and Lisa Rickert at Photoflex for sharing their innovative products; to another Penn Stater at Bushhawk, John Hedrick; and to Brian LeCompte at the Montreal Tourism Department.

To David Maser my gratitude for his fine black-and-white line art.

Finally, my profound gratitude goes out to my agent, Mark Frisk, for his belief in this book.

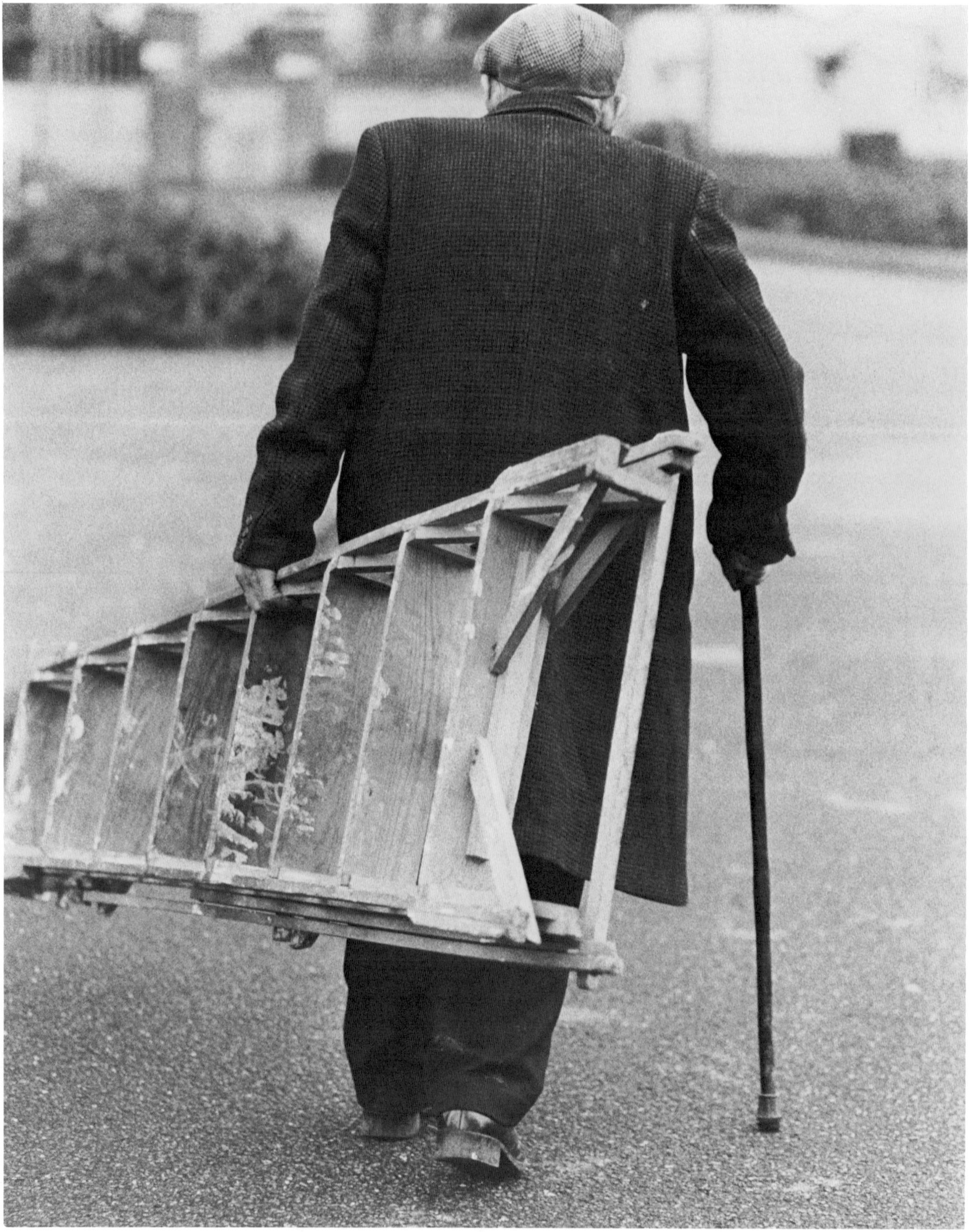

Introduction

During the last few years, the marketplace has seen an explosion of new film types from the major film manufacturers. Ever more discrete niches are being created, resulting in more and more specialized films with specific characteristics designed to serve particular applications. All this means the consumer has more decisions than ever to make.

Film selection is often the only available creative choice remaining to millions of compact point-and-shoot and other automatic camera users. On the advanced amateur and professional front, automation is rapidly becoming the norm, so film choice will become even more critical to these image-makers than it was just a decade ago. To make these critical and creative choices, it is important to understand the nature of the material, its limitations, and the available options.

The following seven chapters contain all of the vital technical information that any photographer, professional, or amateur could want. I have gathered data on a myriad of film types now available and organized it into a usable format. Included are detailed evaluations of the creative characteristics of film arranged by general type, specific brand, and descriptive designation. Throughout this book, I have emphasized both traditional and unorthodox applications to image making, with the goal being to answer the question of which film or films are most appropriate in a specific situation to achieve a desired result.

The first three chapters provide a solid base from which to work, answering the questions, "What is film?," "What are its characteristics?," and "What types of film are out there?" Chapter 4 discusses filters and their effect on the final image. Chapter 5 shows how the choice of film impacts the creative process. Chapter 6 provides the detailed data on almost 120 different 35mm films. Chapter 7 provides insight into film processing. Finally, numerous appendixes give addresses and information that will further enhance the photographer's creative control over the medium. The photographs have also been carefully chosen to illustrate a particular characteristic of a certain type of film.

Please note: For the purpose of this book, the designations "slide," "transparency," and "reversal" are synonymous and I use them interchangeably. Also the designations "print film" and "negative film" have identical meanings.

Figure I.1 "The Dignity of Work," Marc Levey. Agfa Pan ISO 400.

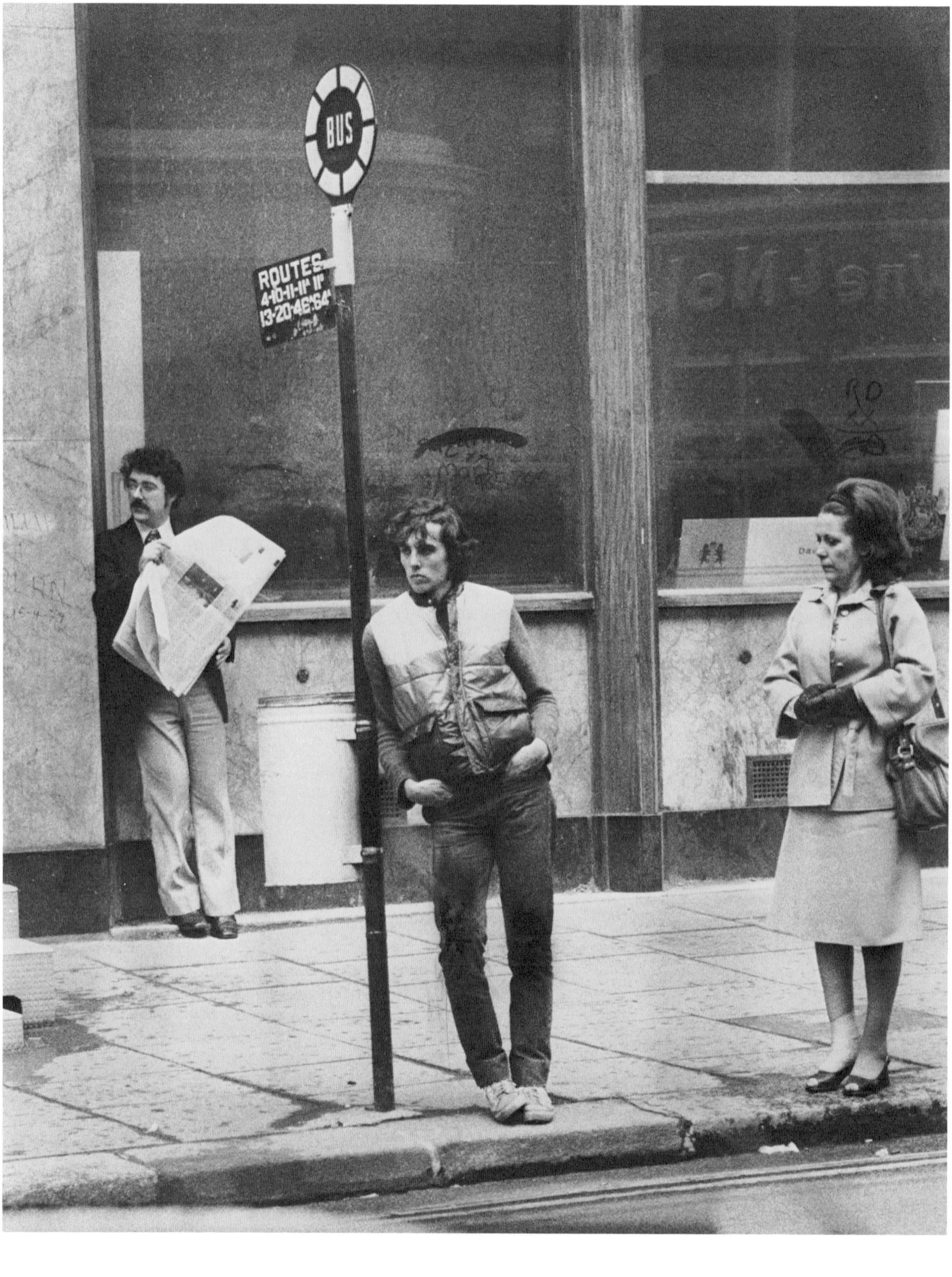
BUS
ROUTES

Chapter 1

What Is Film?

In this chapter you will find in-depth descriptions of the major characteristics of photographic film. The characteristics are as follows:

grain
sharpness
acutance
exposure latitude
contrast
color saturation
color balance
film speed (ASA/ISO)
reciprocity

We will explore why each quality is important and how they relate to each other. Further, we will suggest how each characteristic can contribute to successful and imaginative picture-making.

Grain

Grain is defined as the subjective salt-and-pepper appearance of a negative, slide, or print resulting from clumps of metallic silver or clouds of color dyes congregating together in the image. Generally, grain is most apparent in the clear mid-tones that lack much pictorial detail.

Granularity

Grain is an observable structural aspect of a negative, slide, or print. It is the measured variation in the distribution of an otherwise uniform silver deposit or dye cloud. Grain growth is linked to film speed. As film sensitivity increases, and corresponding ISO numbers go up, so usually does the appearance of grain. We can generalize that faster films are grainier than slower ones. (See "Film Speed," within this chapter.)

The issue of what is acceptable or objectionable grain becomes important only when an image is magnified sufficiently for the grain pattern to become visible. Color films, print or slide, contain no residual silver in their finished forms because the dyes forming the color image create dye clouds at locations previously occupied by developed silver clumps. Depending on the nature of the former silver inhabitants and the composition of the new color dye clouds, what appears as grain in black-and-white images can be seen in color negatives and transparencies as well.

Grain or even dye clouds are not necessarily visible to the naked eye. What we see are not individual grains but groups of grains or dyes at different strata within the emulsion layer of the film or print. The degree of visible graininess is also a function of how regularly or randomly the grains of silver are distributed. An observer may not perceive grain in a certain image if it is arranged in a regular, repeating pattern. However, the same observer might see a second image as grainier than the first even though it contains fewer silver particles by actual count—simply because its architecture scattered the silver grains randomly.

It has been a long-standing goal of film designers to create emulsions and papers with regular, uniform, and repeating grain patterns composed of identically shaped crystals in an ever-thinner emulsion substrate. It is important to point out that in film design, as in fashion design, thin is good. The thinner the emulsion, the more it can potentially resolve. Also, since thin emulsions tend to cause less light scatter than thicker ones, they can contribute more readily to the perception of fine grain.

Earlier we stated that color films and prints display what appears to be grain in much the same way that black-and-white material does. However, the manner in which grain-like structures are produced in color films and prints is quite different from black-and-white, but the resulting image-degrading effects are similar for both.

Why Is Grain Important?

Grain is an important consideration for the following four reasons:

1. Grain can set practical or esthetic limits to the degree of enlargement of a photographic print or projection of a transparency.
2. Grain can limit the amount of visual information (detail) produced in an image.
3. Like audio noise that distorts music, grain can stand in the way of viewer's perceptions of the visual message.
4. As maligned as grain is, it can be desirable and esthetically pleasing.

Relationship Between Grain and Resolution

As with resolution and other dimensions of film performance, it is possible to quantify grain. Describing it numerically involves using the RMS granularity number. This number is often accompanied by a descriptive verbal classification of each film. Following are the general classifications both in RMS numbers and their equivalent descriptive classifications:

Diffuse RMS Granularity Numbers	*Descriptive Classifications*
5 or less	Micro-Fine
6–10	Extremely Fine
11–15	Very Fine
16–20	Fine
21–24	Medium
26–30	Moderately Coarse
33–42	Coarse
45–55	Very Coarse

Grain and resolution might be categorized as separate entities but in reality they are rather interdependent. While there are no absolute international standards by which to measure graininess, (RMS) Diffuse Root Mean Square granularity values are the ones most commonly quoted. Comparing RMS numbers across film categories, such as between color print and slide film, is much like comparing apples and oranges. It is not a meaningful measure in any sense. In order to get any useful information about relative granularity, you must compare apples with apples and oranges with, you guessed it, oranges. Therefore, compare only slide films with other slide films, color print films with other print films, and so on.

Virtually anyone should be able to note the difference between an image classified as coarse grain from one classified in the micro-fine range, provided magnification and contrast levels for the two images are comparable. It is another matter finding visual grain differences in negatives or slides where a relatively small RMS granularity spread exists. For example, two negatives produced from films with RMS numbers of 6 and 8, respectively, should not display noticeable grain differences. It is only when RMS numbers differ by more than *four points* that finer or coarser grain becomes apparent. Having said this, I hasten to add that recent dramatic strides in photo technology have resulted in several new crystal and dye configurations totally revolutionizing thinking about grain structure in film. No longer is it enough to think of grain merely in terms of mass. We must also consider the shape and orientation of the crystals. Grains are no longer homogeneous or monofunctional. They may contain several layers, each performing a separate function. It is important, then, to examine the new grain technology in some detail.

The New Grain Technology

Historically, film technology has been locked into a rigid relationship among film speed (sensitivity), sharpness, and grain structure. Alter one and you surely change the other two in the opposite direction. Demands for ever faster films led to the introduction of high-speed films, especially of the color print variety. Indeed, we do get to use higher speed film. Emulsion speeds of ISO 200, 400, 1000, and even 3200 are now on the shelf. There is a price to be paid for all this speed. Initially, high-speed film was very grainy, with huge, rough-edged crystals. These films were not particularly sharp. We had speed, but it wasn't useful for anything except small snapshot-size prints. If you produced anything over a 5 × 7-inch enlargement, the quality just wasn't there. Critical photographers were not content to settle for mushy, ill-defined, low-light images and they voiced their disapproval with their wallets. High-speed-film sales didn't go anywhere.

About ten years ago, several major producers of photographic film around the world launched a major research effort to change the traditional film-quality triangle relationships. They asked one question: Could one design a film that was faster (more sensitive) while at the same time limit or eliminate the adverse effects that such speed increases had on image sharpness and the appearance of grain? The first positive answer to this question came in 1983 when Eastman Kodak announced a new method of making silver halide crystals. Until this time, crystals were formed in a more or less pebble shape and were randomly distributed within the film emulsion layer. To change this, Kodak researchers did two things. First, they computer-designed the now-famous pioneering T-Grain (tabular), a crystal whose thinner, flatter shape presented more surface area and thus was able to capture more light for the same amount of silver than conventional grains. To increase their efficiency, T-Grains were also designed to align their surface parallel to the emulsion surface automatically.

What Kodak accomplished in one decisive stroke was to make a grain structure that improved on previous crystals in a number of highly significant ways. First, the new structure was more regularly shaped with fewer large crystals that had heretofore been the cause of visible and objectionable grain patterns, specifically in high-speed film. Also, the new design was able to limit the number of very small grains that had downgraded sharpness. An additional benefit of T-Grain research is that it allows for a measurably thinner emulsion layer, which in turn reduces light scatter throughout the film strata. Thinner emulsions, as previously noted, lead to improved image sharpness.

First applied to black-and-white films, T-Grain technology, as well as its clones and relatives, is now commonly used to produce both black-and-white and color films of notably higher quality, meaning visibly less grain, more resolution, and higher emulsion speed. This can be illustrated graphically in the following chart on comparative grain appearance:

Today's Film		*Same Type Film (1985)*
ISO 25	=	More visible grain but not objectionable
ISO 100	=	ISO 32 or slower
ISO 400	=	ISO 100
ISO 1000–1600	=	ISO 400

Again, the current crop of film exhibits grain that is not only less obtrusive, but more regular and evenly distributed, thus contributing to sharper-appearing images.

Several film manufacturers have been hard at work on variations and improvements to the basic T-Grain configuration. Table 1.1 is a summary of the approaches numerous film-makers have adopted.

Sharpness

Image sharpness is one of the most discussed and least understood phenomena in photography. In reality, sharpness is not just one thing, it is many. At times, sharpness is measurable; at others, it is merely a subjective impression. Sharpness is complex because so many factors contrib-

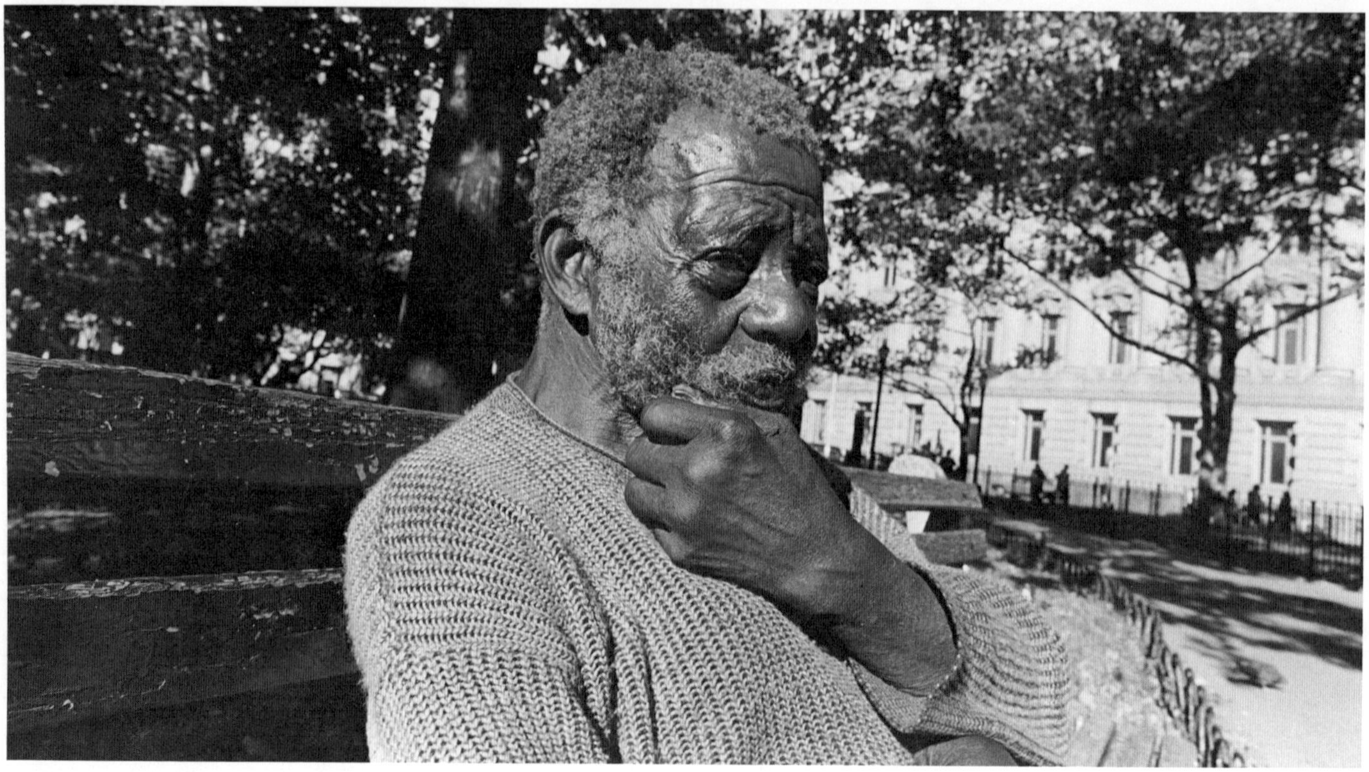

Figure 1.2 Paul Duda. Kodak T-Max ISO 100.

Table 1.1 The New Grain Configurations

Film Maker	*Designation of Technology*	*How Different*	*Stated Advantages*
Agfa	Twin Crystal	Both large, flat crystals and pyramids. Uniform size. No overly large, irregular, or small crystals to affect adversely grain growth and sharpness.	Better sharpness at all speeds. Smaller apparent grain size.
Fuji	Sigma Crystal Technology (twin crystal)	Two grain shapes in one crystal core. High light sensitivity outer shell designed to give more efficient development. Grain speck technology results in smaller, more uniform, dense grains. Thinner emulsion (15%).	Higher effective film speed and lower grain.
Kodak	T-Grain	Thin, very flat and uniform, densely packed grains. Automatically orient themselves parallel to emulsion layer.	T-Grains capture more light than pebble shapes. T-Grains more sensitive because of increased surface area.
Konica	Clean Multistructure Crystal	Large outer shell to gather light energy. Small inner core to develop into very fine grain with more uniform shape.	Smaller grains. Visibly higher sharpness.
Scotch	Composite Structure	Flat, button-shaped crystals automatically orienting to emulsion layer. Thin crystal with internal fine-grain core for finer grain.	Better film speed/grain relationship. Sharper overall.

ute to or distract from it. The impression of sharpness can be affected by the following:

film grain
acutance
subject contrast
subject movement
film resolution
film speed
film contrast
inherent lens contrast
use of contrast-altering filters
color contrast
picture-taking techniques (focusing, exposure settings, etc.)
method of viewing (slide projection, direct print inspection, etc.)
the physical viewing environment
viewer's physical or psychological state

While most of these factors will also affect other qualities of an image, impressions of sharpness are most at risk. If given a choice, most observers (even sophisticated ones) will choose a sharp image over an unsharp one.

You may ask, "How about images that portray motion, soft focus, local sharp, unsharp differences, or ones featuring very limited depth of field?" When the rationale for the use of these techniques is clearly evident to the viewer, the image will be accepted for what it is. If it is difficult to comprehend the reasons or motives for unsharpness in an image, most people will reject it out of hand and will view it as nothing more than a fuzzy, poorly executed picture of dubious merit. Let us now turn our attention to the major factors influencing overall image sharpness.

Film Resolution

By definition, *film resolution* is a measure of a film's ability to record distinctly very fine and closely spaced details in subjects recorded on the emulsion. Thus, resolution is a quality of film that generates detail. Usually, the more detail that is observable, the higher the film resolution. Conversely, low-resolution film generates relatively fewer image details. In just about all instances, pictures shot in high-contrast light will generate substantially more lines of resolution than under low-contrast light. Resolution figures for film are quoted for both high- and low-contrast lighting. The figures for high-contrast are as much as 60 percent greater than those for low-contrast targets. As with measures of granularity, resolution is described numerically and verbally. Table 1.2 illustrates the commonly held measures of film resolution.

Acutance

Another objective physical measure of sharpness is *acutance*. It is not, as some assume, the same as resolution. Acutance is a measure of the ability of film to delineate critically between adjacent tones contained on a negative or transparency. The clearer, finer, and more precise this delineation is, the higher the acutance of the film. The actual process of measuring a film's acutance level involves bringing an opaque knife edge in direct contact with the film surface being tested and then exposing the film to a well-collimated light source. The resulting image on film should be an easily observable edge between dark and clear portions of the film.

Comparing edge widths of different films of the same general type produced under identical conditions gives us some idea of different films' relative ability to distinguish edge boundaries sharply. High acutance does not always go along with high resolution. In the case of some films, there actually seems to be an inverse relationship. Some high-resolution films appear to exhibit substantially less acutance than similar film of lower resolution. New thin-emulsion films (films that are thinner from front to back) are able to generate higher levels of acutance than their conventional counterparts.

Logically, one would assume that always choosing a high-acutance film that also displays high resolution is the thing to do. This isn't always possible. Film design involves a series of compromises, and creating a film that is fast but still has all the desired sharpness qualities may not be within the film-maker's ability. It is a tribute to the state of the technology that we even come close to the ideal combination of sharpeness, acutance, and speed.

Exposure Latitude

Exposure latitude is defined as the amount of over- and/or underexposure a certain film can tolerate with normal processing and still produce acceptable image quality. Several factors govern a film's latitude parameters. A film's type, inherent design, speed, subject brightness ratio, and, in some instances, the method or specific chemistry used in processing could all be important determinants. Here are some guidelines with respect to film latitude:

1. Slow films generally exhibit narrower exposure latitude than do medium- or high-speed emulsions.
2. Black-and-white and color negative material offer significantly more latitude than do transparency films.
3. Negative materials show much more overexposure latitude than on the underexposure side. For some negative films, this can amount to two or three times as much overexposure latitude.

Table 1.2 Resolving Power for Films

1000:1 Luminance Ratio (High-Contrast Target Only)

Category	*Resolution Numbers*
Ultra High	600 ll/mm plus[a]
Extremely High	250–500 ll/mm
Very High	150–200 ll/mm
High	100–150 ll/mm
Medium	60–80 ll/mm
Low	50 ll/mm or less

Resolving Power of Selected General-Purpose Films

Film	*ISO*	*Low-Contrast Subject*	*High-Contrast Subject*
		Black-and-White Film	
T-Max 100	100	63 ll/mm[a]	200 ll/mm[a]
T-Max 400	400	50 ll/mm	125 ll/mm
Neopan 400	400	50 ll/mm	100 ll/mm
Agfa Pan 400	400	50 ll/mm	110 ll/mm
Ilford HP-5+	400	50 ll/mm+	100 ll/mm+
		Color Negative Film	
Kodacolor Gold 100	100	100 ll/mm[a]	
Ektar 25	25	160 ll/mm	
Realia	100	125 ll/mm	
HG 400 Professional	400	100 ll/mm +	
Optima	125	100 ll/mm +	
Ultra	50	150 ll/mm +	
Super SR 100	100	100 ll/mm	
Super SR 400	400	100 ll/mm	
HR 100	100	100 ll/mm	
One Film	200	98 ll/mm	
Ektar 1000	1000	80 ll/mm	
		Color Transparency Film	
Kodachrome 25	25	100 ll/mm[a] +	
Ektachrome 100 HC	100	100 ll/mm +	
Ektachrome 400 HC	400	80 ll/mm	
Velvia	50	160 ll/mm	
100 RDP	100	125 ll/mm	
P1600	1600	70 ll/mm+	
RS 100	100	100 ll/mm+	
RS 1000	1000	60 ll/mm+	
Scotchchrome 400	400	89 ll/mm	
Scotchchrome 1000	1000	67 ll/mm	

Source: Eastman Kodak Co.
[a] ll/mm = Lines per millimeter.
Author's note: What is shown here as ll/mm is the familiar designation. The accurate technical description is line pairs/millimeter (lp/mm).

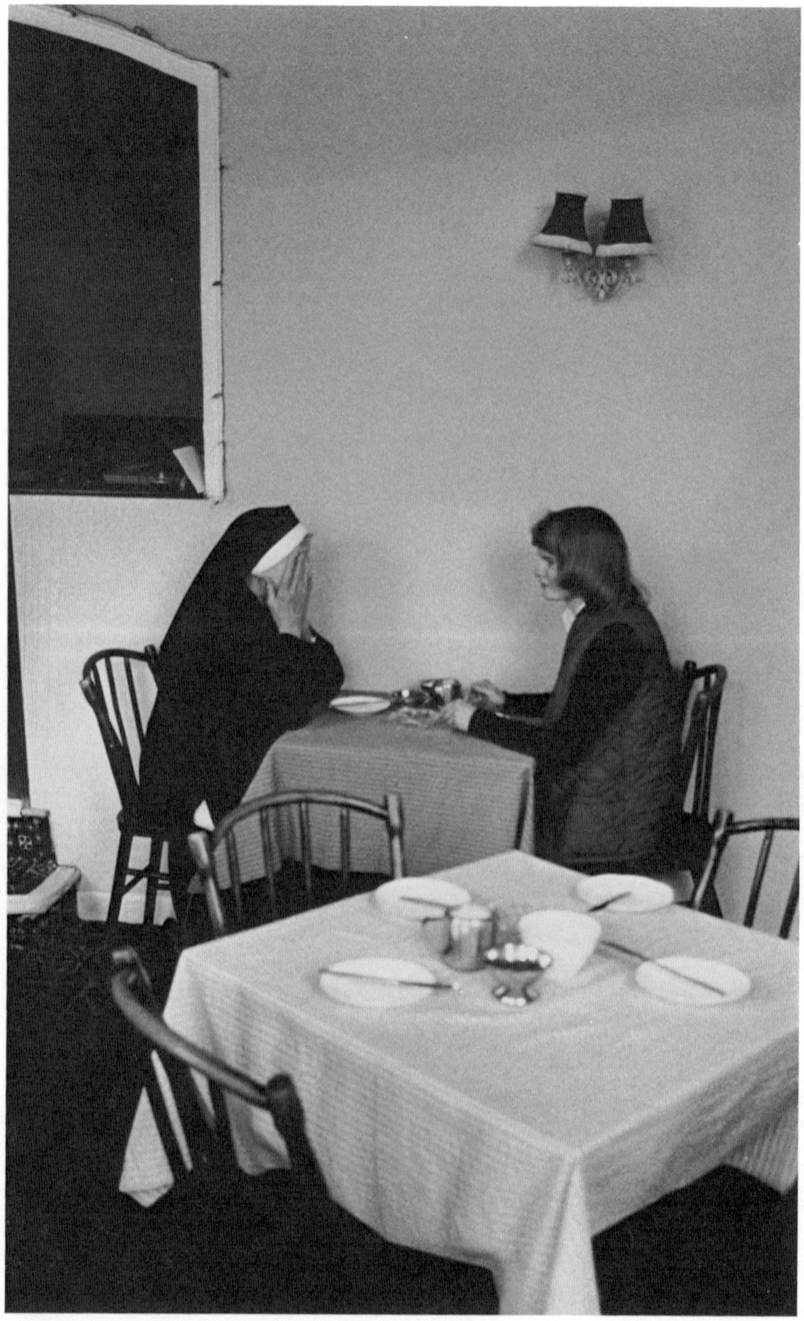

Figure 1.3 Marc Levey. Fuji Neopan ISO 1600.

4. Transparency films are much more tolerant of underexposure than of overexposure. They exhibit at least twice as much latitude on the underexposure side. Even so, transparency films have the least exposure latitude of any film type.
5. Exposure latitude is greatest when film is used to photograph subjects under low-contrast lighting. It follows that as subject contrast increases, exposure latitude decreases.
6. Exposure latitude is controversial because:
   a. There is no objectively correct exposure;

b. What appears to be a well-exposed frame to one observer is over- or underexposed to another;
c. Published exposure latitude figures are judgments and represent consensus at best.

7. Exposure latitude is up to 50 percent greater when shooting under even, diffused light (as on an overcast day) than under bright, direct sunlight or nondiffused electronic flash.
8. A majority of experienced photographers indicate the best starting point for exposure is to rate the film at its manufacturer's recommended ISO. Numerous published tests confirm this judgment.

Table 1.3 provides some general guidelines for exposure latitude by film type.

Table 1.4 details working exposure indices for selected popular color films. The ratings were arrived at by taking an average of ISOs used by the panel of expert photographers working with the author on the preparation of this volume. While many felt the rated ISO was optimum, especially in low-contrast light, numerous films were rated at ISOs other than their designated speed. It is of interest to note that almost all panel members used personal EIs (Exposure Indices) different from manufacturer recommendations when using slide film and were more likely to accept nominal speed ratings for print film. This is probably due to the wide exposure latitude inherent in print film as compared to slide material. In practice, applying corrections to established ISO numbers sets a personal exposure index most suitable for your own equipment, applications, and taste. Even though stated ISOs are used by many photographers, they should not represent an absolute value.

Contrast

Contrast is one of those photographic terms used constantly, and more often than not with little (if any) notion of its meaning. Since there is more than one type of contrast in photography, the term needs to be defined more precisely. There is more than one kind of contrast, you say. Yes indeed! Here are a few: subject contrast, tonal contrast (as in black-and-white), color contrast, emotional contrast, luminosity contrast, and so on.

Table 1.3 General Guidelines for Exposure Latitude by Film Type

	Latitude	
Film Type	*High-Contrast Subjects*	*Low-Contrast Subjects*
Black-and-white	1 stop under and up to 2 stops over (varies with processing materials and techniques).	1½ stops under and 3 stops over (varies with processing materials and techniques).
Color negative	1 stop under and 2 stops over.	2 stops under and 3 + stops over.
Color transparency	½ stop under and 0–½ stops over.	1 stop under and ½–1 stop over.
Chromogenic black-and-white (Ilford XP-2)*	1 stop under and 2 stops over.	2 stops under and 3 stops over.

*This film can be exposed at different ISO's and processed normally.

Table 1.4 Optimum Exposure Index by Selected Individual Film

Film	*Low-Contrast EI*	*High-Contrast EI*
Ektar 25	Rated speed	ISO 32
Gold 100	Rated speed	ISO 64–80
Ektar 125	Rated speed	ISO 100
Ektar 1000	Rated speed	ISO 800
Vericolor III	Rated speed	Rated speed
Ultra 50	Rated speed	Rated speed
XRS 100	ISO 125	Rated speed
Optima 125	Rated speed	Rated speed
Portrait 160	Rated speed	Rated speed
Realia	ISO 64–80	Rated speed
Super HG 100	Rated speed	Rated speed
160S pro.	ISO 125	Rated speed
Super HG1600	Rated speed	Rated speed
Super SR 100	Rated speed	ISO 160
HR 100	Rated speed	Rated speed
SRG-3200	ISO 1600	Rated speed
One Film	Rated speed	Rated speed
Kodachrome 25	ISO 32	Rated speed
Kodachrome 64	ISO 80	ISO 80
Kodachrome 200	ISO 250	ISO 250
Ektachrome 50 HC	Rated speed	ISO 64
Ektachrome 64X	ISO 80	Rated speed
Ektachrome 100 Plus	Rated speed	ISO 125
Ektachrome 400	Rated speed	Rated speed
P 800/1600	Rated speed	Rated speed
Velvia	Rated speed	Rated speed
Fujichrome 100 RDP	ISO 125	ISO 125
Fujichrome 400 RH	Rated speed	Rated speed
Fujichrome 1600II	Rated speed	Rated speed
Agfachrome RS 100	ISO 125	ISO 125
Agfachrome CT-200	Rated speed	ISO 250
Presentation Chrome	Rated speed	ISO 125
Scotchchrome 100	Rated speed	ISO 160
Scotchchrome 640-T	Rated speed	Rated speed

Contrast also exists in the physical structure, emotional content, and impact of an image. Some critics go so far as to argue that contrast is everything in a photograph. They assert that without significant contrast, either physical and/or emotional, there would be precious little for viewers to look at. How dull, they say! As the noted photographer and writer Freeman Patterson puts it in his best selling book, *The Joy of Photography*, "Contrast of any sort produces tension and makes a composition dynamic."

There is little doubt that contrast in all its forms constitutes a major visual element in the appeal that an image makes for attention. Contrasts drawn boldly or subtly can make or break a picture. This is reason enough to take a detailed look at contrast as one of the principal design elements in photography. As film selection and use go a long way toward defining the quality of an image, it is incumbent upon us to understand all that goes into making an image visually appealing. Clearly, film contrast is an important element in the quality equation. Given the preceding, let us now turn our attention to the subject of image contrast in detail.

Contrast in Black-and-White

In a black-and-white photograph, *contrast* is defined as the difference in brightness from one part of the negative and print to another. The measured brightness ratio difference between the lightest and darkest portion of an image is expressed as a ratio of f-stops, such as a 7:1 brightness ratio.

Scale is also a term used to describe tonal contrast range. *Short-scale* black-and-white negatives are those containing relatively few distinct gray tones from the darkest to lightest portions of the image. *Long-scale* black-and-white negatives contain many discernible individual tones of gray between the darkest and lightest parts of the negative.

The Characteristic Curve

Characteristic curves for film and other light-sensitive materials are a graphic representation of how light reacts on these materials and produces density in the exposed and developed image. Stated differently, the characteristic curve can, for any given film, developer, and time combination, show us the density value resulting from a specific exposure.

The curve has several important components highlighted in the accompanying illustration (see Figure 1.4). The curve plot describes exposure as a function of density and begins with

Point A—*Base Fog Density Level,* which is the area where there exists no visible exposure but only a gray mist obscuring detail.

Point B—*Threshold Exposure,* which is the point on the curve at which density just begins to appear out of the base fog level.

Point C—Known as the "toe," it is the area of least usable density and is just beyond the base fog density area. In practice, this is an area that without special treatment suffers from serious underexposure. Some photographers refer to it as "D-Min." Note that as exposure is increased from this threshold value, density grows with more details in a visible image.

Point D—Not really one point but the straight-line portion of the curve, it is on this diagonal that negative density increases at a more-or-less uniform rate. From the toe region to the point on the straight-line portion of the curve where it reaches its maximum density is usually no more than 7 f-stops. The slant of a normal contrast negative presents a straight-line slope of roughly 45 degrees. The angle described by the straight-line portion is termed ***gamma***. A more active gamma than the normal 1.0 describes a more acute slope and is one that produces more contrasty nega-

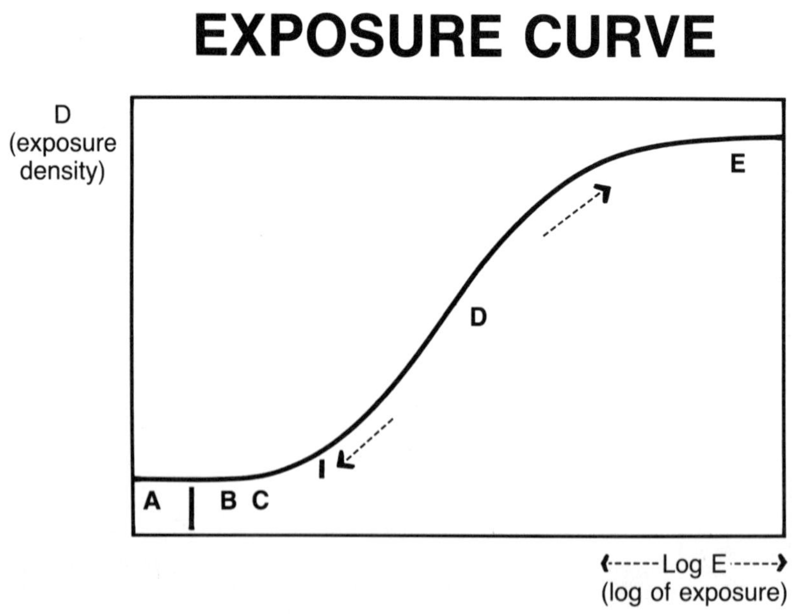

Figure 1.4 Typical characteristic curve.

tives. Gammas of 3.0 and beyond are not uncommon for the highest contrast lithograph black-and-white films.

Point E—As the curve begins to transition from diagonal to horizontal at the top of its travel, we reach the portion known as the *shoulder.* It is in this area that maximum density is achieved. Termed a film's "D-Max," it is also called the area of "Max. Black."

All films respond differently to various combinations of developers, temperatures, developing times, and agitation techniques to produce curves with individual personalities. Particular combinations of these and other variables expand and contract shoulder and straight-line portions of the curve, thus producing differences in density or contrast. Likewise, the toe portion can also be affected.

With color negatives and, more acutely, color transparency processing, creative variations in contrast are much more severely limited, if not totally unavailable, to the darkroom worker. For the color photographer, overall image contrast control is usually only possible when first choosing one film over another by its inherent contrast qualities. And while some contrast variations are possible during actual exposure or by employing contrast enhancing filters, most contrast control decisions need to be made when choosing which specific color to use. It can be illuminating to examine and compare the curves of several color films. It makes sense to compare color negative films with one another and not with transparency films. The same holds true when choosing transparency films—don't compare them with color negative films.

Since most color films have at least three light-sensitive layers, they exhibit three separate and not identical curves, one each from the magenta, yellow, and cyan layers of the film. These three curves do not match exactly because no color film responds to all colors in an identical manner. Such discrepancies can indicate color-balance problems or particular over-or undersensitivity to certain areas of the visible spectrum representing particular colors.

Contrast in Color Films

Unlike black-and-white film, where contrast is initially determined by exposure but can be later modified to a considerable degree during processing and printing, color contrast is almost solely a product of the inherent contrast designed into the film by the manufacturer. Other things affecting contrast in color photography include scene contrast, exposure, and color qualities of each film. Bear in mind that overall contrast in color materials cannot either be controlled or modified with the ease of such changes in black-and-white films and papers. See Table 1.5 for the relative overall contrast qualities of selected films.

Often, color contrast and color saturation are confused with one another. *Color saturation,* or *chroma,* is a quality of color film that identifies a color as a pure hue, undiluted by black, gray, or white. Compare this to *color contrast,* which is defined as the relative visual weight of two or more colors in close proximity to one another on the color wheel. It is of further interest to note that because the three color dyes contained in most color films are combined in different ratios and amounts, certain colors produced by particular color films will appear more color-saturated than others. Thus, a film's saturation qualities can have a bearing on its overall color contrast.

As a general rule, films that are judged to be more contrasty than average are also seen as having higher levels of color saturation. This is particularly so with films exhibiting high color saturation in the warm (red, orange, yellow) end of the spectrum, an area of high visual weight.

Table 1.5 Relative Overall Contrast Qualities of Selected Color Films

Low Contrast	*Normal Contrast*	*High Contrast*
	Color Negative Films	
Fuji Realia	Agfacolor Optima 125	Kodak Ektar 25
Agfacolor Portrait 160	Kodacolor Gold 100	Agfa Ultra 50
Konica SR 160	Konica Super SR 100	Fujicolor Super HG 100
Kodak Vericolor III S	Kodak Ektar 125, 1000	
Scotchcolor 200	Kodak Ektapress 100, 400	
Fujicolor 160 S		
Fujicolor HG 400 Professional		
Konica SRV 3200		
Kodak Ektapress 1600		
	Color Transparency Films	
Kodak Ektachrome 100	Ektachrome 100HC	Fujichrome Velvia
Agfachrome 50, RS 100+	Ektachrome 100 Plus	Fujichrome 100
Agfachrome RS 1000	Scotchchrome 100, 400	Fujichrome P1600
Fujichrome 50	Kodachrome 25	Kodachrome 64
Kodak Ektachrome 64 Professional	Kodachrome 200	Polaroid Presentation Chrome
Kodak Ektachrome 100 Professional	Fujichrome 400	
Scotchchrome 1000		
Scotchchrome 640-T		

The Film Speed–Contrast Relationship

Traditionally, slow films showed greater contrast than did faster emulsions. This was acutely so in color material. Slow black-and-white film, although more contrasty than faster films, did not present as many problems because contrast could be controlled or altered in either processing or printing. For those who preferred to work in color, controls were just not available, especially if the film they chose was slow. Until very recently, contrast control of color material has been strictly limited; for the nondarkroom worker, nonexistent.

Now, conditions are somewhat different. Film-makers have begun to offer emulsions that are custom-designed with a nontraditional mix of characteristics.

No longer do slow color films need to be more contrasty than their faster siblings. Presently, although it is still not possible for Kodak or Fuji to produce an ultra high speed ISO-1600 color film with the contrast qualities of an ISO-100 emulsion, the gap has been narrowed considerably. Fortunately, we are no longer saddled with high-speed color film that betrays its user by delivering tell-tale, muddy black-brown shadows and detail-less highlights. The old axiom of "faster film, less shadow detail" is far less true today, as it is harder and harder to tell ISO-100 negatives or slides from ones generated from ISO-400 material. Still, as Table 1.5 demonstrates, there are films that display high or low contrast by design.

Color Balance

What we call *color* exists in a very narrow portion of the electromagnetic spectrum. Visible light waves placed approximately in the middle of the spectrum have wavelengths of between. 000016 and .000028 inch and are designated in units called ***nanometers*** (nm) and range from 400nm to 750nm. Figure 1.5 is a representation of the array of visible light along the entire electromagnetic spectrum.

Color Temperature

Color temperature is a means of measuring the energy level of color and is expressed in degrees Kelvin (K). Each color on the visible spectrum is assigned a Kelvin temperature or, more accurately, a range of temperatures, so knowing the color temperature of a light source at any one time gives us important information about the quality of the light. Examine Figure 1.5. At the warm red end, small shifts in color temperature will result in large changes in color rendition on film. At the cool end, however, small shifts in color temperature result in comparably less change in color rendition. Thus, color films that are designed and balanced for daylight are able to produce acceptably accurate color under a wide range of natural or electronic-flash lighting conditions, while films balanced for tungsten illumination only respond and deliver acceptable results within a rather narrow band of color temperatures. It is worth noting that the

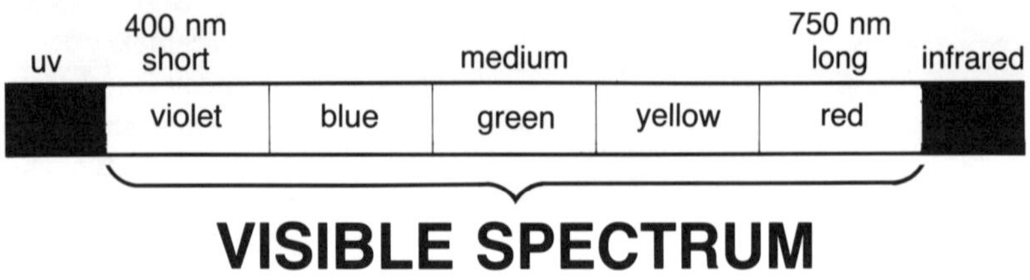

Figure 1.5 Electromagnetic spectrum.

range of acceptable color with tungsten film is far smaller than when using daylight-balanced film under its designated light sources.

Remember that there are separate films made to be used with 3200K and with 3400K. Only one group of films is necessary to cover the entire daylight spectrum from about 5000K to over 10,000K. Although we have poor color memory, our psychological tolerance for certain colors that do not appear natural in a photo is very limited at best. Skin tones, blacks, whites, and grays are critical hues in this regard.

Not only are color films balanced for daylight or tungsten illumination, most individual films by brand are made to favor a particular cast from warm (red, orange, and yellow) through neutral, where colors appear more or less as we expect them to look, to the cool end of the spectrum (colors ranging from neutral to those with a bluish cast).

Film-makers go so far as to adjust color dye ratios for different film batches intended for sale in different parts of the world. Therefore, a film intended for the U.S. market might reflect a discernibly different color balance than the same brand and type sold in Europe or Asia.

What makes all this so difficult is the concept that acceptable color is a purely subjective decision on the part of each observer. For instance, what I consider pink, you might very well perceive as red. There are no objective, universal colors everyone agrees upon. There are only ones that a consensus points to and says, "These are colors that most agree fit the arbitrary names we have assigned to them." Thus, colors such as red, blue, green, or purple are only psychologically acceptable and by no means absolute.

There are methods and tools at our disposal to modify the color balance of film so that the images they produce exhibit acceptably natural colors. Filtration is the primary color-altering technique. (For detailed information of filters, see chapter 4). It should be noted that even with perfect technical balance between film and light source, visible color imbalances still occur when lenses are coated with microscopically thin layers or coatings meant to contain aberrations, increase light transmission, or improve contrast. The effect of lens coatings on color balance and color rendition is variable even from lens to lens within the same brand lineup. Be aware that hues become contaminated by color reflected onto subjects from adjacent objects. Large areas of painted walls reflecting light on any photographic subject will invariably alter the overall color of the subject. If, for instance, one photographs a subject dressed in a red sweater standing near a yellow wall and the light from that wall reflects onto the subject, then the red sweater is likely to be recorded on film not as a pure red but as a somewhat orange, less intense hue.

The following chart details color temperatures for numerous common light sources used in photography. Although not all-inclusive, it does demonstrate the range of temperatures with which we must contend.

Color Temperatures of Common Light Sources

Degrees Kelvin (K)	*Light Source*
1800–1900	Lit candle
2450	25 watt lightbulb
2600–2700	60 watt lightbulb
2800	100 watt lightbulb
3200	Tungsten photo lamp
	500 watt projector bulb
3400	Studio photo floods
3700–4000	Clear flashbulbs

Color Temperatures of Common Light Sources (con't)

Degrees Kelvin (K)	*Light Source*
4000	Warm white fluorescents
4400	2 hours after sunrise, 1 hour before sunset
5500	Typical daylight
	Noon daylight (few or no clouds)
6000–6500	Open shadows
	Some flash units
6500–8500	Overcast skylight
6500–18000+	Blue, clear sky

Color Saturation

Until very recently, films that featured high color saturation were relatively rare and almost always of the color transparency variety. For technical reasons, film-makers found it extremely difficult to produce color negative films with the same intense, vibrant, saturated color as they managed to build into color slide film. Now, technological innovation in photo chemistry at the molecular level has led to new dye formulations and methods to generate more intense and pure color, which in turn has led to a mini-revolution in the design and manufacture of color print films.

Although color slide film still offers generally higher levels of color saturation, contemporary color print films are capable of delivering rich, striking color that is far superior to previous print emulsions and, in some cases, approaches the saturation of slide film. While much credit for these improvements goes to the aforementioned innovations in photo film chemistry, several improvements in print paper and processing chemistry have also led to a better finished product.

Thus far, we have examined color saturation's effect, but what exactly is it? One way to think of it is as a film's ability to render colors that exhibit rich hues that are at the same time intense and close to what the eye and mind accept as realistic renditions. I did not say *accurate* renditions because, like color balance, not enough or too much color saturation is subjective. Comparisons under somewhat controlled conditions, however, do offer us some evidence about differences in color-saturation levels from brand to brand, within a brand, by specific type, and even from emulsion batch to batch (a batch being a serially numbered production run of a certain film). It is enough today that the current trend is to design and market color film that yields high color saturation, which the buying public seems to want in ever-increasing amounts. A collateral effect of high color saturation is increased *color contrast.* In the world of highly saturated color, adjacent and even complementary colors stand out more vividly and vie for attention with greater urgency than in the more subdued world of muted color. While this is not at all a bad thing, it does mean that we need to exercise more care in film selection and use.

To an extent, color saturation can be modified by adjusting exposure. For example, underexposing slide film to be used in sunlight by about one-third to one-half f-stop can increase color saturation while still retaining much if not all of the shadow and highlight detail. Likewise, overexposing print film a like amount under the same conditions raises color saturation visibly.

There are reasons for the trend toward higher and higher levels of saturation. First, the buying public asked for it, particularly in color print film which, by the way, accounts for the bulk of all film sales worldwide. Many industry surveys confirmed these beliefs. They demonstrated quite conclusively that, deep, rich—but not necessarily accurate—color was what the picture-taking public wanted. Of the several principal qualities of color film, such as grain, sharpness, contrast, exposure latitude, color accuracy, and color saturation, it is hyper-color saturation of the "knock your socks off" variety that photographers seem to want most. Because of this, we find film-makers such as Germany's Agfa replacing their entire line of slide and print films once noted for their soft, relatively muted, and to my mind, beautiful color palette with much more aggressive, contrasty, electric color, more in line with what seems to be in fashion. I, along with many, many color film users, lament the passing of films such as the classic Agfachromes if only because their demise shrinks the creative color options to a precious few. The irony of the present situation is interesting to ponder. In the recent past, the goal of many color adherents was to find ways and means to increase color saturation. Now we see photographers employing techniques and devices to desaturate today's supersaturated color emulsions. Granting that almost all current color films offer higher levels of color saturation than in the past, there is still some spread in the range of saturation they present. To paraphrase George Orwell's *Animal Farm*, all films are saturated but some are more saturated than others. The following comparison chart details some of our findings regarding color saturation. Remember, these comparisons are relative. One final point. We are beginning to see film-makers offer individual color emulsions with different application-oriented saturation characteristics incorporated as their principal *raison d'être*.

Color Films Exhibiting High Color Saturation

Negative Film	*Transparency Film*
Kodak Ektar 25	Kodachrome 25 Professional
Fujicolor Super HG 100, 200	Kodachrome 64 Professional
Konica Super SR 100, 200	Fujichrome Velvia
Agfa Ultra 50	Fujichrome 50 D
Kodak Kodacolor Gold 100 Plus	Fujichrome 100 D
Konica Impresa 50	Ektachrome 100 Plus Professional
	Polaroid Presentation Chrome
	Agfachrome RS 50 Plus
	Ektachrome 64X, 100X, 400X

Color Films Offering Relatively Muted Color Palette

Negative Film	*Transparency Film*
Fujicolor Realia	Agfachrome CT-200
Kodak Vericolor III	Agfachrome RS 200 Professional
Kodak Ektapress 100	Kodak Ektachrome 64
Scotchcolor HR 100, 200	Ektachrome 100 Professional
Agfa Portrait 160	Scotchchrome 400, 1000
Agfa XRS 100	
Fujicolor HG 400 Professional	
Konica SRG 160 S	
Scotchcolor 160	

Film Speed

Film speed is the term often used to describe an emulsion's sensitivity to a given amount of light. Sensitivity, in turn, is a product of the amount and configuration of silver halide crystals coated on the film's emulsion layer. This is true even for color and chromogenic black-and-white films where the final result is devoid of silver, but instead is composed of dye clouds. Even these two film types begin with the exposure of light-sensitive silver.

Two scales are used to denote film speed. ISO stands for International Standards Association, the body setting many industrial standards for precision measurement. DIN (a German designation meaning Deutches Industrie Norm) serves the same purpose as ISO does for the remainder of the photographic world. The two scales, ISO and DIN, are different in one respect. ISO is arithmetic; the doubling of an ISO number indicates an exact doubling of film speed or sensitivity (e.g., ISO-200 film is twice as sensitive as an ISO-100 emulsion). The DIN system, on the other hand, is not an arithmetic progression but is instead logarithmic. On the DIN scale, moving up three numbers denotes a doubling of film speed (e.g., going from DIN 18 to DIN 21 denotes doubling film speed). The equivalent ISO numbers are ISO 50 to ISO 100. Film speed is also denoted by a commonly accepted verbal classification system:

Slow	ISO 6–50 DIN 9–18
Medium	ISO 64–160 DIN 19–23
Fast	ISO 200–800 DIN 24–30
Ultra-Fast	ISO 1000–3200 DIN 31–36

In addition to the intensity of the light striking a film's surface, there are several other factors affecting film speed. Many black-and-white films are not sensitive to all portions of the spectrum equally and because of this have different EIs. As an example, if you would choose to use a black-and-white film whose red sensitivity was low indoors under red-rich tungsten illumination, the film's effective speed will be lower than its rated speed. In daylight, or using electronic flash (which is not as red-rich), the normal speed rating will be restored.

Both black-and-white and a number of color films' speed ratings can be successfully uprated to provide more than normal sensitivity. An ISO-100 film can be exposed at ISO 200 or more, often with little sacrifice in picture quality.

The following films have proven to be especially good performers when uprated.

Black-and-White Film

Agfa Pan 400
Fuji Neopan 400, 1600
Ilford HP5 Plus
Ilford XP-2 400
Kodak T-Max 400
Kodak T-Max P 3200

Color Negative Film

Agfacolor XRG 400
Kodak Ektapress 400, 1600
Fujicolor HG 400 Professional

Note: Standard machine C-41 processing defeats much of the effects of uprating print film. Uprated print film must be custom-processed for best results.

Color Transparency Film

Agfachrome RS 100 Plus
Agfachrome RS 1000
Fujichrome 100 RDP
Fujichrome P 1600II Professional
Kodak Ektachrome 160
Kodak Ektachrome 400 X
Kodak Ektachrome P 800/1600
Scotchchrome 400
Scotchchrome 640-T
Scotchchrome 1000

Note: Standard machine processing will result in darker than normal slides if exposure adjustment is not made at the time of exposure. Normal film density can be achieved only if film is unrated at time of exposure and film-processing times are appropriately recalibrated.

Film Speed–Quality Relationship

There seems to be a definite relationship between film speed and other important film-quality dimensions. In the past, this relationship was fixed and profound: as film speed increased, image quality in other dimensions decreased—and markedly so. Current films still operate under the same inverse speed-quality formula but technological developments during the past decade have reduced the quality tradeoff one had to make to achieve higher film speeds. No longer must we put up with ISO-400 film that is a low-contrast, relatively unsharp, and (in the case of color) low color saturation material with coarse, image-distorting grain. Today's high-speed films are phenomenally good compared to those of a few years ago. Later, we will examine how much of a tradeoff is necessary to achieve higher and higher film speeds.

Film-Sensitivity Scale

The sensitivity or speed of all photographic film is designated by a number on a scale containing a series of ISO numbers. As mentioned earlier, ISO is the organization charged with the responsibility of setting such standardized systems. Often, film speed is expressed in both ISO and DIN numbers. In some older photo texts, you will find film speed quoted as an ASA number (ASA stands for the now disbanded American Standards Association). In point of fact, ISO and ASA number systems are identical. The ISO scale is as follows:

ISO 3, 6, 8, 10, 12, 16, 20, 25, 32, 40, 50, 64, 80, 100, 125, 160, 200, 250, 320, 400, 500, 640, 800, 1000, 1250, 1600, 2000, 2500, 3200, 4000, 5000, 6400.

ISO–DIN Equivalents

ISO	10	12	16	18	25	32	40	50	64	80	100	125	160	200	250	320
DIN	10	12	13	14	15	16	17	18	19	20	21	22	23	24	25	26

ISO	400	500	640	800	1000	1250	1600	2000	2500	3200	4000
DIN	27	28	29	30	31	32	33	34	35	36	37

ISO	5000	6400	12500	25000
DIN	38	39	42	45

ISO speeds are rounded off to the nearest ⅓ of an f-stop. Because of this, ISO speed in the preceding table is stated to a tolerance of + or– ⅙ f-stop. Every third DIN number denotes one full f-stop increase in film sensitivity.

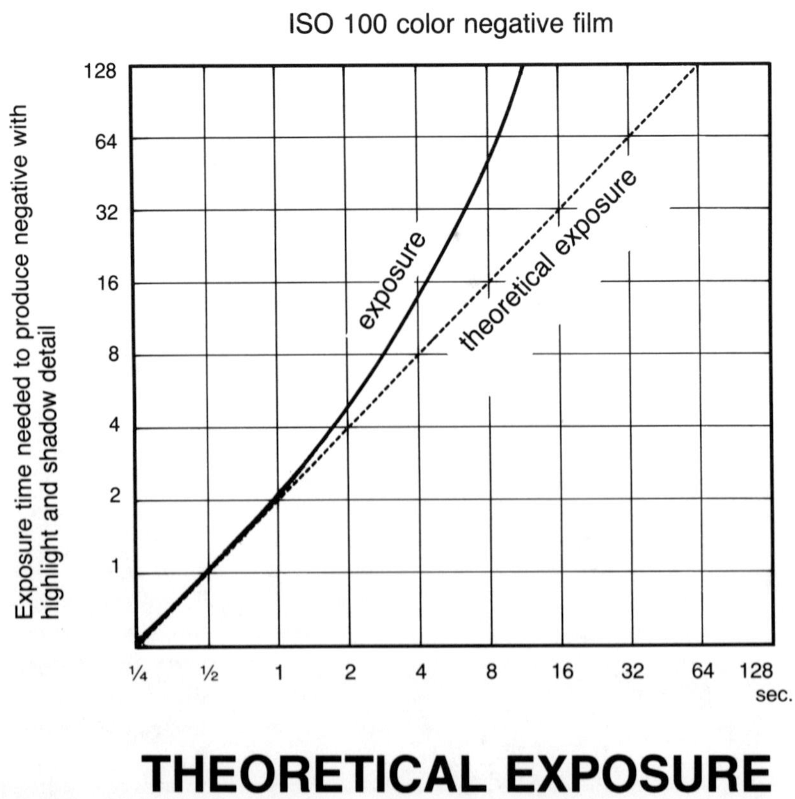

Figure 1.6 Typical exposure curve plotted against theoretical.

Reciprocity

Photographic film responds to exposure in a predictable, linear manner from about 1/10th second up to 1/10,000. Within this shutter-speed range, exposure remains the same, if when you decide to change f-stop or shutter-speed settings, you halve one while at the same time doubling the other. Stated differently, exposure consists of a reciprocal relationship between shutter speed and lens aperture. This, then, is the so-called Reciprocity Law. When this law fails, we incur a condition known as *reciprocity failure* and adjustments to shutter speed and/or filtration are required. Without adjustment, film appears underexposed and, with color emulsions, serious color shifts also become evident.

In Figure 1.6, a typical ISO-100 color negative film responds to exposure between specific shutter speeds in a linear fashion. You can see that theoretical exposure begins to deviate from its predicted path at extremely slow shutter speeds. Modern color films have been designed with extended straight line-response at high shutter speeds to a maximum of 1/10,000 second. For practical purposes, reciprocity failure is not much of a problem at the high shutter speed end of the range.

For specific adjustments to counteract the adverse effects of reciprocity law failure, consult chapter 6 for each film.

DX Coding and Freshness Dating

Almost all current 35mm film cartridges (except 35mm Polaroid instant films) are coded with a pattern of silver and black squares that electronic sensors inside the camera film chamber or within automated film-processing machines read. Also, the outside labels of film cartridges contain a Universal Product Code (UPC) readable by electronic scanners similar to

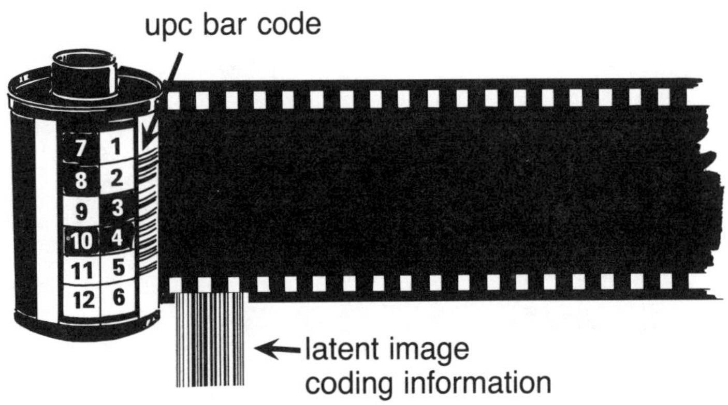

Figure 1.7 DX coding and freshness data.

the ones in most supermarkets and department stores. Along the perforated edge of the film itself is an electronically retrievable latent-image bar code. A third bar code can be seen alongside the black-and-silver DX pattern on the outside of the film cartridge. What vast information do these sets of codes impart? First, the autosensing coded patterns of up to twelve individual sensors inside camera film chambers sense patterns for individual film type and signal this information to the camera's onboard computer central processing unit. The twelve discrete squares that constitute a pattern of information match up and make contact with respective sensors in the camera. As is illustrated in Figure 1.6 these are arranged from top to bottom, right to left. Positions 1 through 4 designate the film's particular ISO by broad category—slow, medium, fast, or ultra-fast. Positions 5 and 6 indicate ISO more precisely; 7 through 10 tell the camera the number of exposures on the roll, and 11 and 12 the film type, negative, transparency, etc.

The UPC symbol on the cartridge allows automatic sorters at processing plants to identify correct film type for processing. Last, film edge coding on many color negative films assists in optimizing automatic printing exposure.

Even if you use a top-of-the-line professional-caliber autofocus single-lens-reflex (SLR) camera, you will not have to worry about setting proper film speed or programing the camera automatic rewind for proper film length because the coding and onboard camera electronics do all this and more.

What if you do not wish to use the film at its rated ISO? Your camera might contain provision for setting the ISO manually or for some other form of exposure compensation. If not, you may wish to affix little 5.5mm × 7mm pieces of black and metalized silver tape over existing DX patterns to signal a different ISO to your camera. You will only need to alter positions 1 through 6, but you must be precise. The pattern guide in Figure 1.8 was furnished by *Popular Photography* magazine.

Freshness Dating

Contained on the outside of the box or cylinder holding the actual film cartridge is a "use by" date. This indicates the date by which the film should be used to ensure best results. Beyond the date printed on the box, film sensitivity and/or color balance may change. Film does deteriorate over time but film a few months out of date and stored in a dry, relatively cool place will retain its optimum performance. Beyond a few months, nasty things begin to happen. You will find a more thorough discussion of freshness dating, storage, and extended film preservation in Appendix E.

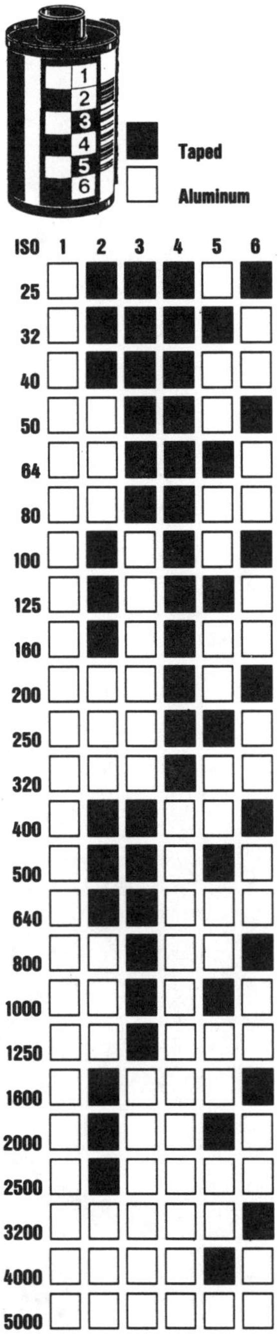

Figure 1.8 DX coding.

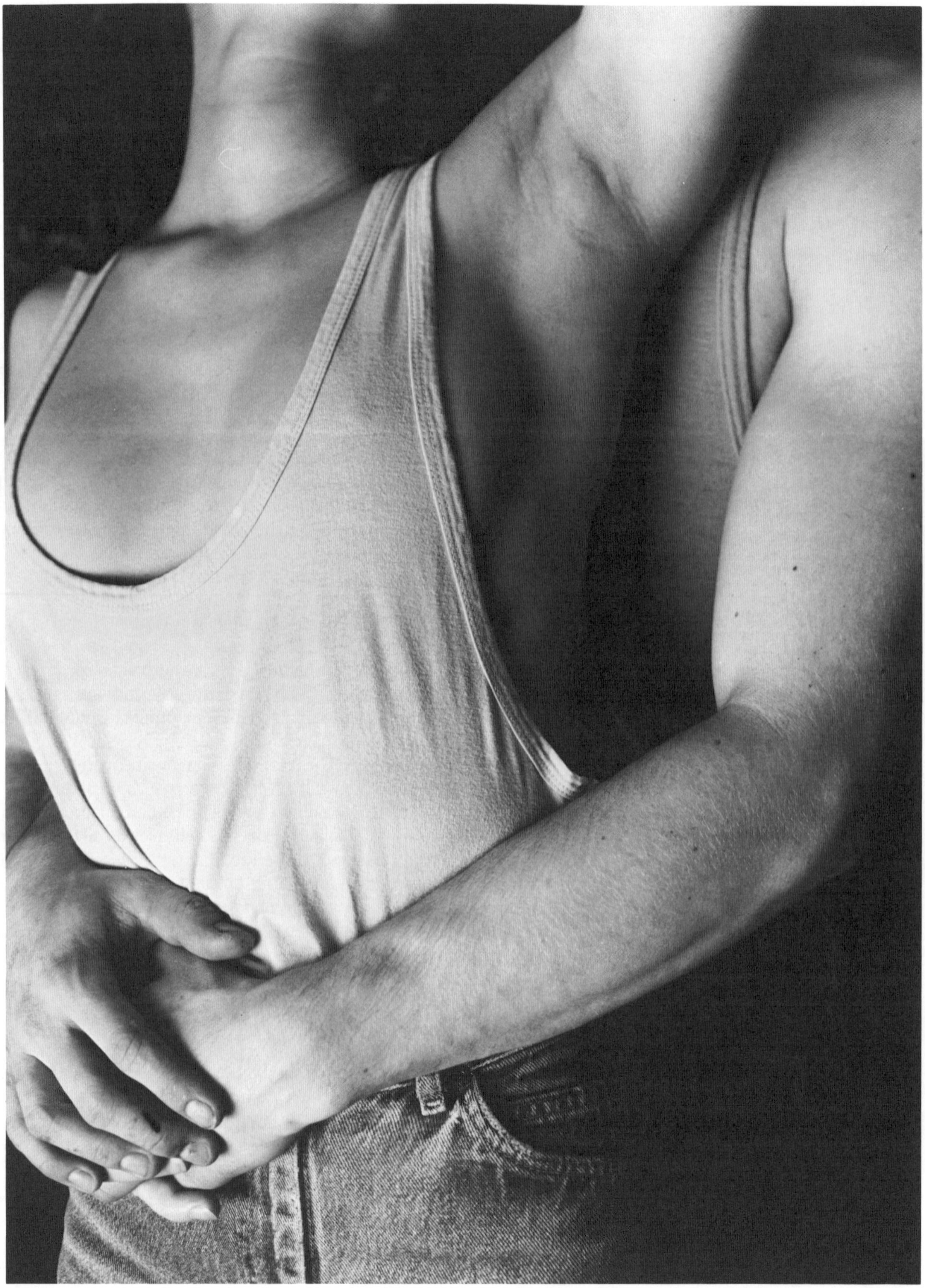

Chapter 2

General-Purpose Films

35mm film is grouped into four categories:

1. Black-and-white film that produces a negative, which in turn is projection- or contact-printed to make a positive print. Black-and-white transparencies can also be made.
2. Color negative film that produces a color negative that can be projection- or contact-printed to make a positive print. Negatives can also be made into color transparencies.
3. Color transparency film that generates positive slides as finished products. Transparencies can be printed directly by projection or contact-printed. Internegatives can also be made from slides to produce high-quality prints. (Note that the terms *transparency* and *slide* are used interchangeably and are identical in meaning, at least for the purposes of this book.)
4. Numerous special-purpose films in black-and-white, color negative, and color transparency formats. These films will be discussed in detail in the next chapter.

This chapter will devote itself to an examination of general-purpose films. Beyond the broad categories mentioned above, each film type is further subdivided by ISO number as being slow, medium, fast, or ultra-fast. Ordering film in this manner allows us to further divide 35mm film into sixteen distinct categories. Using this system of categorization, differences in the design, intended purpose, advantages, and liabilities of film in the general-purpose group can be studied and compared in more detail. Also, the discussion in this and the next chapter set the stage for in-depth evaluations of each and every one of the more than 100 films currently on the market.

Black-and-White Negative Film

Today's black-and-white film user has more than sixteen separate emulsions from which to choose. These range from slow, ultra fine grained emulsions to extremely fast, reasonably sharp, available-light film, permitting photographers to shoot scenes inaccessible in the past. A shared characteristic of all these films is their vastly improved grain structure. This in turn has led to improvements in adjacent edge sharpness. The result—sharper negatives from faster film.

Although black-and-white film use only accounts for only a small percentage of total film sales, it is presently enjoying somewhat of a renaissance. For many, it still remains the most persuasive form of photo expression because, say its adherents, its aesthetic potential cannot be matched by the realism of color. Whether or not this is a defensible position is not the point.

Figure 2.1 Paul Duda. Plus X Pan ISO 125.

What is obvious is the fact that black-and-white forms much of the body of photo expression that is regarded as the highest form of the art. Black-and-white film has some notable advantages over color. First, its use reduces all the visual elements in a scene to levels of gray. Skillful manipulation of the camera and, even more so, skillful darkroom work give careful and dedicated users almost infinite control over the look of the finished print. The range of control is far greater than with any color process.

At one time, local commercial photo labs routinely offered extensive black-and-white processing and printing services. With the ascendancy of color, especially the color print as the medium of choice for the masses of amateur snapshooters and advanced amateurs, these black-and-white services have dwindled alarmingly. Significant numbers of photographers who would like to work in black and white but don't live near large metropolitan areas are finding it increasingly difficult to find custom black-and-white facilities. And even if a nearby lab does black-and-white work, turn-around time can be three to five times longer than color printing.

Aside from the difficulty some experience finding dependable and professional black-and-white labs, the medium draws to it those individuals who see the world as some of the greatest figures in our craft did and still do. Names like Bresson, Lange, Smith, Weston, Adams, and Brassi come to mind. All evoke memories of symbol-laden imagery. Form, shape, line, light, and contrast were content. Today, those traditions are carried forward by the likes of John Sexton, Eva Rubenstein, Jerry Ulsemann, and many others. For them, black-and-white makes their visual statements immediate and clear, unobstructed by the imposition of color. It is the medium of choice because of its ability to distill a message down to its basic elements. With no concerns about elements such as color balance, black-and-white photographers are free to concentrate on the composition and meaning of the image itself. Whether this is true or not is a matter of great debate. Black-and-white is not for everyone. The fact remains that many photographers, myself included, simply do not see in black-and-white. We prefer color for numerous reasons, some of which we don't quite understand ourselves.

The wonderfully gifted, color photographer Ernst Haas suggested in his writing that each has within us a color and compositional bias that he called our "key." By that he meant that we all have a predisposition to seek out specific identifiable color combinations and compositional patterns. When I analyze my images, I find diagonals and triangles appearing over and over: strong, warm saturated colors—oranges, yellows, and reds predominating. Following Haas's reasoning, and from my experiences over more than two decades as a teacher, it seems fair to assume we also have a built-in predisposition toward or away from black-and-white image-making. However, there is no way of knowing this unless you experience black-and-white. If you haven't experienced the medium yet, it might prove illuminating to experiment a bit. Lurking inside you might very well be the next black-and-white giant.

General-Purpose Black-and-White Films Currently Available

Film	*ISO*
Agfapan APX 25	25
Agfapan APX 100	100
Agfapan 400 Professional	400
Forte Pan	100
Forte Pan	200
Forte Pan	400

General-Purpose Black-and-White Films Currently Available (con't)

Film	*ISO*
Fuji Neopan 400 Professional	400
Fuji Neopan 1600 Professional	1600
Ilford Pan F	50
Ilford FP-4 Plus	125
Ilford Delta 400	400
Ilford HP-5 Plus	400
Ilford XP-2	Variable (50–1600)
Kodak Technical Pan	25
Kodak T-Max. 100 Professional	100
Kodak Plus -X	125
Kodak T-Max 400	400
Kodak Tri-X	400
Kodak T-Max P3200	3200

Slow Black-and-White Film: ISO 25–32

This category of film is the sharpest, finest grained of all black-and-whites generally in use. These films feature high contrast that can be fine-tuned in film processing. They are also capable of resolving more image detail than almost any camera or enlarging lens can reproduce. A really first-class camera lens should be able to resolve 100ll/mm shooting a high-contrast subject; slow black-and-white films developed in fine-grain chemistry can resolve 200+ll/mm. If it's just so much resolution overkill, why bother with the slow films at all? It is the overkill capability that makes slow black-and-white films so useful. The reserve capacity of these films allows them to deliver excellent results in less-than-optimum lighting conditions. Most can deliver 200ll/mm shooting a high-contrast subject and are still able to resolve 90–100ll/mm in low contrast—more than enough to record every line of resolution a good lens is able to transmit to the film. Additionally, it should be obvious by now that everything in photography involves tradeoffs. For slow-speed black-and-white films it has been speed for grain. Less speed—less grain.

Improvements in grain and overall image technology have made ISO 100–speed film the new standard. However if you desire to produce exquisitely detailed, grainless enlargements larger than 11 × 14, you should master the use of one of the slow black-and-white emulsions. With care in the darkroom, you should be able to make all-but-grainless, super-detailed 24 × 30-inch or bigger prints.

Medium-Speed Black-and-White Film: ISO 100–125

By far the most popular black-and-white films are the "do everything" medium-speed films. If not quite as sharp or free of visible grain as slower films, they are more useful because of their higher sensitivity. Not only do they permit hand-holding but they also offer greater flexibility in terms of shutter speed/aperture selection. As is the case with their slower cousins, very large, impressively detailed prints can routinely be produced from carefully developed medium-speed negatives. Up to about 16 × 20 inches, most viewers will be hard-pressed to identify prints made from slow- as opposed to medium-speed negatives. Clearly, if your goal is to generate enlargements containing the utmost detail and minimal grain, use a slow film. On the other hand, if you only need moderate enlargement sizes (not exceeding 11 × 14), today's medium-speed black-and-white films should do very well. Two additional facts support the use of medium-speed film over traditional slow emulsions. The new film technology has given us quantum improvements in grain structure and attendant developing chemistry. The result? Medium-speed film negatives of exceptional quality that surpasses that of the slow films of only five years ago. Using a medium-speed film allows us the option of using higher shutter speeds, smaller f-stops, or both. Being able to select higher shutter speeds should improve the chances of achieving sharp available-light images by several fold. I also note that contrast is far more controllable when using medium-speed film.

High-Speed Black-and-White Film: ISO 400

Like other black-and-white films, emulsions in the ISO-400 range have been markedly improved. Some older films, such as Kodak Tri-X, are still on the market mainly because of their proven qualities and singularly desirable performance characteristics, very different from the newer competitors like Fuji Neopan 400 or Kodak T-Max 400.

Old or new, all high-speed black-and-white films display more grain and somewhat lower image resolution than medium or slow material. As with all other types of film, these differences are much less pronounced than they once were. Before, it was obvious when a print was produced from high-speed film; that is not necessarily so anymore.

High speed still remains the film of choice among photojournalists, sports, and black-and-white nature photographers, where success is dependent on high shutter speeds. Also, these films can be more readily "push-processed" to higher effective speeds than other classes of black-and-white films by using high-energy developers intended for the purpose. Admittedly, pushing does increase grain, but not enough to make the two to four times speed increase visually too expensive.

One final thought: Grain is not always undesirable. Under the right circumstances, it has a role to play, underlining powerful emotions at the heart of an image. Again, it's another case of a photographic condition having opposite aspects, on the surface seeming negative but when turned around in a creative manner, the result can be positive.

Ultra-High-Speed Black-and-White Films: ISO 1000-3200

A few years ago the introduction of Kodak T-Max P3200 caused a real stir in black-and-white circles. It took all the advancements in film design, T-Grains, and enhanced chemistry, and applied them to the problem of producing a superior ultra-high-speed film that not only controlled grain growth but was reasonably sharp and delivered decent blacks (D-Max). Kodak has succeeded admirably. P3200, developed in its special T-Max developer, can offer us usable EIs up to and sometimes exceeding ISO 6400.

For decades preceding T-Max P3200, the only ultra-fast film purposely designed for the marginal lighting some of us call "available darkness," was Kodak Data Recording Film, a not-so-wonderful emulsion whose grain structure looked like grotesquely enlarged salt and pepper. With fortune smiling on you, you could find some recognizable detail, but just barely. Since it did give a sometimes usable image under dreadful conditions, it was tolerated. What an improvement P3200 is!

There is even a third choice now. Fuji has refined its ultra-high-speed film, Neopan 1600 Professional. Nominally rated at ISO 1600, it can be push-processed to ISO 3200 for more speed or downrated to 400 or 800 for finer grain. Like P3200, it takes advantage of Fuji's advances in the photo chemistry craft to deliver results with remarkably better grain and sharpness than one could reasonably expect from a film this fast. In some respects, 1600 Professional is a more remarkable achievement than P3200, for, at ISO 800, it performs just like a medium-speed emulsion.

35mm Color Negative Films

During the calendar year 1990, photographers took over 15 *billion* pictures. Of this astronomical total, fully 95 percent were color prints made from negatives. That's nearly sixty color prints for every man, woman, and child in the United States and that in just one year.

What this immediately implies is that the color negative film, processing, and printing business is really big, with huge potential profits yet to be made. It is little wonder, then, that the color negative segment of the business is so hotly contested. Currently, we have over forty films in this category from which to choose. Many of these films claim to be the sharpest, finest grained, and most color-saturated on the scene. In truth, the quality of most of these products is so high that finding meaningful distinctions among them is difficult, if not impossible. This does not mean there are no differences between them, but those differences are, for the most part, subtle. The newest strategy in drawing performance lines between various print emulsions is seen in the way the recent introductions target specific and rather narrow segments of the market. Where before the goal was to achieve a reasonable share of a much larger general market, these all-but-custom-designed films seek to dominate a smaller portion of the market. Numbers of films such as slow, super-sharp high-saturation emulsions, soft-contrast moderate-speed, and even high-resolution high-speed films now dot camera-shop shelves. This trend to target buyers more narrowly will continue as research identifies even more fertile avenues for sales.

Why do print films so dominate the market? First, they are very well suited for the immensely popular point-and-shoot autofocus cameras. Offering as they do an exposure latitude of up to five f-stops, precise exposure ceases to be a major issue. Exposure can and has been totally automated. Second, snapshooters seem to prefer finished prints over color slides by a wide margin. Last, the proliferation of one-hour automated-processing minilabs has made for nearly instant gratification. Shoot a roll, drop it off at the one-hour, do a little shopping, and presto! Finished 4 × 6s.

The major problem with the preceding scenario is that most of the creative decision-making that powers photography as an intense and intimate experience for so many of us is turned over to a technician or, even worse, to a machine. This is not an indictment of machines per se. It is true, however, that decisions about density and color balance are no longer the photographer's. Of course, one could have negatives processed and printed by a custom darkroom, but for most it is too expensive, inconvenient, and time-consuming. No instant gratification here! The result of all the automation is a general lowering of expectations on the part of the color-negative user. Because automated processors and printers usually follow a program that shoots for average or middle ground, results, while good technically, often seem dull or commonplace. Most of the billions of prints serve as records of events, places, or people. This is fine; recording is one of the legitimate purposes of our craft. It is just that today's color negatives and resulting color prints offer so much unrealized potential, especially when making larger-than-snapshot-size enlargements.

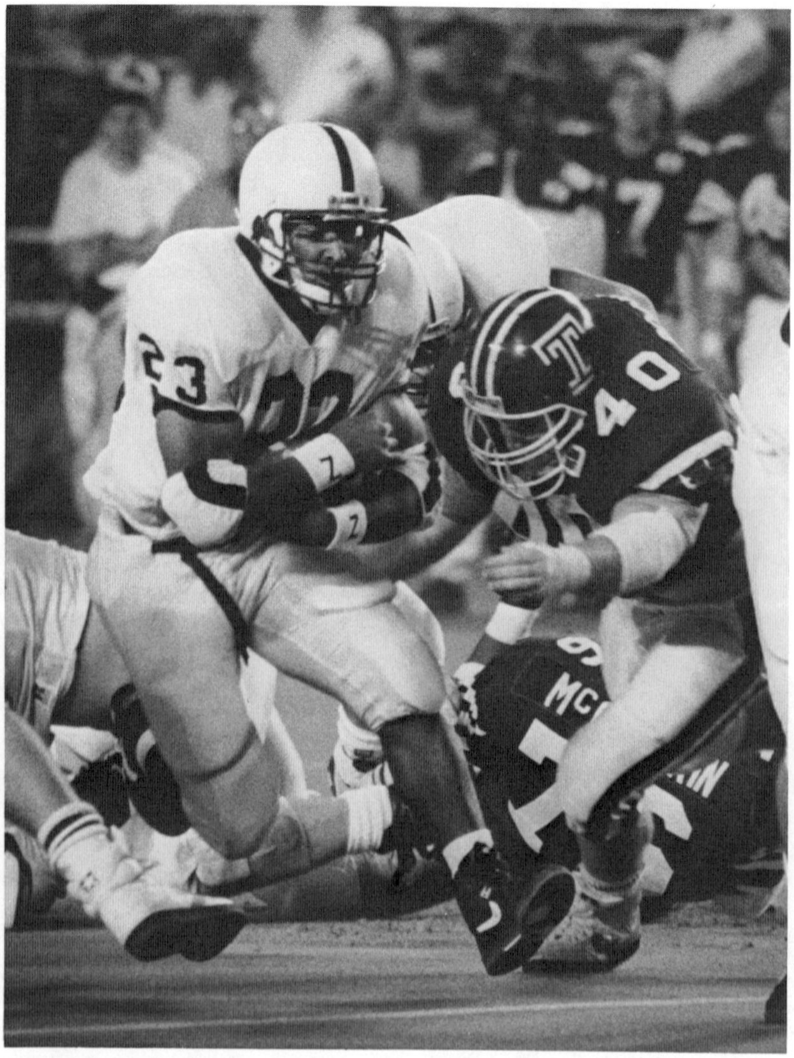

Figure 2.2 Steve Manuel. Kodak T-Max ISO 3200.

A word about nomenclature. Color print film can be instantly identified by the word *print* in the film's name. Therefore, all Kodacolor, Agfacolor, Polacolor, Fujicolor, and Scotchcolor films produce color negatives from which color prints are made. Be aware that there are a few print films with non-"color" names, such as Fuji Realia, Kodak Ektar, Ektapress, and all Konica film imported into the United States, that are print films. Incidentally, with very few exceptions, slide films contain the word *chrome* in their names.

The following are a few of the stated advantages of color print film:

1. Very wide exposure latitude—much more so than color slide film.
2. Custom printing from a negative presents opportunities for great creativity.
3. Print film is available almost everywhere.
4. Prints from color negatives offer more pleasing results from nonstandard (tungsten, fluorescent, mixed) lighting. Better than prints from slides or original slides themselves.
5. Color negative film responds well to overexposure. Overexposure lati-

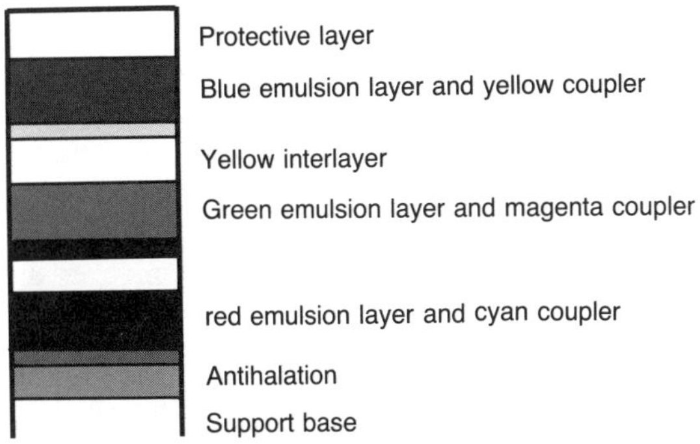

Figure 2.3 Cross-section of a typical modern color print film.

tude is so great that acceptable prints can be made from color print film overexposed by as much as 3½ f-stops.

6. Snapshooters and other amateurs seem to derive more satisfaction from color prints in hand than they do from color slides needing projection.
7. New automated color printing fosters almost instant gratification.
8. Color print film controls contrast and color balance better than does color transparency material.

How Color Print Film Works

Color print film is composed of eight or more microscopically thin layers, three to four of which contain embedded silver halide crystals and adjacent color dye couplers. The silver, at first, and eventually the dye couplers alone, work together to reproduce the full spectrum of color that we see in a fully processed color negative or color print.

When color film is exposed to light, silver halides undergo a chemical change. They are sensitized by the light, in much the same way black-and-white film is, and they form an invisible or latent image. When the film containing this latent image is subjected to immersion in a chemical color-developing agent, each of the color layers containing silver halides undergoes further chemical changes. Initially, halides are converted to black silver. Next, color couplers form dye clouds in one of three subtractive primary colors (yellow, magenta, or cyan).

Parts of the original scene photographed that were blue are represented on film in the yellow dye layer. Wherever green existed in the scene, that is now represented in the magenta layer, and red subject matter becomes part of the cyan dye layer of the negative. In sum, white appears black, blue records as yellow, green as magenta, and red as cyan in the bottom dye layer. It is important to note that dye clouds are not solid opaque blocks of color. They are color specks of varying densities, representing chromatic differences in specific colors. Dye clouds exist at different depths from top to bottom and, when viewed via transmitted light, combine to present a full color replication of the original scene. Eventually a fully processed negative is placed in an enlarger or other printing device, light is passed through the negative, and a full spectrum of color exposes sensitized color-printing paper. Before getting to the printing stage, several additional steps are necessary to complete film processing.

The Bleach-Fix Stages

These steps are necessary to complete color print film processing. After processing in the color developer, a bleach is introduced to remove all silver in the emulsion since the silver, having done its job of attracting and instructing color dye clouds, is no longer needed. After bleaching removes the rest of the silver, all that remains are individual color dye clouds in the several layers that now contain the image itself. A thorough, controlled washing then removes remaining unused halides and color couplers. The image on film is then chemically fixed for permanence. A number of washes and a chemical stabilizer are applied to complete processing. The film is then dried in a dust-free environment. What emerges from the chemical metamorphosis is an emulsion consisting of three (four with some films) semitransparent dye layers, each exhibiting varying densities of color. All the dye layers combine to produce a visible full-color image.

After the film is completely dry, it can be printed. The color negative is then inserted into a film carrier. Enlarger light is passed through a set of calibrated gel or dichroic filters of yellow-, magenta-, and cyan-colored material. This light goes through the negative and is then projected onto a piece of color printing paper. The paper's construction is very similar to that of the film itself. There are some differences—principally, the paper's top, middle, and bottom color-record layers (which are light-sensitive) respond to the colors red, green, and blue, respectively, compared to the film's top-to-bottom sensitivity of cyan, magenta, and yellow. As with color print film, the paper's three color-record layers, once light-sensitized and chemically processed, combine to make a full color image that mirrors the color nuances and tonal range of the original scene.

35mm General-Purpose Color Print Films Currently Available

Film Type	*ISO*
Agfacolor Ultra 50	50
Agfacolor Optima 125	125
Agfacolor Optima 200	200
Agfacolor XRS 100	100
Agfacolor Portrait 160	160
Agfacolor XRG 100	100
Agfacolor XRS 200	200
Agfacolor XRG 400	400
Agfacolor XRS 400	400
Agfacolor XRG 400	400
Agfacolor XRS 1000	1000
Fujicolor Realia	100
Fujicolor Super HG 100	100
Fujicolor 160 S Professional	160
Fujicolor Super HG 200	200
Fujicolor Super HG 400	400
Fujicolor Professional HG 400	400
Fujicolor Super HG 1600	1600
Ilfocolor 200	200
Kodak Ektar 25	25
Kodak Ektar 25 Professional	25
Kodak Ektar 100	100
Kodak Kodacolor Gold 100 Plus	100
Kodak Ektapress Gold Professional	100
Kodak Ektar 125	125
Kodak Vericolor III Professional	160
Kodak Kodacolor Gold 200 Plus	200
Kodak Kodacolor Gold 400 Plus	400
Kodak Ektar 1000	1000
Kodak Vericolor 400	400
Kodak Ektapress 400	400
Kodak Kodacolor Gold 1600	1600
Kodak Ektapress Gold 1600	1600
Konica Impresa 50	50
Konica Super SR 100	100
Konica SRG 160 Professional	160
Konica Super SR 400	400
Konica Super SR 3200	3200
Polaroid One Film	200
Scotchcolor HR 100	100
Scotchcolor 160	160
Scotchcolor HR 200	200
Scotchcolor HR 400	400

Technical Aspects of Modern 35mm Color Negative Film

Improvements in Color Film Design

Photo chemists in the United States, Europe, and Asia have succeeded in leaping ahead with huge quality gains in just about every measure of color print film performance. In about fifteen years, research and design teams around the world have managed to halve grain size while at the same time reconfiguring it into more uniform and efficient shapes. The research teams of the major players in the game—Kodak, Fuji, Konica, and Scotch—all made the same assumptions about the grain size–film speed relationship, namely, that it is highly desirable to reduce grain size while maintaining or, even better, increasing film speed. Several novel approaches were developed and all succeeded to one degree or another.

Kodak adapted their flat, emulsion surface–oriented T-Grain structure first developed for black-and-white film. In its current form, it allows color print film designers to use crystals extremely uniform in size and shape. Since T-Grains present more surface area, more dye clouds are formed, leading to higher levels of color saturation. These innovations have found their way into several films in Kodak's current lineup, including the Ektar and Ektapress films. Fine grain, typical of slow films of the seventies and eighties, is now seen in medium-speed films. For example, Ektar 100 now displays finer and more uniform grain than did any color negative film on the market in 1985.

These practical improvements result in more visible detail, smoother texture especially in open highlight areas, sharper adjacent edge separation, less color smearing, and the appearance of higher overall image sharpness. Not to be outdone, other film-makers introduced new emulsions with design innovations of their own. In Japan, Konica and Fuji perfected their own versions of the flat crystal. The dual-crystal design they use is essentially an outer crystal shell encapsulating an inner core, each with a specific and quite different function. Using the outer/inner technique, the larger outer shell gathers much light while the smaller inner core is eventually developed to yield a finer, reduced-size grain. At the same time in Germany, Agfa, with their twin crystal, and Scotch in the U.S. rushed ahead with variants of new smaller, more uniform crystal shapes that enhance the fine-grain appearance of their current film offerings. And the battle rages on. As long as it does, we photographers win with more and better film performance.

Aside from grain structure, the other major technological advance in color print film involves changes in the chemical compounds remaining to form a color image after light-activated converted silver is washed away in processing, that is, the color couplers. Previously, when activated by the color developer chemistry, color couplers grew at random, more or less unchecked by any other physical or chemical activity. Large dye clouds resulted and a phenomenon known as "color clumping" led to serious deterioration of the color image. Color clouds became larger and larger, joining together to form indistinct masses of color, reducing visible fine detail. As the clumping increased, grain grew more visible. To reduce these undesirable effects, photo scientists devised a new coupler mechanism, the Developer Inhibiting Releasing Coupler. DIR, unlike its predecessor, does not produce uncontrolled dye clouds but instead generates many more smaller uniform clouds that intermingle with one another and form a much finer grained negative. DIRs and their newest incarnation, DIARRs, help control unwanted interlayer contamination between the several color-record layers. This results in higher levels of color purity and color saturation. Other improvements have resulted from reductions in overall emulsion thickness, which in turn reduces light scatter. When light scatter is decreased, image sharpness increases proportionally. Thinner emulsions also lead to noticeable enhancement of edge effects and these in turn produce more discernible boundary contrast (acutance).

It is instructive to examine color print film by categories, as each of the four, slow, medium, fast and ultra-fast, has unique peculiarities and strengths.

Slow Color Negative Films: ISO 25–50

At present, there are only three films in this category: They are Kodak Ektar 25, Agfa Ultima 50, and Konica Impresa 50. All three are intended for the advanced amateur or professional, as almost no compact point-and-shoot autofocus camera now in the hands of millions of snapshooters is capable of setting either ISO 25 or 50, and so will expose the film incorrectly.

For those who own cameras able to accommodate slow films, the quality potential is enormous. None of the films claim to be a general-purpose type; for all-around use, they are just too slow. But in the hands of photographers who prefer to work carefully with static subjects such as still-life, landscapes, macro or formal portraiture, slow color negative films afford an opportunity to make negatives of such quality that they rival or surpass larger formats in such important performance dimensions as grain and sharpness.

The three films in this category have a very wide exposure latitude. Agfa's Ultima features what they claim is the highest color saturation of any print film. Our findings verify this claim. High color saturation comes with a price—significant increases in contrast. You will be faced with some shooting situations where the film will be too contrasty to retain detail in both shadows and highlights.

Ektar 25 is also a highly saturated film, but not as much as Ultima. Further, as one might suspect, it is somewhat sharper than the slow Agfa and displays lower contrast. The cost is one full f-stop less in film sensitivity compared to Ultima. Be aware that the Agfa film is one of a new family of color print films, each offering different and very specific performances. Konica's new Impresa 50 is a film combining extremely fine grain with moderate contrast and remarkably accurate color reproduction.

If image detail, high color saturation approaching the best transparency films, lowest possible grain level, and greatest total quality in 35mm negatives are among the most important film selection criteria for you, *and* you are willing to sacrifice a degree of picture-taking flexibility, then the slow films in this category could be an excellent option.

Medium-Speed Color Negative Film: ISO 100–200

By far the most numerous and best-selling of all the color print films, medium-speed negative emulsions account for more than 75 percent of all film sales. Within this broad category are almost all of the amateur-designated films for mass consumption. Not surprisingly, it is these films that have benefited most from the tech revolution we reviewed earlier in this chapter. To say that these films are an improvement over their forerunners is an unmitigated understatement.

Because these films need to print well even if they are grossly misexposed, all display ample under- and overexposure latitude, perhaps the most of any one film category, color or black-and-white. The latitude can be as much as seven f-stops, with over four stops on the overexposure side. Our experience indicates the exposure latitude figures are a bit optimistic. Two f-stops over and 1½ to 2 stops under are the practical limits if you expect any decent quality. It's true that you can get a recognizable image outside these limits, but not without serious image degradation. And since many automatic metering systems contained in popularly priced automatic cameras aren't remarkably accurate, you may very well need all the exposure latitude you can get.

Over twenty individual films can be listed in this category, but not all are intended for the mass market. At least five emulsions are specifically designated for professional use. They feature lower contrast and softer, more subtle color gradation than the majority of films in the group. Mainly they are intended for studio and portrait work where very accurate and pleasing softer skin tones are demanded. A disadvantage of professional films is their lack of widespread availability. Found only at large, well-stocked photo specialty shops, they command a dollar premium, often being priced as much as 30 to 35 percent higher than other same-speed films.

Laboratory and field tests confirm the long suspected belief that several films in this class are indeed faster than their marked ISO ratings. By deliberately underrating their film, some film-makers build in an automatic exposure cushion of one-third to two-thirds of an f-stop overexposure, thus insuring a normally processed negative that appears especially color-rich and saturated. In fact, many experienced color print film users routinely overexpose their film by this amount. Refer to chapter 6 for those films whose published ISO we believe are overstated because, if the film is already overrated, one needn't do it manually.

High-Speed Color Negative Film: ISO 400

At one time thought to be the ultimate in imaging technology, ISO-400 color print films have steadily improved. They now are thought of and used as general-purpose films. After all the improvements of recent years, these films are capable of producing negatives qualitatively superior to many conventional ISO-100 films. Noticeable gains in color saturation and D-Max (maximum black), two of the weak points of previous 400-speed films, make the current crop formidable image-makers.

It is in grain that we find the most impressive gains. With today's ISO 400–print films, grain is very well behaved, even at highest degrees of magnification, as in 11 × 14 or 16 × 20 enlargements. Obviously, ISO films are not fully competitive with ISO-100 emulsions but the gap between the two groups is narrowing and at times is difficult to identify. Several ISO-400 films are custom-designed to function well under nondaylight tungsten or mixed light sources. Among these, Fujicolor Realia is notable. By adding a fourth color-record layer to the usual three, Fuji has been able to align the film's color response more closely to that of the human eye. In addition to the normal blue-, green-, and red-sensitive layers, Fuji added an additional blue-green sensitivity layer. This new layer generates DIRs in the film's bottom red-sensitive layer. The result is a film that responds much better to green-rich fluorescent light. An extra benefit of the added layer is more true-to-life colors in the important skin tones. Konica offers us an ISO-400 color print film featuring very low granularity, high color saturation, and excellent response to tungsten illumination, even though it is a daylight-balanced film. Kodak also includes T-Grains in its popular Kodacolor Gold Plus 400 film, and we can see fundamental improvements in grain-size sharpness and acutance.

Ultra-High-Speed Color Negative Film: ISO 1000–3200

Film in this group is diverse, including as it does some emulsions that are 1½ to 3 times faster than those we once considered high-speed material. These are truly available-darkness types. That is, films capable of producing an image in light so low the photographer is barely able to discern the subject. And because they all contain large, numerous silver halides, no one expects them to perform like an ISO-100 film. They do surprisingly well, with the most advanced among them able to deliver negatives sharp enough to produce 8 × 10-inch prints acceptable by professional standards.

Hand-held available-light color photographs once thought to be impossible are now within easy reach of almost any photographer. As an example, average street lighting at night calls for an exposure of 1/15 second at f-2.8 using standard high-speed ISO-400 film. Switching to an ISO-3200 film would allow an exposure of 1/60 second at f-4. Multiple-burst fireworks calling for 1/30 second at f/2.0 using ISO-400 film can be made at 1/125 second at f-4 with ISO 3200. If the tradeoff in grain and sharpness is acceptable, ultra-high-speed films can be real problem-solvers.

35mm Color Transparency Film

Even though 35mm color transparency film accounts for a mere fraction of total film sales, it still enjoys great popularity among advanced amateurs and professionals because it remains the medium of choice for magazine and book publishers. Further, most color photo competitions insist on transparencies and in some cases have absolute prohibitions against prints. Some discerning photographers insist that the color slide provides the greatest impact. Viewed in a darkened room, slides come to life. They can reproduce contrast ranges of better than 200:1, where a color print viewed by reflected light does well to display a contrast range of 50:1.

The laws of physics and human vision favor the color slide. When viewed in their intended form—that is, projected—most observers judge the slide to be superior to the color print. As we shall soon discover, several additional factors enter into the equation.

As of this writing there are over forty individual 35mm color slide films from at least six major suppliers. These range from the very slow but wonderfully sharp Kodachrome 25 rated at ISO 25 to Fujichrome P 1600II at ISO 1600.

Why so many different films, even by the same manufacturer? One answer can be found in market economics. Film-makers feel they must have a complete lineup of films to match the competition, type by type. Brand loyalty plays a role. Substantial numbers of photographers simply will not use anything but their accustomed brand of film. To some degree, this accounts for Kodak's dominance of the U.S. market. Other companies find it difficult to dent Kodak's marketshare, no matter how innovative their own emulsions might be.

Recently, however, there have been some signs of increased activity on the part of Kodak's chief competitor, Fuji. Green Fuji boxes are beginning to show up in places such as supermarkets, formerly the exclusive domain of the giant from Rochester. Whether this represents a breaking of the virtual monopoly Kodak has enjoyed, only time will tell. But in the film business it is the consumer who benefits from renewed competition. Let us hope the shelves become crowded with yellow, green, silver, blue, and black boxes or cylinders.

There is yet another reason for the proliferation of color slide films. Color print material can be successfully manipulated when it is printed in the darkroom—for example, color can be modified or contrast increased or decreased. A color slide, on the other hand, is a finished product after processing. What you get is already on the film as a positive image. You may choose to make enlargements from slides, but the range of control is more limited for the home darkroom worker. For the person interested in working with slides, creativity takes place chiefly in-camera. Film-makers provide us with many emulsions from which to choose; each has a unique identity or personality. Therefore, one film might prove to be exceptionally adept at producing accurate color, while another may work significantly better in shadows than the rest. Yet a third could be the most adept at producing believable earth tones. Choice gives us a very diverse palette. In my view, the more the better—there can never be enough films.

Historically, color transparencies displayed finer grain and higher color saturation than color print film. This disparity exists today. Transparency film adherents claim several other advantages for their emulsions, such as:

1. Picture cost per shot is lower using slide film.
2. If you compare films of the same ISO, slide film is finer grained and sharper than print film.
3. Slides are more compact and are easier to sort and locate.
4. Slight underexposure enhances color saturation.
5. Slide film lends itself to push-processing more readily than does print material.
6. There are a number of slide films balanced for shooting in tungsten light. There are no tungsten-balanced general-purpose print films.
7. Prints from slides, especially those made using Ilford Cibachrome paper and chemicals, will last far longer than any negative print material.

How Color Transparency Film Works

Color transparency film is similar in construction to color negative film. Processing most slide film can be accomplished at home using process E-6 or similar chemistry. Just about any local processing lab is E-6 capable.

The E-6 Processing Sequence

1. Film is first exposed to light and is thus sensitized. Reaction of light and silver halides contained in several strata within the film create what is in effect three separate latent images. These are called color-record layers.
2. A black-and-white developer is used to convert the three record layer silver halides into black metallic silver. Color couplers adjacent to the metallic silver in the record layers are at this point totally unaffected. The primary colors, blue, green, and red, are represented in the three color-record layers of the film as a black-and-white silver image only.
3. Following the black-and-white developer, usually referred to as the first developer, the film is subjected to a reversal bath that chemically fogs the remaining unexposed silver halides and changes them to a silver positive. (The old now-discontinued E-4 process required physically re-exposing the film to a weak light source to reverse the image from negative to positive.)
4. Next, a color developer is introduced that activates color dyes from the

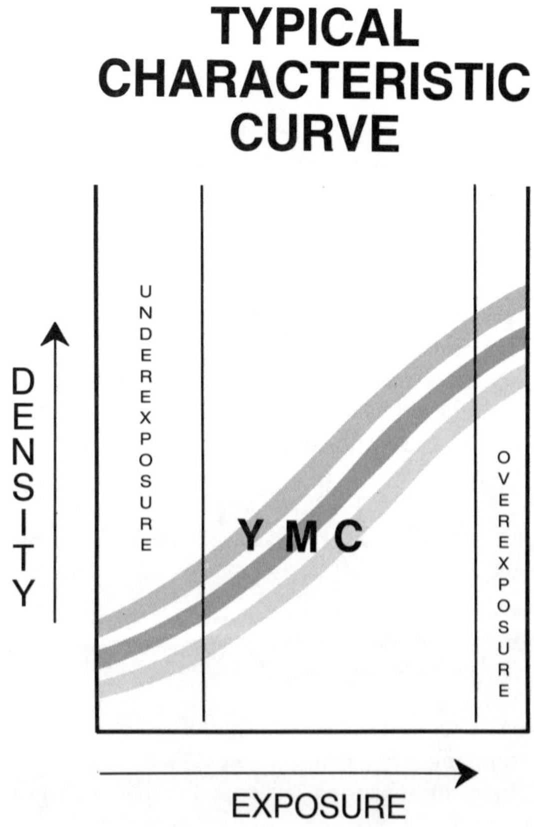

Figure 2.4 As exposure increases, color saturation and negative density also increases.

couplers linked to the positive silver image generated in the reversal step only. The couplers that were attached to the negative silver image in the first black-and-white development step are not activated by this color developer.

5. The final three steps—bleach to remove all silver in the image, leaving only color dye clouds in the three record layers to form a positive image, a fix step to achieve image permanence, and finally a stabilization bath to neutralize residual chemicals—are essentially the same as in color negative processing.

After drying in a dust-free environment, the full color positive is now ready for projection or printing. The processing sequence just described is applicable to all films using the E-6 processing methodology.

Nonsubstantive Films

There is a small but important group of transparency films containing color couplers in the developer. These are the Kodachromes. This family of high-performance films is processed in a very complicated and precise K-14 chemistry and need a great degree of precision only possible under stringent industrial control. Kodak has limited the number of labs where Kodachrome processing takes place in order to assure strict quality control.

Nonsubstantive Kodachrome films add color couplers during development. Although Kodak does not recommend it, a few labs across the country will push- or pull-process Kodachrome. Why does this admittedly less convenient and more expensive film and processing exist when there are so many other fine E-6 films from which to choose? For one thing, Kodachrome has gained a fiercely devoted following and for good reason. The slowest of the family, Kodachrome 25 and its professional stablemate Kodachrome 25 Professional, are the finest grained, sharpest color films of any brand or type. Additionally, all Kodachromes exhibit rich, well-saturated color and excellent contrast. The addition of an ISO-200 film to the Kodachrome family further enhances their appeal. Many would claim there is something different about the way a Kodachrome looks, particularly at high magnifications. Colors seem to remain discrete and better defined than with other films. Also, there is a positive predisposition toward Kodachrome on the part of editors, art directors, and ad people. In the ad business, the term often used for slides of any kind is "Kodachrome".

Slow Color Transparency Films: ISO 25–64

In this category, two films, Kodachrome 25 and the newer Fujichrome Velvia, are the ones by which all others are judged. Both offer micro-fine grain, with Kodachrome being a hair sharper. In practical terms, the differences in sharpness are hardly noticeable. Each is capable of recording better than 125 lines per millimeter, which is far more than most camera or projector lenses can resolve. The two films are contrasty, but not so much that they obscure important shadow detail. When it comes to color rendition and saturation the two films part company.

Each has a unique personality. Kodachrome 25 and its 3400K tungsten-balanced derivative Kodachrome 40 display rich well-saturated colors and tend to be neutral/cool. Velvia's color saturation can only be described as spectacular, perhaps the highest of any film. Fuji's engineers and chemists created a minor miracle in that the film's image-shadow detail is maintained exceptionally well in spite of the high color contrast. Even on cloudy overcast days, color jumps off a Velvia slide. Kodachrome

35mm General-Purpose Color Transparency Films Currently Available

Film	*ISO*
Agfachrome RS 50 Plus Professional	50
Agfachrome RS 100 Plus Professional	100
Agfachrome CT-100 i	100
Agfachrome RS 200 Professional	200
Agfachrome CT 200	200
Agfachrome RS 1000 Professional	1000
Fujichrome Velvia Professional	50
Fujichrome 50	50
Fujichrome 50 D Professional	50
Fujichrome 64 T Professional	64
Fujichrome 100	100
Fujichrome 100D Professional	100
Fujichrome 400	400
Fujichrome 400D Professional	400
Fujichrome 1600II Professional	1600
Kodachrome 25	25
Kodachrome 25 Professional	25
Kodachrome 40	40
Ektachrome 50HC	50
Kodachrome 64	64
Kodachrome 64 Professional	64
Ektachrome 64 Professional	64
Ektachrome 64(T) Professional	64
Ektachrome 64X Professional	64
Ektachrome 100 Professional	100
Ektachrome 100X	100
Ektachrome 100 Plus Professional	100
Ektachrome 160 (T) Professional	160
Ektachrome 100 HC	100
Kodachrome 200	200
Kodachrome 200 Professional	200
Ektachrome 200	200
Ektachrome 200 Professional	200
Ektachrome 400X Professional	400
Ektachrome 400 HC	400
Ektachrome P 800/1600 Professional	800/1600
Konicachrome 100	100
Polachrome CS	40
Polachrome HC	40
PolaPan	125
PolaBlue	8
PolaGraph HC	400
Polaroid Presentation Chrome	100
Scotchchrome 100	100
Scotchchrome 400	400
Scotchchrome 640T	640
Scotchchrome 1000	1000

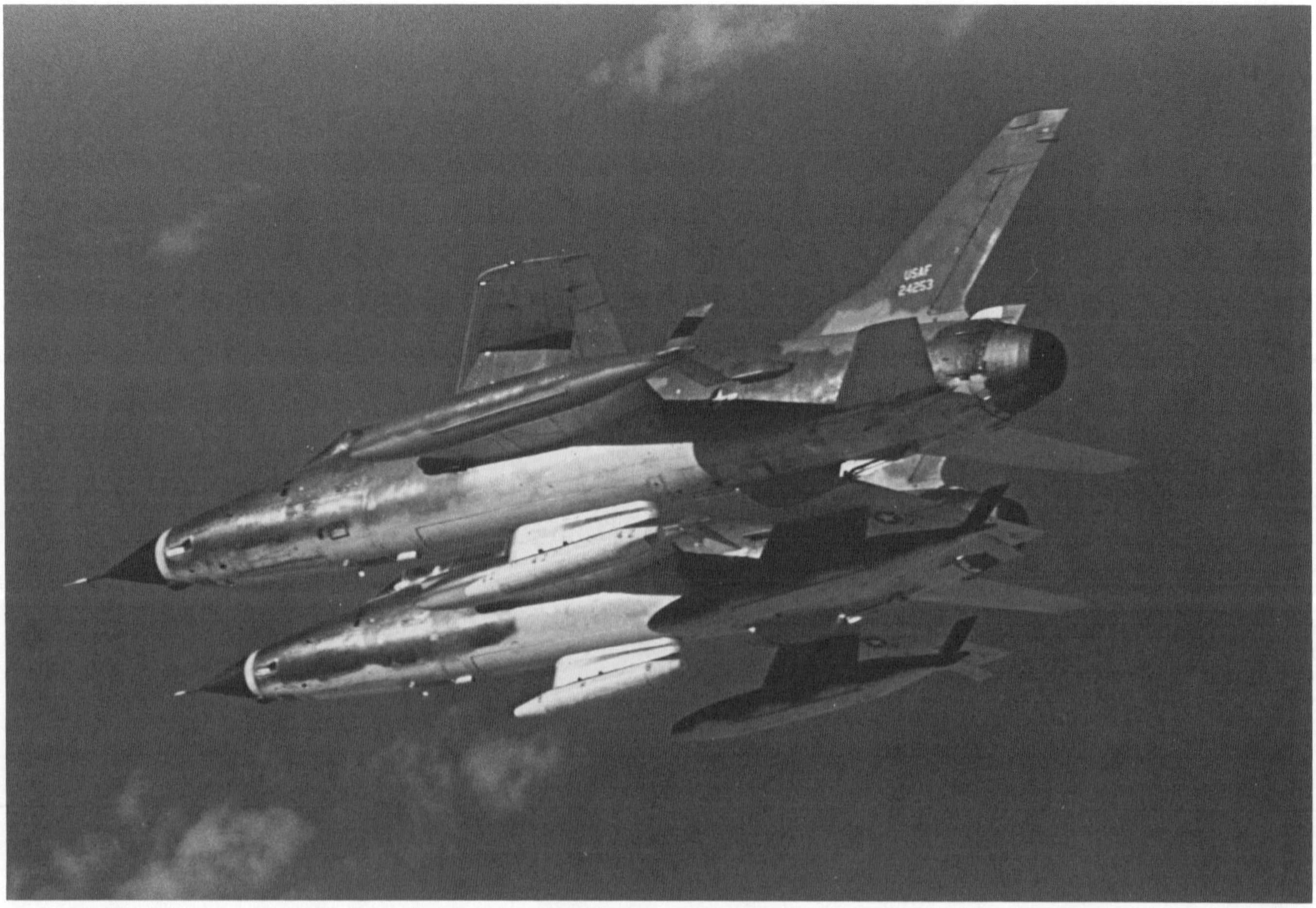

Figure 2.5 Marc Levey. Ilford Pan F ISO 50.

Even on cloudy overcast days, color jumps off a Velvia slide. Kodachrome seems subdued by comparison.

When faced with two such superb films, creative decision-making comes into play. Is super-hot color rendition always desirable? Does Kodachrome 25's slight sharpness edge and superior grain structure justify giving up one full f-stop film speed to Velvia? Tough questions to answer definitively. I suggest there really is no yes or no answer. These and other questions can be answered situationally, that is, in terms of photographic intent.

There are other excellent films in this category, each having a distinctive role to play. Agfachrome 50RS Plus is an excellent all-around emulsion, if not quite as sharp as the first two on the list. It features a rich but more subdued color palette, especially when compared to Velvia. Agfa has shifted color saturation, abandoning its long-held and much cherished soft, lower contrast color mix, favoring as it did pastel renditions and earth tones, for the more in-vogue high color saturation motif. To their credit, Agfa didn't abandon their heritage completely with their latest slide films. The slower ones exhibit restrained color next to their competition. Once again we have a choice!

Another very important film in this category is Kodachrome 64 and its Pro variants. It is a film popular with professional photographers who make their living from color. Being more than one full f-stop faster than Kodachrome 25, it is a logical alternative to the slower film for those who routinely shoot nature, sports, landscapes, or those who feel they need the extra speed, but insist on the distinctive look provided by Kodachrome. There is no denying Kodachrome 64 is visibly grainier than either K-25 or Velvia. However, its grain is still minimal, and when exposed at ISO 80 it gives a slide that is dramatic, with deep, rich color especially in the red, orange, and yellow portions of the palette. To demonstrate to yourself how popular this film really is, you only need to glance through the film credits of any photo book or magazine and count the number of pictures shot with Kodachrome 64.

Japan's Fuji gives us yet another choice in the slow color slide film category, namely Fujichrome 50 and Fujichrome 50D Professional. Before the introduction of Velvia, these two E-6 films displayed the highest saturation and sharpness outside the Kodachrome camp. Along with high saturation came the mixed blessing of high image contrast, good in low-contrast light but difficult to control on bright, sunny days. Contrast problems aside, it has enjoyed a steady increase in popularity among knowing photographers who have come to appreciate its ability to render colors with extreme accuracy.

Kodak's counterpart to Fujichrome 50 is Ektachrome 50HC. It is an exceptional film, exhibiting all the good traits desired in a user-friendly emulsion: high color saturation, excellent sharpness, tight uniform grain pattern, reasonable controlled contrast, and decent exposure latitude. Next, we have one of Kodak's long-standing Ektachromes, Ektachrome 64 Professional. It represents an alternative to the higher saturation films in the category. It is sharp with good color definition, but is more soft-working in terms of color gradation. The film is an interesting choice for high-contrast light situations where the goal is to wring out as much tonal separation from shadows to highlights as possible. Last, we come to one of the newer films in the group, Ektachrome 64X Professional. Not a totally new emulsion per se, it does nevertheless give us another option. Based on Ektachrome 64, it has been optimized for overcast or shade. By increasing magenta and yellow dyes, this film will warm up scenes shot under cool light without the use of a warming filter. In other situations, it provides a very warm palette.

Medium-Speed Color Transparency Film: ISO 100–200

Over eighteen films constitute this largest category among slide films. As with their color negative brethren, medium-speed color slide films appear under seven major labels and countless private-house brands. They benefit from all the latest technology and constitute the bread-and-butter of the slide lines.

Flat or multiple grain structures are rapidly replacing conventional, randomly shaped arrangements. Mixed size and shape grain structures are also prominent among contemporary medium-speed slide films. The net result of the new grain patterns is film that is sharper and exhibits less apparent grain. Throughout this group, there are improvements in color couplers, some borrowed from color negative film research. New DIRs and DIAAs allow for cleaner, less contaminated dyes in each color-record layer. Thinner overall emulsions reduce light scatter, which in turn improves sharpness. Improvements in edge effects—sharply etched boundaries between area of color or contrast—contribute further to the appearance of sharpness.

Medium-speed films, in particular ISO 100, present us with a burden of riches. Almost all are very, very good, offering as they do an excellent balance of speed, grain quality approaching that of slower films, first-rate color reproduction, moderate contrast, pushability with little if any loss in quality, and varied color interpretations.

Other virtues of this group include the ability to absorb elevated ISOs for more color saturation without any other harmful effects. And, with all this, they are more than two full f-stops faster than Kodachrome 25.

Which film is best ? Again, it comes down to a matter of personal taste in which the subjective character of each film, such as the appearance of grain (e.g., smooth and creamy with Fuji emulsions or the more aggressive, sharp-edged Kodak structure), becomes the deciding factor.

No matter how much objective numerical data we accumulate, it is your eye and mind that should remain the ultimate judge. This is not to say objective data is not useful, for it most certainly is; it helps narrow down the field. Examine the test information—lines per millimeter, percent contrast, MTF curves, and the like—but pay as much attention to the sometimes conflicting subjective impressions.

High-Speed Color Transparency Film: ISO 400

Several films in this group have recently been completely revised, while others have been steadily improved throughout the years. Films in the ISO-400 range are designed to deliver reasonable images under hand-held, available-light shooting conditions. For the most part, until very recently, ISO-400 color slide films have represented a compromise at best. To deliver high film speed or sensitivity, they needed to employ large, irregularly shaped crystals. What resulted was somewhat objectionable grain, readily visible in the finished slide. Even though one person's objection is another's satisfaction, most agreed that 400 films were not capable of large magnification projection or enlargement to anything beyond 8 × 10 inches. At any rate, ISO-400 slide film had been looked upon with distain by purists, exhibiting large clunky grain, gray-brown murky blacks, low contrast, and mediocre color reproduction. Yet these films were the only way to shoot many difficult low-light subjects. Things have gotten better! Today's ISO-400 slide films are very different from the ones we were forced to use a few years ago. They project well and can be made into excellent 8 × 10-inch or even bigger enlargements. They are an improvement in every way. And, as a bonus, they are able to deliver more-than-acceptable quality at several times their rated ISO. For instance, Scotchchrome 400 is specially designed to be push-processed up to ISO 1600. Conversely, Fujichrome 400 is optimized for outstanding results at its rated ISO. It has very good color fidelity and surprisingly low grain, often being mistaken for an ISO-100 film. One suspects that when the full technology is applied to this class of films, we will witness even more dramatic improvement.

Ultra-High-Speed Color Transparency Films: ISO 1000–1600

The "Ultras" are a group of four films, two of which are of the variable ISO type. Fujichrome 1600II and Ektachrome P 800/1600 are designed to be used at several ISOs, ranging from 400 to 3200. In testing these films, we have discovered the useful ISO range to be no more than 1600 if one has any hope of preserving reasonable grain and tolerable color. Both films, but Fujichrome in particular, do very well when the ISO rating is kept to 800. Note: these films project better than they print.

The third member of this group is Agfa 1000RS. It is a true ISO-1000 film exhibiting high sharpness, good color, and a pronounced but tight and even grain pattern. It is wonderfully moody when used to photograph on overcast days. Try finding splashes of color or candid portraits on the street.

The last film in this group is in some ways the most interesting. It is unique in that it claims to be the world's most grainy color film. The film is none other than Scotchchrome 1000 and its slightly slower tungsten-balanced clone, Scotchchrome 640T. Rated at ISO 2000 or higher, the film produces images where physical scene elements are broken down into patches of color and shade much like nineteenth-century Impressionist paintings. With the right light and subject, the film can contribute to the making of a powerful message.

It is not too difficult to see that there is a very wide choice among slide films. Negative film selection is a bit easier because there is opportunity to fine-tune the image during printing. Using slide film restricts that to some extent; it is up to the slide-film user to make creative decisions before and during shooting. In the next chapter we will examine specialized films, many of which are transparency materials.

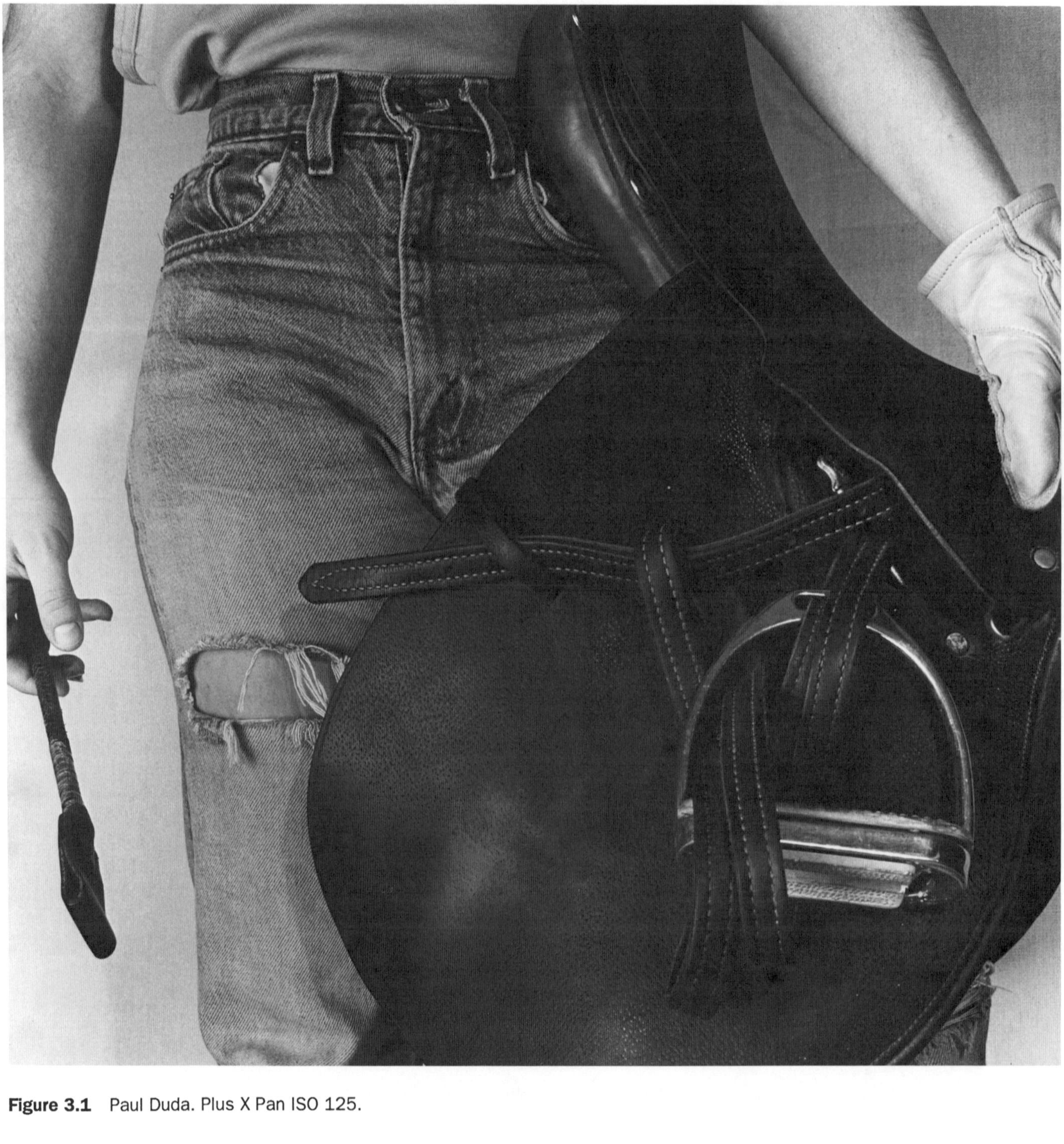

Figure 3.1 Paul Duda. Plus X Pan ISO 125.

Chapter 3

Special-Purpose 35mm Films

Thus far, we have concentrated on the films used day in and day out to produce the vast majority of images made worldwide. These black-and-white and color emulsions, in both negative and transparency versions, are the bread-and-butter tools of professionals. Besides these many films meant for general use, there are a number of specialized emulsions designed to serve very specific functions. In addition to their intended role, special film types are used by photographers to create all kinds of startling visual effects unforeseen by the emulsions designers. Used with imagination, films such as infrared or photomicrography deliver images pushing creativity to its limit. Let us take a closer look at these specialized films in the following ten categories:

1. Black-and-white and color infrared
2. Photomicrography
3. Data recording
4. Chromogenic black-and-white
5. 35mm Polaroid Instant black-and-white
6. 35mm Polaroid Instant color transparency
7. 35mm Polaroid high-contrast transparency (E-6)
8. Lithography-line
9. Orthochromatic
10. Duplicating

Because of their special properties, many of these films require extreme care in handling, use, and processing. Some are very sensitive, others extremely slow. Several must be loaded in total darkness. Some call for the use of special filters, while yet a few need nonstandard processing chemicals and/or techniques. Last, the storage requirements for a number of the films in this chapter are quite different from ordinary emulsions.

Are these sometimes hard-to-find films worth all the effort and special care they demand? The answer must be a resounding yes, only if they are used as part of a total image-design concept and not just as something different employed to create a visual oddity.

Black-and-White Infrared Film

Several manufacturers make infrared film. In this country, the most readily available is Kodak High-Speed Infrared Type 2481, packaged in 36-exposure, 35mm cassettes. Infrared film is sensitive to the portion of the electromagnetic spectrum measured for about 700–900nm. It is also sensitive to some of the same visible light sources as are normal black-and-white films. To achieve a true infrared-only image, untainted by visible light, special filters must be employed over the taking lens. Kodak generally recommends a Wratten Gelatin filter No. 25 (Red). Virtually all filter-makers offer a No.

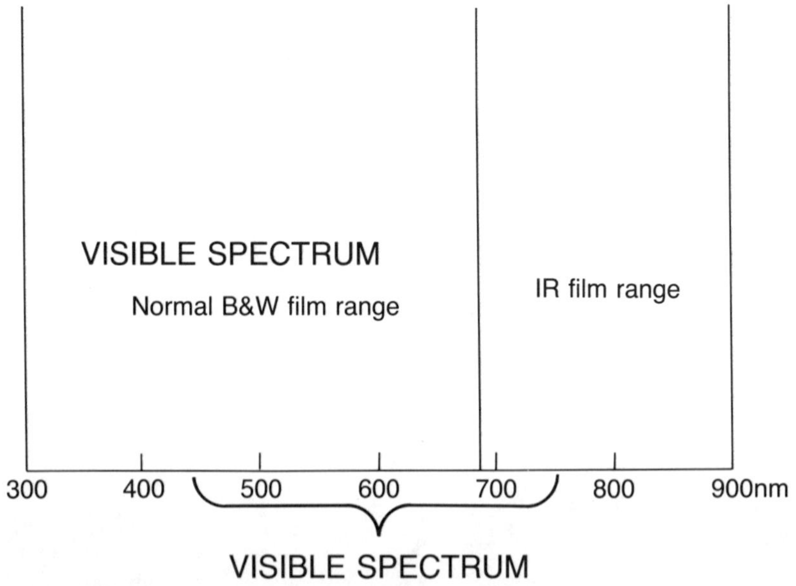

Figure 3.2 Infrared visible spectrum relationship.

25 or its equivalent. Think of it as the basic beginning infrared filter. A number of filter-makers offer a series of infrared filters to achieve varying effects. You will find a detailed discussion of them in chapter 5.

How Infrared Film Works

Because black-and-white infrared film is sensitive to almost all visible light, plus ultraviolet and infrared, a deep red filter is employed to absorb almost all but the infrared portion of the spectrum. If a total infrared image is required, you will need to use an infrared filter made for the purpose. These opaque filters pass infrared only and no visible light. Because they are very dense, viewing the scene to be photographed through an SLR viewfinder is not possible with the IR filter in place.

Infrared radiation is present in a variety of light sources. All the common ones photographers encounter—tungsten, strobe, and sunlight—emit varying amounts of infrared. Be aware that infrared behaves differently than visible light. It is most pronounced at sunrise and sunset and shortly thereafter. Some subjects reflect infrared more strongly than others. Human skin and green chlorophyll-bearing live plants are good infrared emitters. Shooting these subjects with IR film produces surreal shimmering white skin tones or vegetation. On the other hand, clear sky and other low IR-sensitive portions of a scene will be rendered very dark or even dead black. Very dramatic!

There are a few hitches using black-and-white infrared film. To begin with, there is no set ISO number for any IR film. This is because IR radiation does not correspond to that of visible light, so ordinary light metering is virtually useless. Also, IR radiation varies wildly depending on reflecting surface, light source, and any number of other factors. In sunlight, seasoned IR users will start off at 1/125 second at f-11 for distant scenes and 1/30 second at f-11 for close-up subjects less than 100 feet from the camera. Even with these guidelines, most users bracket up to four stops up and down.

The second hitch involves focus shift. Because IR wavelengths are longer than visible light, focus must be adjusted when using IR film. A diminishing number of camera lenses include a red dot or line indicating the

IR compensating point to use. In practice, if shooting at f-8 or smaller f-stops, the slight difference between visible focus points and IR can be discounted.

Infrared film's high contrast and pronounced grain is a curse for some and a blessing for others. In any event, do not expect fine-grain, mural-quality negatives. What you can expect is a negative with good tonal qualities and distinctive rendition of high IR radiating subjects. One thing is sure. Results are difficult to predict accurately. One last caution about handling IR film. All IR emulsions are very sensitive to visible light and *must* be loaded and unloaded in complete darkness. A darkroom is highly recommended for these operations.

Black-and-White Infrared Applications

Historically, black-and-white infrared film has been used for several scientific, forensic, and environmental applications. Many years ago, photographers discovered the esthetic and creative potential of this material. They found they could literally create images that could not be seen or experienced by any other means. Thus, they could turn the invisible into the visible. The images were unlike anything else in our visual experience. Infrared film lends itself to portrait, still-life, abstract, and particularly landscapes. Black-and-white IR prints are unmistakable, exhibiting an unworldly tonality with biting, highly visible grain and an overall textured, contrasty appearance. Handled with care and intelligence, black-and-white IR images let us see the world in a way unimagined by anyone fifty years ago.

Infrared Color Transparency Film

Color infrared film is supposed to create false colors. It is intended for use in agricultural surveys, military intelligence gathering, and many other commercial and scientific ventures. At the present time, Kodak Ektachrome Infrared Type 2236 is the only IR-sensitive color film produced. Structurally, color IR film is similar in several ways to normal color transparency material. There are noticeable differences, however. While normal slide film contains three layers sensitive to visible blue, green, and red light, color IR film's top layer is IR- rather than blue-sensitive. Also, the color layers in IR film are shuffled. Cyan dye is found in the top layer, then yellow, and on the bottom magenta.

Kodak recommends the use of a deep yellow, or better still, a No. 25 Red filter. With no filter in place over the lens, resulting slides will appear overly blue or magenta. When you use a red filter, color will not be realistic. There will be wild color shifts, with skin tones taking on an ashen zombie-like look, a whitish color, tinged by blue or purple. Overall, the slide will have a yellowish cast. If you shoot this film outdoors, most healthy vegetation will be rendered in shades of red. Diseased or dead vegetation will show up as blue-green. These are general expectations and results vary greatly depending upon factors such as the amount of IR present, the filter in use, and the nature of the reflecting surface. Orange, purple, or any other color filter can be tried with IR color film. In addition to semitransparent color filters, several filter-makers offer opaque IR filters usable with black-and-white and color IR film. These special filters absorb all visible light and only pass IR radiation onto the film for a true IR image. Be aware that Ekta-chrome Infrared Film must be processed in E-4 chemistry. This process had been discontinued and may be difficult to locate. For a list of some labs that still do E-4 processing, refer to Appendix C of this book.

Infrared in Use

These guidelines apply to black-and-white and color infrared film.

1. Try to shoot subjects side- or front-lit. This will produce the strongest IR effect.
2. Use infrared compensation in focusing if your lens is so marked.
3. Process color infrared film in Process E-4 chemicals. Use Kodak 76 full strength for black-and-white infrared film.
4. Test film/filter combinations under a wide range of lighting conditions and exposure settings.
5. Both IR film types penetrate haze very effectively.
6. Handle both films in complete darkness. Store in unopened containers; black-and-white infrared film in the refrigerator, unopened color IR film in the freezer. Load film in the camera in a changing bag or, better still, in the darkroom.
7. Cover the clear film identification window found on the back of many recent model cameras with two or more layers of heavy black card stock. If you fail to take this precaution, you risk IR light leakage and eventual fogging.
8. Plastic developing tanks are not IR light-proof. Use stainless steel tanks of the Kinderman or Nikor type to develop black-and-white infrared film.
9. Even though Kodak recommends ISO 50 as a starting point for exposing IR, you may discover (as have numbers of long time IR users) that higher starting ISOs are more suitable.
10. For detailed information about infrared photography, consult the Kodak publication, *Applied Infrared Photography.*

35mm Polaroid Instant Slide Films

Polaroid instant slide film and associated processing systems give easy access to several finished and ready-to-use slide films without the necessity of a darkroom or commercial processing. The system is comprised of five different 35mm films and manual or motorized "instant" processing and slide-mounting equipment.

The film itself is loaded in the camera, exposed, and rewound like any other 35mm film. Each Polaroid instant cartridge is supplied with its own sealed plastic container of processing chemicals that Polaroid has dubbed its "processing pack." After film is exposed in almost any 35mm camera, it is loaded into the auto processor and in a minute or two is ready for visual inspection. A few minutes more to mount the slides in an easy-to-operate slide mounter and you are ready to project or print the finished images.

The color film is different from most other conventional emulsions in that it is based on the additive color principle rather than normal subtractive color alignment of other color slide films. Polachrome CS contains a microscopically fine lined additive color screen consisting of a repeating pattern of red, green, and blue color filters at about 1000 color triplets per inch, with each triplet comprising a single red, green, and blue filter element.

Over the filters is a very thin color screen protective layer and over that a positive image-receiving layer. It is in this layer that the final silver image will be deposited. Four additional layers are situated above the receiving layer. They consist of a protective layer directly above, which also functions as a stabilizer, a release layer that allows upper film layers to be removed during processing. Below the top antihalation layer is a panchromatic silver-halide emulsion layer.

A color image is formed almost in the same way as is a color television picture. Microscopic examination of the white portions of a finished Polachrome CS slide reveal equal amounts of red, green, and blue. One of the problems associated with the use of an additive color screen is that the base density of the film tends to be much higher than that of conventional slide material. Because of its higher density, measured at 0.7 compared to conventional slide films' 0.2, the brightness of Polachrome is considerably less. This brightness differential is most noticeable in the highlight areas (the toe of the characteristic curve) and least in the midtones (straight-line portion of the curve) plus shadow areas (shoulder of the curve).

Since image formation of the four black-and-white films takes place without an additive color screen, their base densities are very similar to regular black-and-white emulsions.

Polachrome CS Strengths and Limitations

Strengths

1. Virtually immediate results
2. Film can be used in any 35mm camera
3. Accurate color rendition
4. Better reciprocity performance than conventional color film
5. Very good overexposure performance

Limitations

1. Relatively slow–ISO 40
2. Visible grain, especially in shadow areas
3. Off-the-film metering produces incorrect readings
4. Very delicate film surface susceptible to damage
5. High cost per slide
6. Polaroid Autoprocessor must be used to develop slides

Polachrome HC

This film is a high-contrast version of the now-standard Polachrome CS. It can be used in very low contrast lighting to generate slides with visual "snap."

35mm Instant Black-and-White Films

Polaroid's 35mm black-and-white films are similar in construction to Polachrome, with one notable exception: none of the black-and-white films employs an additive color screen. The processing of these films is the same as the color versions, although the black-and-white chemistry is somewhat different.

PolaPan CT

This is a medium-speed, black-and-white slide film rated at ISO 125. The film is a full spectrum, panchromatic material that delivers positive, general-application slides with exceptional tonal range and good resolution qualities. Contrast can be modified by altering processing temperatures, with lower temperatures producing increased contrast and higher temperatures less contrast.

PolaGraph HC

Polaroid's high-speed (ISO 400) film is a high-contrast black-and-white slide film intended to photograph low-contrast subjects for copy work or other assignments. The film is processed in much the same way as the other 35mm Polaroid instant films.

Polaroid Characteristic Curves

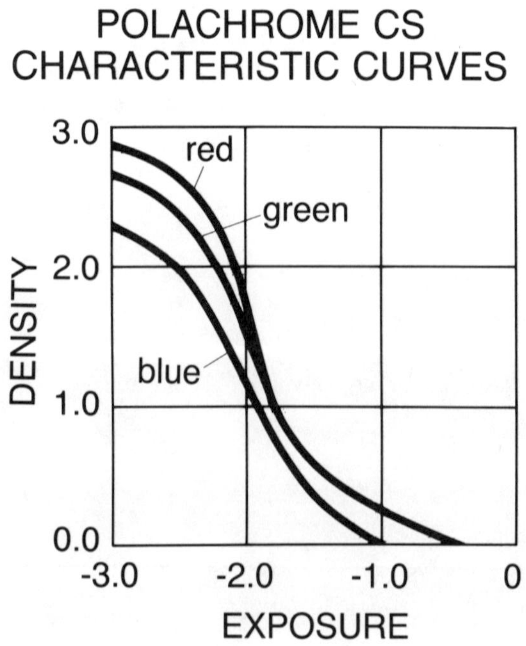

Figure 3.3

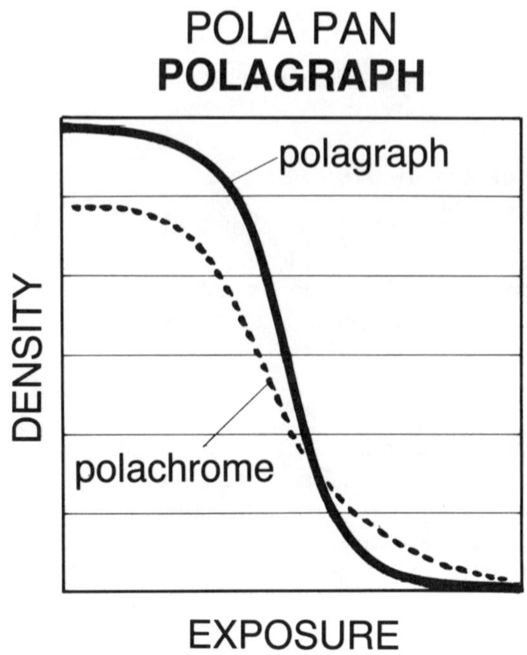

Figure 3.4

PolaBlue BN

PolaBlue is a unique, relatively slow (ISO 40) film. The film creates a blue background, white-lettered slide of typed or printed pages. It is orthochromatic, meaning it is not sensitive to the red portion of the spectrum. Because of this limited sensitivity, red or black subjects are rendered white, while all other colors turn out blue.

ISO Daylight Tungsten

Temperature (°F)	*Daylight*	*Tungsten*
POLACHROME		
100	64	50
90	50	40
80	50	40
70	40	32
60	50	40
50	64	50
POLAPAN		
100	250	160
90	250	160
80	160	100
70	125	80
60	125	80
50	160	100
POLAGRAPH		
100	320	250
90	320	250
80	400	320
70	400	320
60	400	320
50	500	400

Metering Using Polaroid Instant Films

For cameras equipped with off-the-film flash metering, Polaroid Polachrome, PolaPan, and PolaGraph films will give incorrect readings because their emulsion surfaces have different reflectance qualities than conventional films. For normal ambient off-the-film light metering, consult specific camera manuals. For cameras employing off-the-film metering for autoflash operation, compensate using the chart below or the camera instructions.

Guide to Setting ISO Dial on the Camera

Camera Type	*Polachrome Film*	*PolaPan Film*
Minolta X 700	320	1000
Nikon F 3	Use Flash Sensor	
Nikon FE-2	800	2500
Nikonos V	160	500
Olympus OM-4	200	640
Pentax LX	200	640

Polaroid Presentation Chrome Film

Presentation Chrome is a medium-speed (ISO 100), extremely high contrast 35mm color transparency material. It is intended for the reproduction of transparencies off CRT video screens, production of title slides, and copying original two-dimensional art work. It reproduces all color very well and does a specially good job rendering whites.

Unlike most current medium-speed color slide films that possess high color contrast but moderate image contrast, Presentation Chrome is both a high color-contrast *and* high image-contrast emulsion. Great care should be taken when using this film in contrasty light because shadows block up readily, obscuring shadow detail. Also, the film has narrow exposure latitude—less than ½ f-stop over- or underexposure. When used in appropriate situations, results can be spectacular. Presentation Chrome is not available in most camera shops. It is marketed by Polaroid Office Products Group. To find out where you may obtain this product, call Polaroid Consumer Line, 1-800-343-5000.

Tungsten Film

When we view common objects under different light sources, our vision makes subtle adjustments so that white appears white, black black, and other colors their expected hues, regardless of the light source. Color photographic film has no such capability, although Fuji Film and others have introduced special color emulsions that go a long way toward offering us a universal color film that will render color realistically under any light source.

Most of us realize that tungsten lighting in the form of studio photo floods or ordinary house bulbs emit light rich in orange and red. The illumination produced by each of these sources is known as warm light. Contrasted to this, sunlight and electronic flash tubes give off light generically termed "daylight." This type of light is full-spectrum, but at certain times appears blue-rich and therefore cool. When taking pictures under specific light sources, color film must be matched to the light source to produce accurate hues. We can accomplish this in two ways; one involves filtering the light to match the spectral response of the film. This technique will be discussed in detail in a subsequent chapter. The second way we achieve accurate color is to match the spectral response of the color film to the light source.

Our eye adjusts for different light sources automatically. Film does not. For example, shooting with daylight-balanced film under tungsten lighting results in pictures displaying an overall orange cast. An orange color rendition might be perfectly all right and even desirable on some occasions, but usually it isn't. As stated earlier, we can use a filter to address the imbalance or switch to a film matched to the light.

Color film marked "daylight" is intended for use under natural daylight or electronic flash. Film labeled "tungsten" is normally employed for indoor situations where the principal light sources are either floods or other warm tungsten-light sources.

The color quality of light and film is defined by a system where each light source and the color balance of film is identified by its color temperature. What exactly is color temperature? Many years ago Lord Kelvin, an English scientist, invented a method designating color temperatures on a scale that has become known as the Kelvin scale. The specific temperature numbers on this scale are derived from actual wavelengths of light emitted from a dead black, nonreflecting material that physicists term a "black body radiator."

As a black body radiator is heated, it generates light: dull red, then a more brilliant red, orange, yellow, and eventually, an intense blue. Visually matching the color of the black body radiator as it is heated to the color of a specified light source and then measuring the temperature of the black body radiator at this point gives us the color temperature of the light source on the scale beginning from absolute zero. Color temperature is annotated as degrees Kelvin. So, daylight film is labeled 5500K, tungsten film 3200K. (See chapter 1 for the chart listing color temperatures of common light sources.)

Indoor red-rich light has a low color temperature compared to blue-rich daylight. This seeming contradiction occurs because tungsten films designed for use with warm, red-rich lighting have smaller Kelvin numbers than cooler daylight films balanced for blue light. This situation is a constant source of confusion to many but it is the system with which we must live, at least for the present.

We should understand the terms *warm* and *cool* are purely psychological descriptors, while the numerical designations of color temperature as expressed in degrees Kelvin are more correct scientifically.

As we previously learned, color film is most often used under compatible lighting. In order to provide photographers with color films that deliver realistic color under light other than daylight, major film suppliers make a number of color films, almost all of which are transparency materials whose color response is balanced for tungsten light. If you need to match color exactly, you will need a color temperature meter, an instrument that measures prevailing light and reads out its color temperature. After taking a color temperature reading, placing one or more special color-correction filters over either the light source or the taking lens should balance film and light source precisely.

At the cool end of the visible spectrum, color accuracy is a bit easier to achieve because large changes in color temperature do not result in as much of a color shift as at the lower end of the color temperature scale. Aware that color temperature meters and color compensating filters are both expensive and cumbersome, manufacturers make several slide films formulated to give normal color balance under several sources of tungsten light. There are a half-dozen such films that range from ISO 50 to 640. All but two are balanced for studio floodlights burning at 3200K. They are designated Tungsten Type A. One film, Kodachrome 40, balanced for infrequently used 3400K photolamps, is designated Tungsten Type B. Fuji has chosen to balance their tungsten films for a 3100K light source. They believe 3100K is closer to the actual color temperature of photo floods.

Tungsten film is useful under ordinary house lights although 60–150 watt bulbs burn at between 2300–2800K, which is warmer (more orange) than photofloods. The resulting pictures will be warm but not always objectionably so. Tungsten film can be used in daylight with an 85B filter for a more normal color rendition.

Other Specialized Films

Photomicrography Film

Kodak micrography film is normally used for medical and scientific imaging and not usually for general photography. The film does have several intriguing properties that make it a natural for the creative experimenter.

To begin with, it is an extremely fine grained emulsion, finer in fact than almost any other color film. Next, the film possesses very high color saturation of all colors but especially orange and red. Further, the film delivers very high image contrast, making it a natural for stark, graphic effects.

There are a few drawbacks associated with photomicrography film. At ISO 12 it is very slow. When using the film without a filter, you will notice a pronounced magenta color cast in the processed slides. Last, the film remains one of the few color slide emulsions that must be processed in the long-discontinued E-4 process. Chemistry is no longer supplied for home processing, but a few custom labs around the country are still able to produce finished slides.

Using photomicrography film is always an adventure. You can never be sure of what colors you will get. Using a Wratten CC 20 or 30 G color compensating filter will get you in the color ballpark but that's all. On overcast days, colors—and in particular, red—explode off the slide. Even the film's weird color balance without CC filters can give you interesting effects.

Respooled 35mm Movie Film

Several types of 35mm movie film stock are being respooled into standard 35mm cartridges after having a second edge perforated with sprocket holes. The films are all color negative emulsions and are supplied in several speeds ranging up to ISO 400. At least three manufacturers supply these modified movie films. All are characterized by high sharpness, fine grain, moderate contrast, very good color, and neutral rendition. This film, which is the surplus from movie-makers, is quite inexpensive. In the United States, a number of suppliers specialize in selling and processing these films. These companies offer processing with negatives only, negatives and prints, or negatives, prints, and slides. From what we can judge, the best of these services will give you competitive results: sharp, highly saturated prints along with professional quality slides. I add a word of caution: film stock such as Kodak Type 5247 is almost always surplus. It may be old and out of date with all the attendant problems that produces. If out of date by a significant margin, it is not much of a bargain. Consult Appendix B of this book for our opinion of processors currently supplying this service.

Kodak Data Recording Film Type 2475

If it is black-and-white grain you are after, then data recording film is for you. The film is normally rated at ISO 1000 and at this speed delivers low-contrast negatives. Uprating to ISO 2000 or higher increases grain and contrast. Data recording film will never win any fine-grain or sharpness contests. It is significantly grainier and much less sharp than T-Max P3200 when each is exposed at its rated ISO. With the advent of P3200, recording film has little practical value other than experimentation with abstract exaggerated grain.

Slide Duplicating Films

All duplicating films are supplied as transparency material. They possess low contrast, exceptional grain structure, and wider-than-normal exposure latitude. However, they are also very slow, rated at about ISO 12 or lower. When making high-quality duplicates, be aware that exact reproduction of an original is just about impossible. However, excellent duplicates that are even improved versions are not too difficult to manage. And you don't need sophisticated equipment, either. The actual procedures for slide duplication go beyond the scope of this book, but if you have some means to make 1:1 or greater magnification images and a 3200K light source, you should be able to make good dupes. Granted, higher tech solutions are available; however, they will not always ensure better results. Currently, there are four duplicate films on the market. Three are made by Kodak and one by Fuji. Kodak films differ from one another in several ways. Ektachrome Type 5071 is a film intended to be exposed under tungsten light, and its sibling, Type SO-366, is daylight-balanced. In most other ways the two films are alike. Kodak also makes a duplicating film, Type K/8071, specifically intended to duplicate Kodachrome originals. This film has been found a necessity in dealing with the image-forming dyes used in Kodachrome, dyes very different from those used in Ektachrome and similar E-6 films. Essentially, Type 8071 duplicating film closely matches the spectral response of Kodachrome originals. When duplicating other makes of film, such as Agfa, Konica, Polaroid, or Scotch, a duping film intended for E-6 films is preferable to Type 8071. Fuji Film also offers a slide duplicating film, Type CDU. This emulsion is an E-6 film for use under tungsten illumination, and has an extended slow exposure time tolerance reaching all the way down to twenty seconds. Fuji Film originals and CDU are perfectly matched to one another but any other E-6 film original can be successfully reproduced with this film.

As with all other duplicating films, use of an ultraviolet-absorbing filter, such as Wratten 2B, is strongly advised. Like its Kodak counterpart, Type 5071, Fuji CDU produces excellent quality duplicates, featuring high definition and excellent color purity.

Altering the Color Balance of Duplicates When viewing original slides, you may decide to alter the color balance of the duplicate. In all but very special situations, the overall balance of the dupe must be changed. Use the following table as a starting point to determine the proper filter pack for the results you desire.

If the color balance is too:	*Add these filters*	*or*	*Subtract these filters*
Yellow	Magenta and Cyan		Yellow
Magenta	Yellow and Cyan		Magenta
Cyan	Yellow and Magenta		Cyan
Blue	Yellow		Magenta and Cyan
Green	Magenta		Yellow and Cyan
Red	Cyan		Magenta and Yellow

Vericolor Slide Film 5072/SO-279

Vericolor Type 5072/SO- 279 is an emulsion designed to generate positive color transparencies from color negatives and internegatives. The film is tungsten-balanced at 3200K. It has exceptionally fine grain and it is capable of resolving extreme detail. The film is available in 35mm 36-exposure cassettes and 100-foot-long rolls.

Starting Filter Packs for Printing Negatives onto Kodak Vericolor Slide Film

Film Type	*Starting Filter Pack*
Ektar, Ektapress Gold, Vericolor, Kodacolor Gold	CC20M and CC30Y
Vericolor II Type L	CC20M and CC25Y
Vericolor III Professional, Vericolor 400	CC20M and CC35Y
Vericolor Internegative	CC50M and CC35Y

Chromogenic Black-and-White Negative Film

As of the writing of this book, there remains only one film in this category, Ilford XP-2. Nominally rated at ISO 400, this unique black-and-white film is routinely exposed at EIs of between ISO 50 and 800 on the same roll and then processed normally. It is processed in C-41 color print film chemistry. The difference between XP-2 and other black-and-white films is that with XP-2 you can mix shots taken at vastly different exposure indices and still get very printable negatives from the entire roll. In extreme cases, EIs of ISO 1600 are possible by extending development times. Understand though, that at these high ISOs, there will be some quality loss.

Think of chromogenic film as a color negative emulsion minus color couplers and color dyes. Instead, the film contains a normal silver halide record layer and a special dye coupler that when developed gives a silver less-dye-generated image. Instead of grain producing silver, chromogenic films are composed of smooth monochrome dye clouds that form the image. The image seems less grainy in those areas of the negative where one would normally expect to find high granularity. Overall, it appears less grainy than comparable conventional black-and-white films. Exposing XP-2 at different ISOs changes negative contrast. If you expose the film at ISO 50, 100, or 200, you will get finer grain and less contrast. Conversely, uprating the film increases contrast modestly but apparent grain remains very fine, finer in fact than almost any other ISO 400 black-and-white film.

Is there a downside? Sort of. Chromogenic film must be processed in color print chemistry, C-41 or its equivalent. This chemistry demands much tighter temperature control than conventional black-and-white developers. Also, Ilford makes its own XP-2 developing kit that gives marginally better results but at a higher price and with less convenience. Finally, this film is only to be found in very well stocked photo shops. For reasons I do not understand, this first-rate film has failed to gain universal acceptance. It does have a substantial group of adherents, myself included, who swear by it.

Lithographic and Line Films

Line or "lith" films comprise a group of ultra high contrast black-and-white emulsions capable of delivering stark graphic renditions of full tonal range subjects in which the visual elements of a subject are reduced to pure blacks and whites with all middle grays eliminated. Most film-makers offer one or more lith films, but few are produced in the 35mm format. Lith or line films are employed by graphic artists to give them copies of artwork, as the basis for separations, for silhouettes, or as part of color or monochromatic sandwiches.

One common problem with these films is their tendency to form pinholes in the black portion of the negative. The condition is easily rectified by spotting the pinhole with a paste-like substance called *film opaque.* Line films are almost always orthochromatic, meaning they have greatly reduced sensitivity to red. This does not present any real problems other than when red subject matter constitutes a significant percentage of the image. Also, these films are relatively insensitive to the red portion of the spectrum, so they can be processed by visual inspection under a weak red safelight.

Kodak Technical Pan Film Type 2415

"Tech Pan," as it is known, is credited with the finest grain and highest resolution of any 35mm film, color or black-and-white. Resolution figures of 400 ll/mm are possible. The film has extended red sensitivity resulting

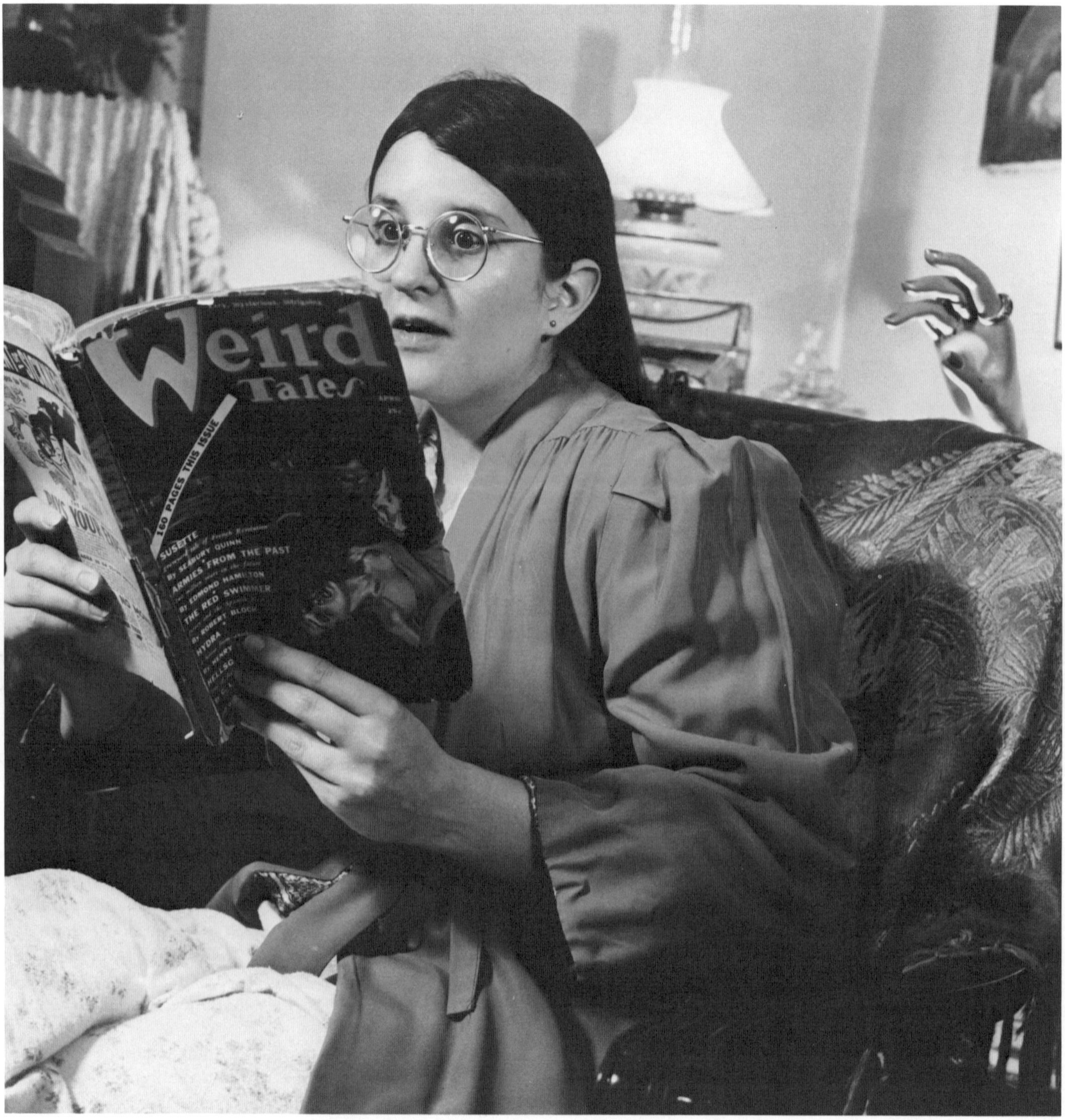

Figure 3.5 Robert W. Bissett. Plus X ISO 100.

in enhanced contrast negatives especially useful for landscape and still-life photography. It is not a film to be used as a general-purpose emulsion for it exhibits very high contrast and has a low emulsion speed of ISO 25. Also, it must be developed in its own special developer, Technidol LC or Technidol Liquid. There are a few other non-Kodak developers that work with the film as well. In use, super-detailed enlargements up to 50× magnification are possible using this film.

If you prefer very high contrast images similar to those made from lith films, develop Technical Pan film in Kodak D-19 or Dektol.

Kodak Direct Positive Pan and Similar Direct Transparency Films

These slow- to medium-speed emulsions are very fine grain films directly generating black-and-white transparencies. Graphic artists and AV specialists use these and the Kodak Positive Film Developing Outfit to make high-resolution black-and-white slides of line art, technical drawings, or text to be used in slide presentations. The films can also be used for general-purpose photography.

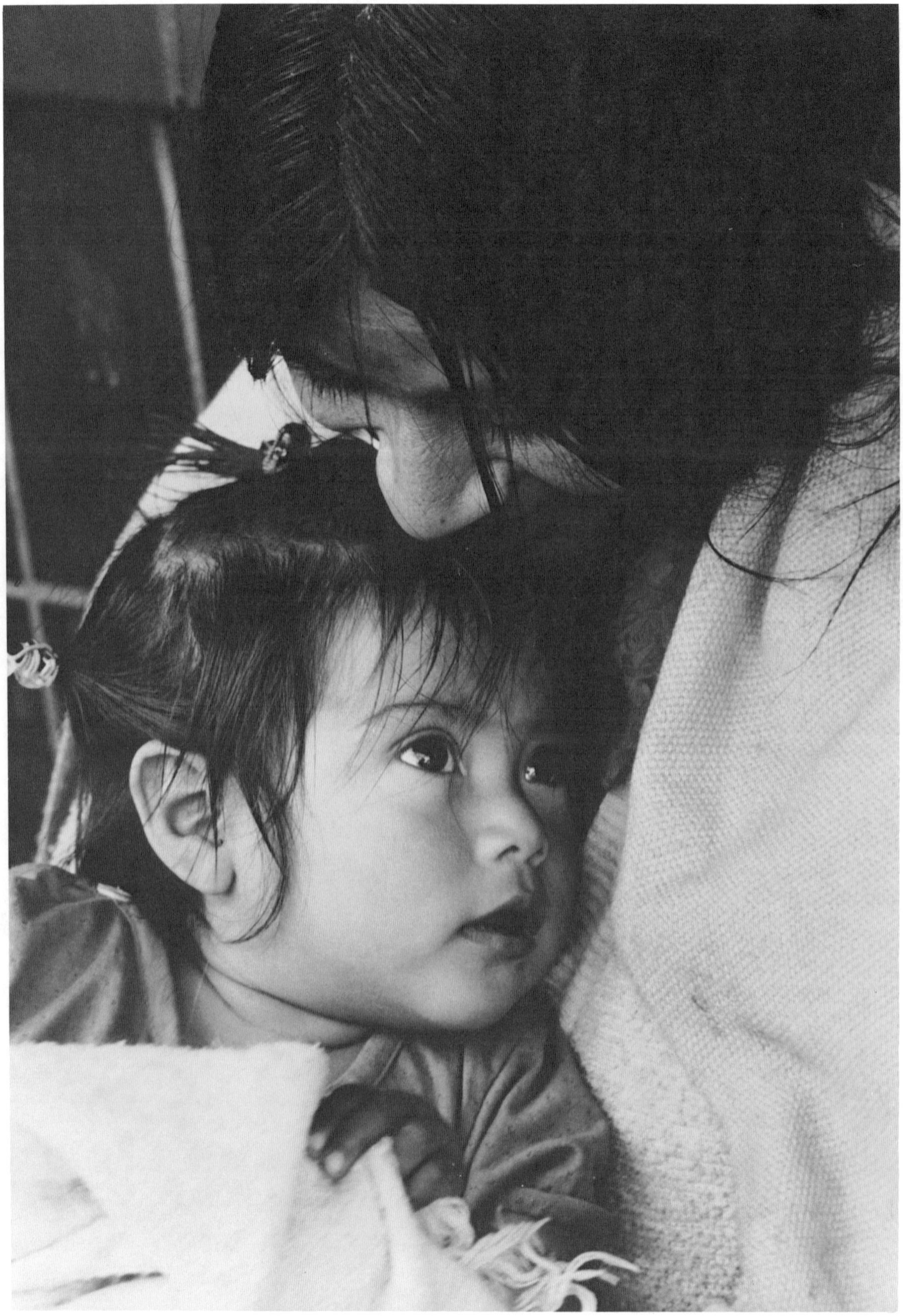

Chapter 4

Filters

All filters employed in color photography alter the color balance of film or the quality of the light. Filters are also used to change tonal relationships in black-and-white images. While most photographers use filters at least occasionally, there remains a debate over their use. Purists argue that one should never place anything between the prime lens and the subject to be photographed. On the other hand, "image modifiers" contend that all is permissible in the name of creativity, even at the risk of an image becoming a cliché. Prudent image-makers fall into the great middle ground between these two extremes. My belief is that one should not try to save mediocre pictures with gimmicks. Gimmicks save nothing! Following this line of reasoning, I suggest the sparing use of filters and that this use should only follow ample forethought. Filters can, however, solve problems, add emphasis, underscore intentional visual statements, and allow for personal interpretation.

How Filters Work

All filters absorb portions of the visible electromagnetic spectrum and pass through (transmit) the remainder. Universally, filters transmit light of their own color. In some instances, they also transmit some of the light of most immediately adjacent colors while at the same time absorbing the other colors.

Colors as we see them are rarely pure. Instead, they are an admixture of two primaries. For instance, yellow is a mixture of equal parts of red and green. If you place a green filter over your lens, it will pass the green component reflected from a yellow subject (such as a lemon) and block the red. Using a red filter when photographing the same lemon will result in red being passed and green being blocked.

It may be useful to go back a bit to examine the phenomenon we call *light* in more detail. There are two main types of light; daylight in its many forms, and artificial light. Sources of illumination run from the primitive light generated by a burning fire to huge and powerful manmade sodium vapor floodlights.

Daylight

As a first principle, daylight is not uniform in color throughout the day. It changes dramatically from hour to hour, beginning as a bluish glow immediately before sunrise, moving to a rich golden hue just after the sun rises above the horizon, then changing once again to a cooler blue as it moves directly overhead, and finally becoming progressively warmer to-

Figure 4.1 Margaret Duda. Ilford XP1/XP2 ISO 400.

ward sunset. Immediately after the sun sets, the sky again takes on an ever-deepening blue cast until all visible light disappears. On a clear day, the atmosphere can look totally colorless. In fact, clear skies contain all the colors of the visible spectrum more or less evenly distributed. In addition to visible colors, the sky reflects infrared and ultraviolet light waves outside the visible spectrum. Even with all this, the sky can still seem totally devoid of color. Everything changes if we alter any component in the color equation. Nature changes this balance as the sun passes through the sky and so can we, with photographic filters.

The Composition of Sunlight

Sunlight is composed of a combination of three visual primary colors; red, green, and blue. Combining two of the three primaries will produce three new colors, the secondaries. Color secondaries are important to photographers in two ways. First, they are frequently found occurring naturally, and second, the secondaries are the colors of the dyes found in all color film.

Secondary colors are derived in the following way:

magenta—equal parts red and blue.
cyan—equal parts green and blue.
yellow—equal parts green and red.

Any color can be generated by altering the proportions of the three primaries. Likewise, photo chemists have the capability to make a color film with any color emphasis simply by altering color dye ratios.

Filter Factors

As stated earlier, practically all filters absorb some light. Absorpsion is not uniform across the entire spectrum as each filter will absorb or pass certain portions of light depending on light source (daylight or tungsten) and color and intensity of the particular filter being used. Bear in mind that using a filter other than one that is clear or nearly clear will alter the color and intensity of the specific filter being used. Also, remember that other than for clear or nearly clear filters (skylight and most UV filters), attaching a filter to a lens will require an exposure increase. Although modern through-the-lens exposure meters automatically compensate for light absorption, the resulting exposure decrease will need to be taken into account. Suppose you attach a polarizer to your lens. The subsequent light loss will be about 1⅓ f-stops. The practical effect is very important. A 1⅓ f-stop loss translates into slower shutter speed, for larger f-stops, or some combination of the two. In marginal light, the new slower shutter speed might preclude hand-held photography. It is also possible that opening the lens to a larger f-stop so diminishes depth of field that important detail is lost to image blur. There are several solutions to the scenario just presented. You may opt for the use of a tripod, switch to a faster film, introduce some form of supplementary lighting, or, as suggested before, split the loss between aperture and shutter speed. The point to consider is that if you plan to use filters at all you should anticipate their effect on exposure and what that means to the look of the final image.

A *filter factor* is defined as a number by which an exposure value should be multiplied to account for the light absorption of the filter in use. Applying a filter factor will result in a correct exposure in cases where a built-in through-the-lens meter is not used. The following chart and Table 4.1 address filter factors and light loss.

Sequence of Filter Factors and Subsequent Light Loss

Filter Factor	*Daylight Loss in F-stops*
1.0	0
1.25	–1/3
1.5	–2/3
2.0	–1
2.5	–1-1/3
3.0	–1-2/3
4.0	–2
5.0	–2-1/3
6.0	–2-2/3
8.0	–3
10.0	–3-1/3
12.0	–3-2/3
16.0	–4
20.0	–4-1/3

Filter Designations

Several years ago, photographic filters were redesignated. Presently, filters bear numbers indicating their specific type. This system has been all but universally adopted. A uniform numerical system became necessary when dozens of new filters were introduced in a short span of time. You may still find old or even dual designations on filters. Refer to the following cross-reference table for the most up-to-date nomenclature. Note: some designations remain the same.

Filter Color	*Old Designation*	*Current Type Number*
Skylight	1 A	1 A
UV	UV	UV
Light yellow	K 1	6
Yellow	K 2	8
Deep yellow	K 3	9
Orange	O 2	21
Green	G	58
Blue	C 5	47
Red	A	25
Polarizer	PL	PL
Neutral density	ND	ND

Filter Configurations

Filters come in four basic forms and can be purchased or cut in round, square, rectangular, or odd shapes on special order. The four configurations are as follows:

1. Thin, square gelatin filters that require a holder and lens adaptor.
2. Number series sized round glass filters needing a retrainer and lens adaptor to fit onto a lens.
3. Threaded glass or plastic filters that screw directly onto the front of a lens.
4. Plastic or resin squares that fit a holder and a screw in the lens adaptor.

Gelatin Filters

Pioneered by Eastman Kodak, gelatin filters are made from very thin gelatin film that is less than 0.1mm thick +/- .01mm. They exhibit excellent optical properties and are uniform from filter to filter. Because of their out-

Table 4.1 Filter Factors and Light Loss for Commonly Used Filters

	Daylight		*Tungsten*	
Filter Number	*Filter Factor*	*Light Loss*	*Filter Factor*	*Light Loss*
1A Skylight	1.0	0	1.0	0
UV-0	1.0	0	1.0	0
3 Light Yellow	1.5	–2/3	1.0	0
4 Yellow	2.0	–1	1.5	–2/3
9 Deep Yellow	2.0	–1	1.5	–2/3
21 Orange	3.0	–1-1/3	2.0	–1
58 Green	6.0	–2-2/3	6.0	–2-2/3
47 Blue	6.0	–2-2/3	12.0	–3-2/3
25 Red	8.0	–3	5.0	–2-1/3
Polarizer	2.5	–1-1/3	2.5	–1-1/3
80A	3.0	–1-2/3	—	—
80B	2.0	–1	—	—
85	1.5	—	1.5	–2/3
85B	1.5	—	1.5	–2/3
Neutral Density	Varies with filter density			

standing optical performance and thinness, the filters take little if anything away from prime lens performance, that is, if they are handled and used with extreme care.

"Gels," as they are usually called, are available in a wide range of colors and in very useful color correction sets we will discuss shortly. Gel filters come only as thin 75 × 75mm, 100 × 100mm, 150 × 150mm, and 350 × 450mm rectangles and will require a filter holder such as a Kodak Gelatin Filter Holder, a filter frame holder, and a threaded ring to which the preceding components can be attached. The entire rig is then screwed directly onto the front of a lens.

Gelatin Filter Sizes

Metric	*Approximate U.S. Units*
75 × 75mm	3 × 3 inches
100 × 100mm	4 × 4 inches
150 × 150mm	6 × 6 inches
350 × 450mm	14 × 18 inches

Gelatin filters are economical because you only need one complete set of attaching components for all the filters you will ever use. Changing from one lens to another involves only one additional component purchase, the specific size threaded lens adaptor.

For all their virtues, gelatin filters have some serious drawbacks. While the entire system *is* economical in the long run, the initial purchase of large numbers of gels adds up to a considerable outlay; each filter costs between ten and fourteen dollars.

Also on the minus side is the realization that gels are the most easily damaged of any filter type. They are coated with a very soft lacquer that abrades easily and is difficult to clean. They do not stand up well to even the slightest mistreatment. If you intend to use gels, plan to replace them at regular intervals, particularly if you use them away from the studio.

Series Sized Filters

This is a filter type usually supplied as round solid glass or gelatin sandwiched between two pieces of optically clear glass. The filters have no threads by which they can be directly attached to a lens. Instead, they are attached by means of a holder/threaded adaptor ring system. Obviously, lenses with different diameters will require separate threaded adaptor rings. Although series filters themselves may be less expensive than the threaded variety, they do require a lot of support hardware that raises the overall cost and is an aggravation, making series filters many times more trouble than they are worth. Also, series filters are supposed to be used only with lenses of specific front diameters. Beyond these ranges, vignetting is likely to occur. Refer to the following chart for series filter coverage.

Lens Diameters Safely Accommodated by Series Sized Filters

Series Size	*Lens Front Diameter*
5	Up to 36mm
6	31 to 47mm
7	44 to 56mm
8	51 to 69mm
9	62 to 85mm

As you can plainly see, there is considerable overlap among filters of different sizes, so if you use lenses with 52, 55, 62, and 67mm front thread diameters, you will only need to employ series 8 filters and appropriate hardware. Attempt to use a lens with a 72mm filter thread and you risk serious edge darkening or vignetting. It is of interest to note that series filters have fallen out of favor, and have been replaced by other, more convenient systems.

Threaded Round Filters

Without doubt, the most universally adopted filter configuration is the round all-glass or glass-laminate, threaded to fit specific size lens diameters. Screw-in filters are available in an endless assortment of types and sizes from 39mm to 82mm. A few special professional filters are also offered in smaller and larger sizes. Filter suppliers produce a full range of products in the following millimeter sizes: 43, 46, 49, 52, 55, 58, 62, 67, 72, 77, and 82mm. Additionally, filter makers like Tiffen and B+W will custom-make filters in almost any size.

Glass filters have the advantage of being very durable, handy to use, and at their best, of high optical quality. They are made in one of two ways. Either they are mass-dyed, or a color is added to a special bonding material sandwiched between two pieces of optically clear glass. Presently, only Tiffen employs the bonding technique in the manufacture of photo filters. In either case, all-glass filters are the most abrasion-resistant and easiest to clean. They are also by far the most popular filter configurations.

Be warned! Screw-in filters may all look alike but vary greatly in a number of significant ways. The best screw-in filters are mounted in sturdy, smoothly threading brass mounts and provide years of trouble-free service. Others that might look the same superficially have mounts constructed of soft aluminum or even plastic. Neither of these materials is as precise or as sturdy as brass.

Another problem arises when filters are not mounted exactly plano-parallel to the front lens element. If tilted ever so slightly, they can distort incoming light rays, causing or exaggerating lens aberrations, and resulting in serious losses in sharpness and contrast. This problem is most often found with inexpensive filters but is sometimes found in the most expensive brands as well.

Some filters will not accomplish what their designation indicates they should be capable of doing. For instance, if the dye used to color a filter intended to be used as an 80A is not uniform throughout the glass, the filter will probably not convert daylight to 3200K evenly enough to give natural-appearing color. Whites, grays, and skin tones could be off enough to create real problems. I urge you not to engage in false economy. Let's use the universal skylight or UV filter as a case in point. It's true that you can purchase a skylight or UV filter for as little as $2.99. On the other end of the price spectrum, you'll need to put out about $10 or $12 for a really first-class filter and a good deal less than $20 for any one of several top-of-the-line brands. For most of us, filters represent a one-time purchase. It is only when we change lens systems or camera formats calling for radically different filter sizes that we buy new ones. To my way of thinking, it makes sense to buy the very best filters you can afford. The difference between

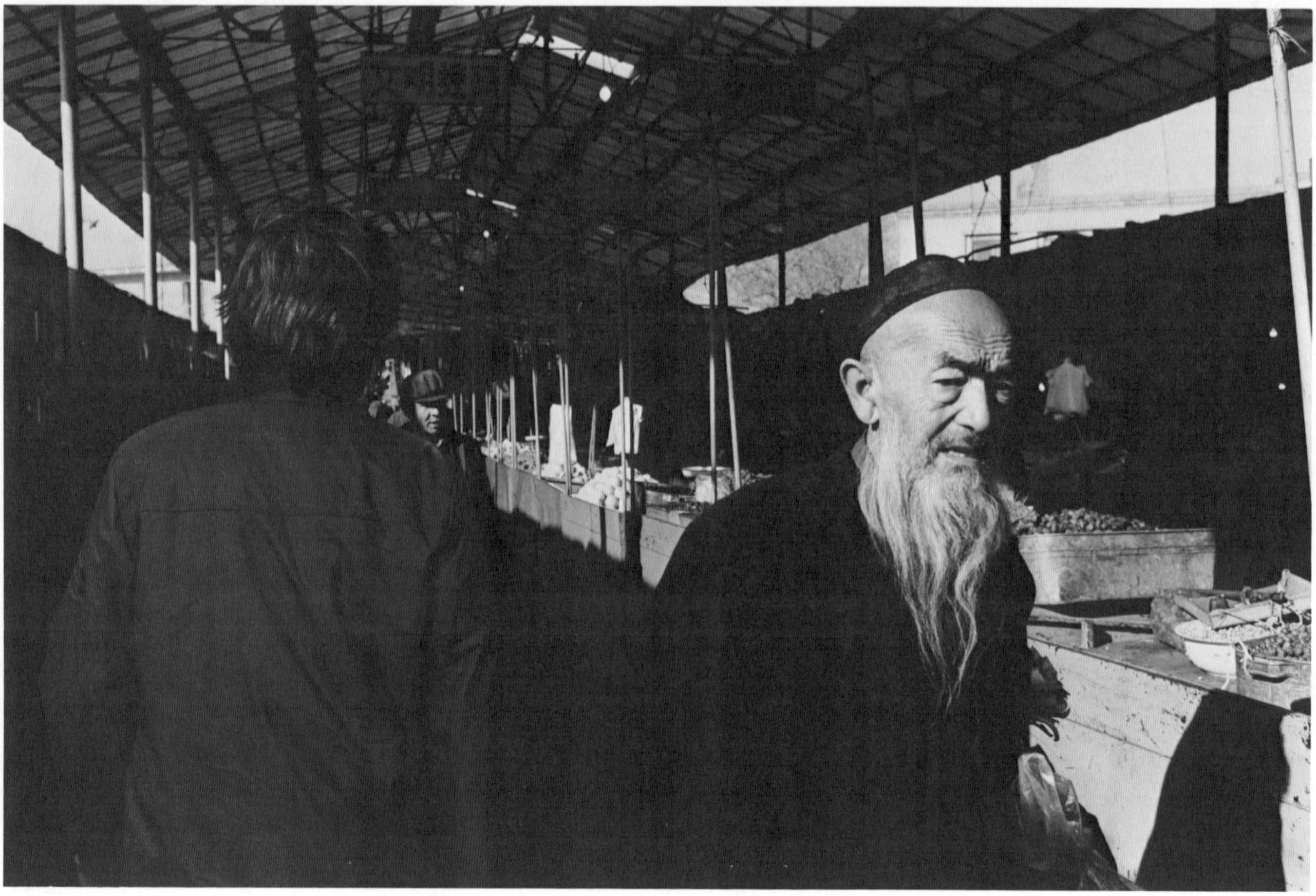

Figure 4.2 Margaret Duda. Ilford XP1 ISO 400.

$2.99 and $12 or even $20 is not so significant. On the other hand, when contemplating the purchase of other popular filters such as circular polarizers, the price range expands greatly. There is a good deal of difference between a decent $20 polarizer and one costing $80. Most of us would consider $80 very carefully even if it's a once-in-a-lifetime purchase.

Again, I advise spending a little extra—to at least the middle price range—on filters you plan to use extensively. This is particularly true for filters designed to hold the lenses' sharpness and contrast. For ones meant to distort, soften, heavily color, or in other ways modify image quality, the issue of filter quality is not so critical. It has been my experience that, for the most part, filter quality and cost go together—to a point. There is a tremendous performance differential between low- and top-end filters. This gap is less significant when comparing performances of mid-range filters and their higher priced competition. What we get when we pay a premium for a top-line filter is a heavier and more precise filter mount and the probability of better quality control in manufacture and assembly. One premium-filter maker states flatly that every one of their products is tested individually before being shipped to distributors.

Step-Up and Step-Down Rings

One problem with screw-in filters is that it might be necessary to own more than one set of filters to accommodate lenses with different front thread diameters. Just as Nikon bayonet-mount lenses won't fit on Minolta camera bodies, so 55mm filters will not fit on lenses with 62mm front threads. Within limits, there is a simple solution to this problem—step-up or step-down rings. These lifesavers are nothing more than thin, double-threaded aluminum rings with a front female thread of one standard filter diameter and a male rear thread of another. Let's say you own a number of 55mm filters and you purchase a new lens calling for 49mm filters. Just obtain a 55mm-to-49mm step-down adapter and all your 55mm filters can now be used on the new lens. The only possible problem with stepping rings comes if you wish to use a smaller filter on a lens with a larger front diameter. Using a 49mm filter on a lens with a 55mm front diameter could lead to vignetting or darkening of the corners and edges of the picture. The problem is pronounced with wide-angle lenses.

Filter Care

No matter how good the lens or the filter attached to it is, dust, dirt, fingerprints, grease smudges, and the like play havoc with picture quality. A really dirty lens or filter will sometimes account for up to a 50 percent loss in lens sharpness and contrast. Also, dirty filters often cause visible focus shifts, again destroying image sharpness.

On more than one occasion, I have observed serious photographers gingerly and meticulously clean the front element of their prized lens, only to reach into their camera bag, grab a loose filter, in one swipe wipe it off with their shirttail, and screw it onto the lens. It is these very individuals who come to me weeks later complaining bitterly about what they consider to be unacceptable images and ask for explanations. The truth is that for some inexplicable reason almost all of us operate as if filters are not part of the optical picture-taking system. We don't treat them in the same way we do prime lenses. We treat filters as we treat our sunglasses—once over lightly. I urge you to adopt the habit of cleaning filters as you clean lenses. Otherwise, don't bother using them at all.

Cleaning, Handling, and Storage of Filters

As a general rule, glass or plastic filters should be treated exactly the same as prime lenses. Only use lens-cleaning tissue expressly made for optical surfaces. Kodak lens-cleaning tissue and fluid or similar products are highly recommended. If you see the lens surface is dirty or dusty, first blow the particles away with an air blower or commercial canned air product expressly made for the purpose. To check the condition of a lens or filter surface, tilt it in various directions so reflections from the surface don't obscure your examination. Never use alcohol! Do not apply the cleaning solution directly to the lens surface; rather, place a small drop on the tissue and use the moistened tissue to clean the glass surface. After applying the fluid to a ragged edge of the tissue, clean in a gentle circular motion until there is no visible trace of the cleaner, dirt, or grease. The key to effective cleaning lies in maintaining a soft touch. Do not apply elbow grease, please. Remember, the cleaning fluid is a mild solvent so it will do its job with little physical effort from you.

If a filter is to remain on the lens, use a lens cap. If not, return it to its storage box or pouch immediately after it is removed. When working in an extremely dusty climate, such as those existing in Africa, parts of Asia, and the southwestern United States, for example, it is wise to invest in a tiny battery-powered microvac. I found it a very good investment on a recent photo trip to East Africa. Another caution: avoid using water to clean filters or lenses. Although some filters are edge-sealed, most are not. Because some filters are made by a laminating process, water can cause swelling, dye spotting, and strains in the glass. When you know you will be traveling or working in a geographic area noted for high humidity, think about acquiring special edge-sealed filters of the Kaesermann type.

Care of Gelatin Filters

Gelatin filters are prone to scratching if mishandled. They should be handled by the edges only. Accumulated surface dust or dirt particles should be brushed off very lightly with a camel-hair brush or blown off with optical-grade compressed air. Do not use lens-cleaning fluid or any other liquid cleaner. Inspect gelatin filter surfaces often for dust and embedded dirt. Gels should be stored in a dry, low-humidity environment of no more than 50 percent relative humidity because the gelatin itself is very susceptible to fungus growth. Kodak states that theirs remain stable up 130° F.

Plastic and Resin Filter Systems

The latest innovation in the world of filters, plastic and resin filters and lens attachments, consists of a modular filter holder capable of accepting up to three filters at once. Through the use of a simple threaded adaptor ring that slips onto the holder, the whole arrangement can be easily fitted to lenses of almost any size. These systems are used in still, video, and movie photography.

The Cokin system, marketed by Minolta, is the most extensive product line. Designed by Jean Coquin, it competes with other major lines, notably Ambico and Multifit. We will use the Cokin system as a model principally because it is the most highly developed. The system itself consists of over 200 individual components, almost all of which are made of high-grade optical-quality plastic or resin material. A few filters are still made of the more traditional glass. Filters come in two sizes, labeled *A* or *P*. *A* filters cover lenses with front diameters between 36mm and 62mm. The *P* filters cover the larger lenses, from 42mm to 82mm. Special holders and adaptors for Hasselblad front-lens bayonets are also made.

At the heart of the system are the filters and many creative front-lens attachments. These range from the familiar UV, Polarizers, and color filters to a bewildering array of filters for special effects. They may be intended to soften, refract, reflect, diffuse, get close, multiply images, create zoom effects, imply motion, mask, multiply expose, and on and on. Any of the new system filters can be broken, but it takes some doing. For instance, dropping one from your hand to the floor should not cause any damage at all. They are also scratch-resistant. A few makers claim their plastic filters are much more scratch-resistant than the glass variety. I cannot verify this claim, but I can say they do resist damage very well. Even when the surface is scratched, the filter may remain serviceable because they are dyed in their molten state at which time color dye permeates the entire filter uniformly from top to bottom. Several tests indicate the best of the space-age plastic filters are of comparable quality to good all-glass filters. The plastic filters cost about the same, weigh less, and are more flexible in use. The wide choice of front-of-the-lens attachments is a two-edged sword. Having many possibilities is a good thing: however, be careful that the effect the filter creates does not itself become the subject matter. Using one or more of the accessories will not always lead to memorable images. In fact, many worthwhile pictures are lost in the rush to create a new effect. Used with discretion, filters and accessories like the Cokin system give us unlimited possibilities to create at a reasonable cost. Examine several catalogs and books put out by system filter makers for ideas.

Selected Filters in Use

Polarizers

Polarizers are among the most versatile filters you can own. They perform five basic functions:

1. Remove most reflections from nonmetallic surfaces. Glass, some foliage, water, plastic, and fabric colors all benefit from polarization.
2. Polarizers are the only filters that darken a blue sky and increase contrast between sky and clouds.
3. They increase overall color saturation by removing blocking reflections.
4. They function much like a neutral density filter.
5. They function to reduce overall scene contrast.

Polarizers appear to be made of a grayish transparent material similar to a neutral density filter. Actually, polarizers are made of a myriad of extremely small (microscopic or smaller) crystals oriented like a Venetian blind. The crystals block light waves vibrating at angles other than parallel to the crystals. Light waves vibrating parallel to the crystal array pass between them. Much as the light passing through the slats in a Venetian blind is controlled by changing the orientation of the blind slots, so turning the rotating polarizer in its mount acts to block the polarized light attempting to pass through. The effect on a blue sky is to darken it at certain angles. The most efficient angle is a 33° arc made by the filter and the reflecting surface. To compensate for the polarizers blocking much of the light attempting to pass through it, exposure must be increased. There is some misinformation about how much light is lost. Authoritative sources on the subject agree a polarizer will impose a 1⅓ f-stop light loss no matter how much polarization effect is seen in the viewfinder.

Polarizers have uses other than darkening blue skies and increasing sky/cloud contrast. They increase color saturation of many objects by reducing diffuse reflections on the subject's surface. Employing a polarizer does not change or add color; its use allows color to be recorded as a richer, deeper, and unmasked hue, all but free from color-robbing reflections. It may seem strange, in view of the increases in color intensity of polarized subjects, but these filters can actually reduce overall scene contrast, particularly in scenes containing major areas of blue sky. By darkening a blue sky, a polarizer diminishes its brightness. The highlights represented by the sky and any shadows in the scene are brought closer together with a net result of a lower contrast range.

An added benefit of this contrast-controlling technique is that shadow detail can be given more exposure without overly lightening the sky. In essence, polarizers act much like a moderate neutral density filter and reduce contrast by about one full f-stop. Polarizers can also remove reflections from glass and water surfaces. The amount of surface reflection that is eliminated depends upon the angle of the polarizer to the subject. There are still further uses for these ubiquitous filters. They cut through light and moderate haze by blocking the haze, allowing polarized light to pass through to the film.

There is no one best way to employ a polarizer. The only clear recommendation is not to overuse it. Even though polarized color is often intense, it cannot save an otherwise weak picture. The best way to judge the effect of polarization is by direct observation. To achieve maximum effect, move the camera 90° to the sun. You should be able to observe the filter effect right in the viewfinder as you rotate the filter in its mount.

In black-and-white photography, the filter is used as you would a red contrast filter. Expect significant increases in contrast between the darkened sky and lighter objects in the scene. Polarizers are brought into use to control overall black-and-white scene contrast on bright, sunny days by absorbing and reducing specular highlights, thus bringing all light values in a scene closer to one another.

To find the part of the sky that will be most affected by polarization, point your forefinger at the light source with your forefinger and thumb circumscribing a 90° arc. As you rotate your thumb, it will describe an arc indicating the polarized portion of the sky.

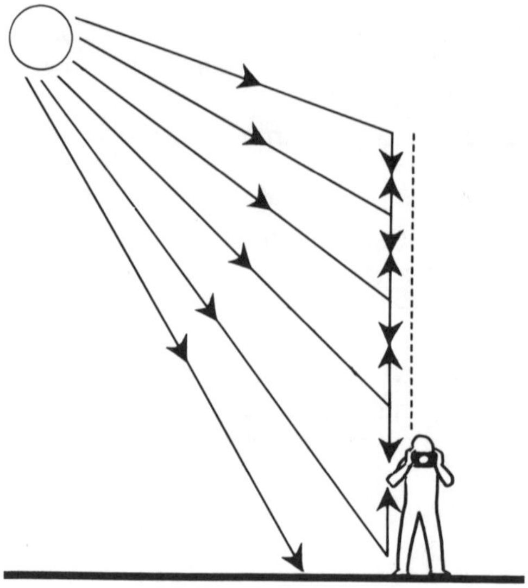

Figure 4.3 With your shoulder pointing at the sun, maximum polarization will be in an arc directly in front and rear of your position.

Neutral Density Filters

Neutral density filters uniformly reduce the amount of light transmitted to the film without having any effect on the color of the image. These filters are calibrated in density units of ⅓ f-stop, with each 0.3 density change equal to one f-stop. They have uses in both black-and-white and color photography. Using N/D filters in black-and-white image-making can alter the distribution of tones and overall scene contrast, while in color work N/D filters allow for the selection of larger f-stops without changing color at all (see Table 4.2).

Specifically, neutral density filters are used:

1. To allow the use of very fast color or black-and-white film under extremely bright lighting conditions such as a sunny day at the beach or in the snow.
2. To reduce required exposure to a point within the range of usable shutter speeds and f-stops.

Table 4.2 Neutral Density Filter Equivalents

Neutral Density	*Light Transmission*	*Filter Factor*	*Exposure Increase*
Blank	100%	0	0
0.1	80	1.25	1/3
0.2	63	1.50	2/3
0.3	50	2.00	1
0.4	40	2.50	1-1/3
0.5	32	3.00	1-2/3
0.6	25	4.00	2
0.7	20	5.00	2-1/3
0.8	16	6.00	2-2/3
0.9	13	8.00	3
1.0	10	10	3-1/3
1.2	6.3	13	3-2/3
1.5	3.2	15	4-1/4
2.0	1.0	100	6-2/3
3.0	0.10	1000	10
4.0	0.01	10000	13-1/3

3. To allow the wider use of slow shutter speeds for such things as image blur.
4. To allow for the use of larger than normal f-stops to increase subject-background separation through shallower depth of field.
5. To allow the use of a constant, preselected f-stop under changing lighting.

One might find the use of neutral density filters advantageous in photographing subjects such as: flowing rivers, streams, and waterfalls. Utilizing N/D filters makes depicting water as a creamy, solid mass much easier. To achieve this effect you will need to use a shutter speed of no more than 1/4 second or, better still, 1/2 second. Panning and zooming are easier with a N/D filter in place over the lens.

A widely accepted use of N/D filters is in candid portraiture and macro photography. Keeping backgrounds simple and unobtrusive helps project the subject toward the viewer. To do this, you will need to blur the background selectively through depth-of-field manipulation. When you are working under bright light, the use of large apertures is severely restricted, especially if medium- or high-speed film is loaded in the camera. One solution to this problem is the N/D filter. Select a N/D filter of the proper density and light will be reduced so you will be able to open up the lens to a point where the background is thrown out of focus.

Neutral density filters are often combined with other filters in split configurations. Usually, the top half is neutral density material and the bottom is any variety of colors or other special-effect filters. Beyond the split configurations, there are also N/D filters constructed to function as variable or graduated units. Graduated N/D filters are useful for controlling overall scene-contrast ratio. This can be accomplished by selectively darkening only one portion of the scene, most often an unusually bright highlight. Bringing the highlight value down compresses the overall brightness ratio (shadows to highlights), thus permitting color film to better record both ends of the brightness scale. A few filters may be stacked together to create a stronger N/D effect but it is far better to use one stronger N/D than several weaker ones. To determine which strength N/D to use, decide how many f-stops decrease in exposure you require and then select the strength N/D filter that will do the job. Refer to the chart on neutral density filter equivalents in this chapter. For example, if you wish to decrease exposure by two f-stops, or use an aperture two f-stops faster than indicated by nonfiltered meter readings but keep the same f-stop, the chart indicates a 0.6 density N/D filter will do the job for you. To confirm the degree of exposure reduction, check the exposure before and after attaching the N/D filter.

Filters for Black-and-White Film

Filters primarily intended to be used with black-and-white film are described by their color. For instance, a perfect blue filter should absorb red and green and transmit blue. Bear in mind, many filters that appear to be a specific color to the naked eye are not necessarily sharp-cutting, that is, they not only transmit their own color but also some of the adjacent colors as well. As a general premise, color filters are used either to maintain natural tonal relationships as the eye sees them or to alter contrast by selectively lightening or darkening tones.

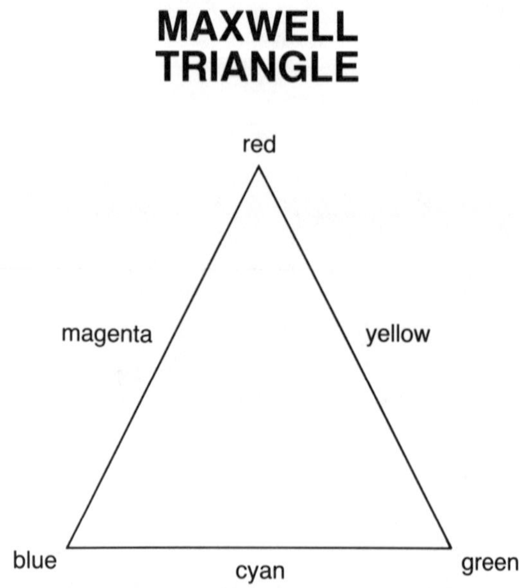

Figure 4.4 The Maxwell Triangle. If you do a great deal of black-and-white phototgraphy, you may wish to make a Maxwell Triangle to keep with you in the field.

Black-and-White Contrast Filters

This group of filters is employed to lighten and darken specific colors in a scene photographed with black-and-white film. The filters are strongly colored and are very efficient in absorbing almost all other colors but their own. Contrast filters are usually supplied in several densities of the same color. The most commonly used contrast-altering filters are yellow, green, orange, blue, and red. There are several ways to predict the effect contrast filters will have on the black-and-white rendition of colored subjects. One method, although not scientifically precise, is close enough to be a useful guide. Labeled the "Maxwell Triangle," it is a graphic predictive tool consisting of the three primary colors—blue, green, and red—and the three additive secondary colors—yellow, magenta, and cyan. In use, the triangle predicts the colors in a scene which will be lightened by a certain filter and which will be darkened. Therefore, the Maxwell Triangle shows that a blue contrast filter will lighten a blue sky as well as having some lightening effect on magenta- and cyan-colored objects in the scene. Using this same blue filter, yellow, green, and red objects will be darkened.

Table 4.3 describes the visual action of color contrast filters by listing their effects on commonly photographed subjects. Table 4.4 gives filter recommendations for specific effects in black-and white.

Filters for Color Photography

There are several groups of filters useful to the color photographer. Previously we have discussed several of the more popular types. Ultraviolet haze, polarizing, and neutral density types are used in both color and black-and-white picture-taking. Additionally, dark blue, orange, red, and yellow contrast-altering filters normally limited to black-and-white photography are sometimes employed to give startling monochromatic color effects. It is also important to note that there are several filters designed for precise color applications.

The three groups of filters used almost exclusively for color photography are:

Table 4.3 Effect of Color Contrast Filters

Filter	*Designation*	*Application*	*Filter Factor*
Light Yellow	No. 6	Darkens sky somewhat; absorbs UV; clouds pronounced; darkens blues.	2×
Yellow/green	No. 11	Foliage and Caucasian skin lightened; renders natural-looking sky.	4×
Yellow	No. 8	Corrects many outdoor tones; increases sky/cloud contrast.	2×
Dark yellow	No. 13	"Minus Blue" reduces haze in aerial work; increases contrast in snow scenes; effect similar to yellow; frequently used with infrared.	2×
Green	No. 13	Lightens foliage, darkens red and orange.	5×
Yellow/orange	No. 16	Reduces effect of skin blemishes.	5×
Orange	No. 21	Blues much darker, especially water; darkens foliage; absorbs UV; reduces distant haze; improves rendition of brick and stone.	5×
Red	No. 25	Dramatic dark blue sky; moonlight effect, especially with polarizer and under-exposure; cuts haze, UV; basic infrared filter gives white foilage, very dark sky.	8×
Dark red	No. 29	Increases contrast copying blueprints; stronger than 25 above, produces black skies. Emphasizes contrast in distant landscapes near horizon line.	16×
Light blue/green	No. 44	Darkens reds to a considerable amount.	6×
Blue	No. 47	Increases haze and fog effect; increases aerial perspective.	6×
Dark green	No. 58	Lightens foliage dramatically.	8×

Table 4.4 Filter Recommendations for Specific Effects in Black-and-White

Subject	*Effect Desired*	*Filter Recommendation*
Blue sky	Preserve normal contrast relationship	No. 8
	Darken	No. 15, 25
	Simulate nighttime	No. 25 + Pol.
Foliage	Preserve normal contrast relationship	No. 8 or 11
	Darken	No. 32
	Lighten	No. 58
Caucasian skin	Natural appearance	No. 8
	Darken	No. 47 or 58
	Lighten	No. 15 or 25

Table 4.5 Conversion Filters for Selected Color Films

	Intended Light Source	
Film Type	*Daylight*	*Tungsten*
All daylight color film	None	80A (- 2 f-stops)
Type A tungsten color film	85 series (-2/3 f-stop)	82A (-1/3 f-stop)
Type B tungsten color film	85 B series (-2/3 f-stop)	None

1. *Light-balancing filters*—filters that correct the ever-changing color of sunlight to match the spectral response of color film. Two series of filters in this group, the 80 and 81 series, add yellow, and the 82 series adds blue. With these sets of filters, you can control excessive amounts of yellow or blue contained in sunlight at different times of the day.

   Some light-balancing filter applications, with daylight-balanced film (5100–5500° K) are as follows:

Lighting	*Filter Application*
At sunrise or sunset	Use 80 series to reduce excess yellow
Noon daylight (no clouds)	None needed
Heavy overcast	Use 81A to reduce excess blue
Open shade (portraiture)	Use 81B to reduce excess blue

2. *Conversion filters*—filters made to affect relatively larger changes in color temperature to allow the use of daylight-balanced film under tungsten light or vice versa (see Table 4.5). Light sources such as photo floods can be used in conjunction with daylight film with the proper conversion filter in place. If, for instance, you have already loaded daylight-balanced Ektachrome 100 Plus, and you wish to get natural-looking color under floodlights, just attach an 85B filter over the lens to convert the film to a spectral response of the lights.
3. *Color-compensating (CC) filters*—these effect small calibrated changes to individual colors. The result fine-tunes color response. They are also used to correct for the odd color rendition of fluorescent lights. CC filters are supplied in calibrated densities of six colors: yellow, magenta, cyan, red, blue, and green. Each carries a designation as follows: CC identifies the filter as a color-compensating type and is followed by a number indicating the specific density of the filter, such as "50," which stands for .50. Finally, a capital letter denoted the particular color of the filter, like *Y* for yellow. Therefore, the full designation of our example CC filter is CC 50Y.

The following chart contains all the CC filters, their density designations, and required exposure increases.

Color-Compensating Filters

Density	*Designation*	*Exposure Increase (f-stops)*
YELLOW		
.025	CC 025Y	—
.05	CC 050Y	—
.10	CC 10Y	1/3
.20	CC 20Y	1/3
.30	CC 30Y	1/3

Color-Compensating Filters (con't)

Density	*Designation*	*Exposure Increase (f-stops)*
YELLOW		
.40	CC 40Y	1/3
.50	CC 50Y	2/3
.70	CC 70Y	2/3
.90	CC 90Y	2/3
1.10	CC 110Y	2/3
MAGENTA		
.025	CC 025M	—
.050	CC 05M	1/3
.10	CC 10M	1/3
.20	CC 20M	1/3
.30	CC 30M	2/3
.40	CC 40M	2/3
.50	CC 50M	2/3
.70	CC 70M	1-2/3
.90	CC 90M	2
1.10	CC 110M	2-1/3
CYAN		
.025	CC 025C	—
.050	CC 050C	1/3
.10	CC 10C	1/3
.20	CC 20C	1/3
.30	CC 30C	2/3
.40	CC 40C	2/3
.50	CC 50C	1
1.10	CC 110C	2
RED		
.025	CC 025R	—
.050	CC 05R	1/3
.10	CC 10R	1/3
.20	CC 20R	1/3
.30	CC 30R	2/3
.40	CC 40R	2/3
.50	CC 50R	1
GREEN		
.025	CC 025G	—
.050	CC 05G	1/3
.10	CC 10G	1/3
.20	CC 20G	1/3
.30	CC 30G	2/3
.40	CC 40G	2/3
.50	CC 50G	1
BLUE		
.025	CC 025B	—
.50	CC 05B	1/3
.10	CC 10B	1/3
.20	CC 20B	2/3
.30	CC 30B	2/3
.40	CC 40B	1
.50	CC 50B	1-1/3

One of the principal uses of CC filters is to balance color film for a variety of intermittent light sources including fluorescent, sodium vapor, and mercury. The chart that follows gives some filtration guidelines for these increasingly popular sources.

Fluorescent Tubes

Tube Designation	*Daylight Color Negative Film*	*Daylight Color Slide Film*
Cool White	CC 30M + 2/3 f-stop	CC 40M + 10Y + 2 f-stops
Warm White Deluxe	CC 60C + 30M + 2 f-stops	CC 60C + 30M + 2 f-stops
Cool White Deluxe	CC 20C + 10M + 2/3 f-stop	CC 20C + 10M + 2/3 f-stop
Warm White	CC 40C + 40M + 1-1/3 f-stop	CC 20C + 40M + 1 f-stop
Daylight	CC 40M + 40Y + 1 f-stop	CC 50M + 50Y + 1-1/3 f-stop
Tube w/no designation	CC 30M + 2/3 f-stop	CC 30M + 2/3 f-stop

Filters for High-Intensity Lamps

High-intensity lamps provide large amounts of light for such applications as street-lighting illumination of indoor arenas, outdoor sports or concert stadiums, parking lots, highway exit ramps, and a variety of other commercial and industrial uses. There are three principal types of high-intensity discharge lamps currently in use: multivapor, mercury vapor, and sodium vapor.

Multivapor Lamps The newest of the three is multivapor. They provide a large volume of light to both indoor and outdoor entertainment or sports complexes. If you use daylight-balanced color transparency film with this light source you will usually get more-than-acceptable color balance and rendition. Light emitted by this source is almost chromatically neutral so the use of CC filters is not generally necessary.

Mercury Vapor Lamps Mercury vapor lamps provide high-volume illumination to commercial areas and are widely employed in modern street-lighting setups. These lights exhibit a bluish-green color cast, calling for some correction in the form of CC Magenta or Magenta-and-Yellow filtration when used as the illumination in conjunction with daylight-balanced color transparency film. Filters are usually not required when using color negative film, since most off-color casts are correctable in the printing stages.

Sodium Vapor Lamps These lamps have found wide use in general outdoor applications. They emit a distinctive yellow/amber color. The warm light yields images that while not accurate, are usually pleasant. Nonfiltered pictures seem realistic because we have come to expect the color shift produced by sodium vapor lamps.

Filter Correction for Use with High Intensity Discharge Lamps

Type	*Daylight Color Negative Film*	*Daylight Color Slide Film*
GE Multivapor	CC 70B + 05C + 3 f-stops	CC 80B + 20C + 2-1/3 f-stops
Deluxe White Mercury Vapor	CCV 40M + 20Y + 1 f-stop	CC 60M + 30Y + 1-1/3 f-stops
Lucalor	CC 70B + 50C + 3 f-stops	CC 80B + 20C + 2-1/3 f-stops
Deluxe White Mercury	CC 40M + 20Y + 1 f-stop	CC 60M + 30Y + 1-1/3 f-stops

Other Uses of CC Filters

As creative tools, CC filters have found much favor among working professionals. Not only do they add precise amounts of predictable color but they can also be combined (up to three at a time) to produce anything from a faint tint to a rich, deep, monochromatic reinterpretation of a scene. They can be used during actual picture-taking or later, when copying slides, to correct or improve the original. CC filters are also widely used by experienced photographers to "tweak" color. Add a CC 40 or 50Y filter to an otherwise dull gray day and the effect is the same as if the sun suddenly broke out from behind the clouds—and without shadows as a bonus. With a number of CC filters you can warm or cool scenes slightly, dramatically, or impart any color cast by merely combining several densities and colors. Further, CC filters can add interest in skin tones by warming pale skin. Be creative by devising new skin colors such as magenta, green, or even violet. Obviously, the possibilities are endless.

CC filters come in two configurations. The original CC filters, Kodak's Wratten series, are available in very thin (.02mm) gelatin squares or rectangles. They are quite delicate and need careful handling. They also require a filter holder and adaptor system to use on a lens. A few companies, principally Tiffen, have developed CC filter lines that are premounted in glass, making them durable and easy to use. Glass-mounted CC filters are supplied in all densities for .025 to .50 and in all six colors.

Special-Effects Filters and Front-of-the-Lens Attachments

Creative filters and attachments number in the hundreds. Photo purists will deplore their use, arguing that anything "doctoring" the image is just covering up the mundane. They go on to assert that gadgets and tricks work against good image-making by emphasizing the effect as subject matter itself. All of this might be true with overused and haphazard trick photography, generating as they do more problems than they solve. Even so, an almost limitless choice of effects available today affords consideration and gives prudent photographers abundant opportunities to use special effects as a tool to give visual emphasis or twist to an otherwise straight photograph.

Several suppliers of conventional glass screw-in filters offer a number of special-effects items. Among the most popular are different types of soft focus, fog, diffusion, center color spot, graduated color, dual image, and many prism-generated alterations, including starbursts, radial color, multiple image, and speed effects. B+W, Hoya, Tiffen, Heliopan, and others sell a good assortment of these attachments and filters.

Cokin, Ambico, and other makers of plastic and resin filters have taken the front-of-the-lens concept and expanded it to a point where the selection almost defies description. As discussed earlier, these filters and attachment systems consist of modular holders , connectors, and lens-attachment units along with literally hundreds of high-impact image-altering devices, usually made from CR-39 plastic resin compounds. This is the very same material used to form lightweight prescription eyeglass lenses and high-quality sunglasses. Ambico brand filters are made of another material, an acrylic coated on both sides with a virtually indestructible resin.

All systems contain conventional filters used for black-and-white contrast modifications, polarizing units, many variations of soft focus, neutral density, diffusion, fog, and so on. Added to these are an array of image-distorting devices or filters too numerous to mention. Beyond filters or image distorters, the systems can accommodate masks that serve several functions, including double exposure, preshaped vignettes, or even thin color gels in jazzy hues such as mauve, tobacco, violet, pink, aqua, or rose. Many of these colors are not available in any other format.

To begin to appreciate the possibilities filter systems afford, stop by your local camera shop and browse through catalogs or illustrated booklets of the system-makers' product lines. Note that most new filter systems include lens-cleaning fluids or special tissue. They also contain lens caps, gelatin filter holders, storage boxes accommodating several filters at once, compartment cases, modular lens shades, and compendium lens hoods. And they continue to expand!

Other Specialized Filters

This section includes several filters made expressly for a particular application.

FL-D: This filter is normally used to balance fluorescent light and daylight color film. It can also be employed to create strikingly warm color that is at once both realistic and warm, and features noticeably brightened highlights.

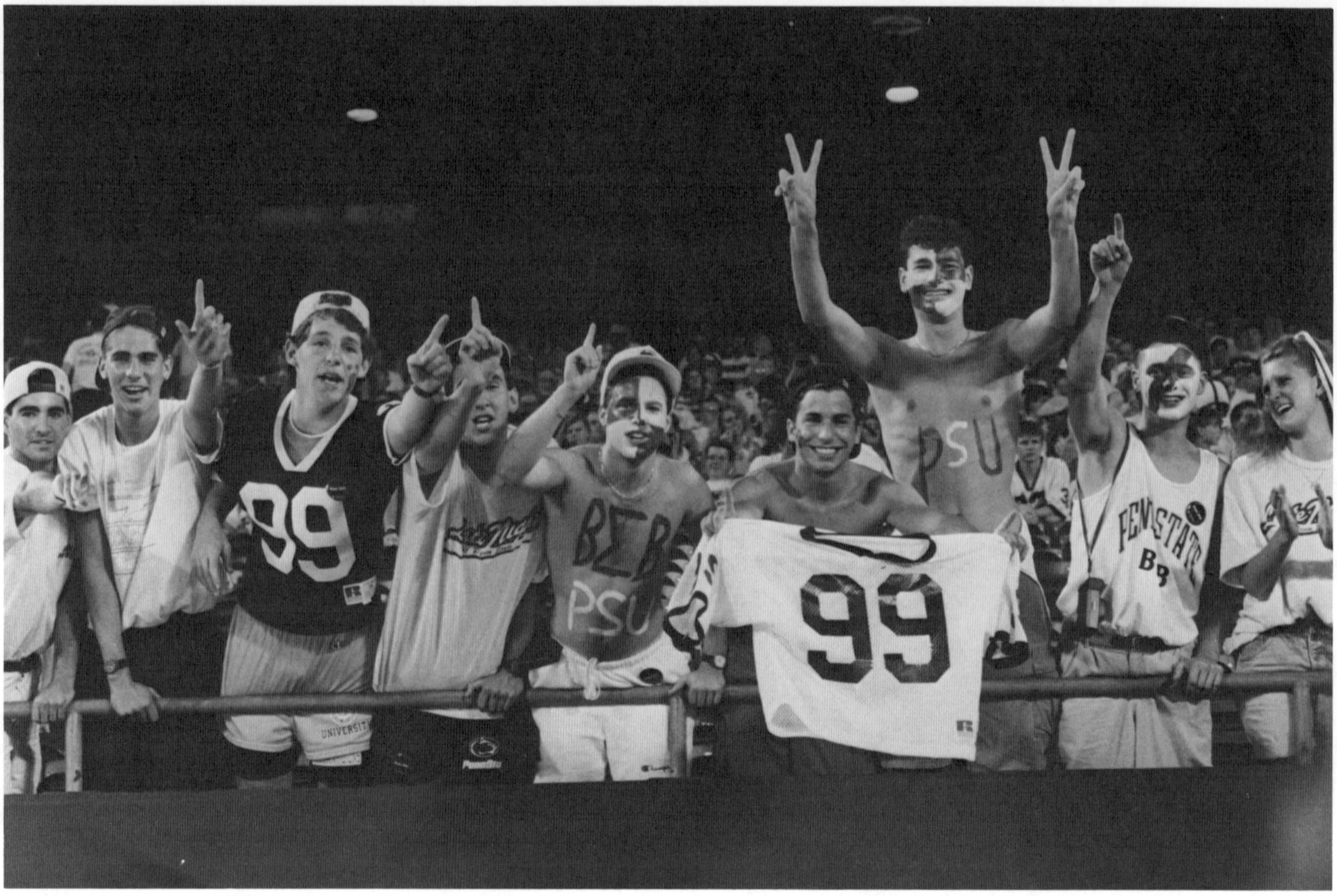

Figure 4.5 Steve Manuel. Kodak T-Max 3200 at ISO 1600.

Tiffen Enhancing Filter: Made of a unique Didyminium glass, this filter was first developed and marketed by a chemist, Howard G. Ross. In its current form, it is supplied in threaded metal mounts in most common millimeter sizes. It can also be had on special order in almost any size necessary. Essentially a very sharp cutting filter, it only affects warm colors, especially red. It seems to have a reduced but still noticeable effect on orange and yellow as well. This filter has little or no effect on the rest on the spectrum. When photographing fall scenes, the effect of the enhancing filter can be and often is spectacular. Its effect is further strengthened by combining it with a polarizer. Sometimes you will need to brighten scenes shot on rainy or overcast days. If warm colors predominate, an enhancing filter will work very well for this purpose. The filter itself imposes a 1⅓ f-stop exposure penalty and, when combined with a polarizer, the loss grows to an almost prohibitive 2⅔ f-stops.

Tiffen 812: Essentially a weak warming filter, the 812 offers just enough warming to turn dull scenes into bright, more cheerful ones that seem more acceptable to a majority of viewers. The filter is also useful as it absorbs excess blue in shade. It is frequently used to add a touch of color to a pale Caucasian face shot on a cloudy day.

Double polarizers or polarizer/faders: The unit consists of two polarizers rotating independently but sharing a common mount. It is very useful to video- or film-makers for fade-in/out sequences since it can go from minimum polarization to full black. The filter is sometimes used as a variable neutral density unit.

Stepless variable neutral density filter: Made up of two N/D filters rather than polarizers, the user dials in the density reduction by turning the filter in its rotating ring.

Wide-angle neutral density filter: This is a normal N/D filter except it is graduated with a stronger N/D effect in the center and decreasing effect toward the edges. It can be helpful to even out the light fall-off present in extreme wide-angle lenses from center to edge.

IR transmission filter: Black, visually opaque filters that only pass IR radiation. There are two types, one to pass IR above 720nm and the other for wavelengths above 900nm.

UV transmission filter: Unlike conventional UV filters transmission filters only pass UV radiation. All visible light is blocked.

Filter Kit Recommendations

An Entry-Level Kit

skylight or UV
81A or Tiffen 812 warming
polarizer
No. 8 Yellow
No. 11 Green
step-up/step-down rings to fit filters to all lens sizes
appropriate lens hoods for all lenses

Intermediate Kit

skylight or UV
81A or Tiffen 812 for warming
polarizer
N/D .50
CC 30R for fluorescents

No. 8 Yellow
No. 11 Green
No. 25 Red for infrared or dramatic sky
several special-effects filters selected to enhance often photographed subjects
step-up/step-down rings
lens hoods for each lens

Advanced Amateur or Professional Kit

skylight or UV
CC 30R or FL filter
polarizer
set of light-balancing 80 and 85 series filters
No. 8 Yellow
No. 11 Green
No. 21 Orange
No. 25 Red
No. 47 Blue

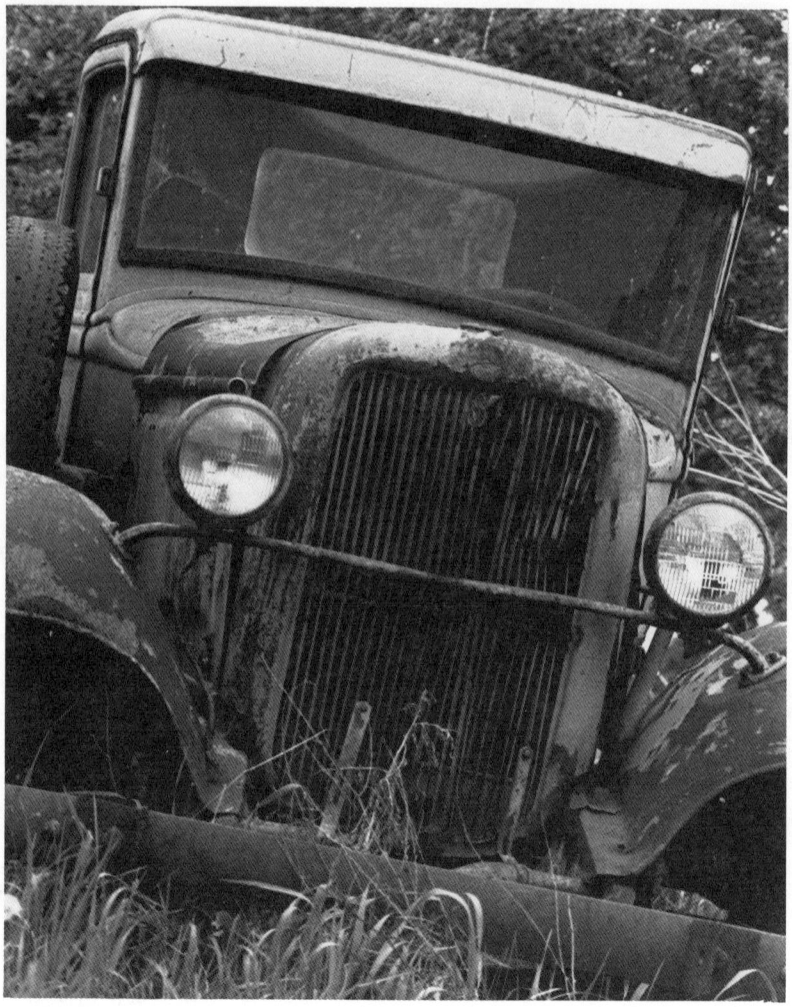

Figure 4.6

No. 58 Green
neutral density filters to 2.0 or a variable N/D filter
numerous special-effects filters, front-lens attachments, and adapters
step-up/step-down rings to fit filters to all lenses
lens hoods for each lens

Filters You Can Make Yourself

Diffusers and Soft-Focus Filters

Beginning with clear glass opticaps or skylight filters, any number of substances may be applied directly to the filter surface or wrapped around it to produce one of several effects. Plastic wrap makes a good diffuser, and the degree of diffusion depends on how tightly you stretch the wrap over the front of the lens. You may want to try old pantyhose as diffusers. Since they come in many shades, the effect will differ from one to another. If you don't want to bother with any of these, simply breathe on the front element of the lens and take the picture. You must be quick about it though, for the diffusion effect dissipates rapidly. You can also make a clear center spot and diffused field filter. Apply petroleum jelly in a circular manner to all but the very center of a UV or skylight filter. Clear polyurethane, lacquer, or nail polish works well. Obviously, the filter used for these modifications will be rendered useless for other tasks.

Again using a skylight filter or clear Opticap as the basis, construct a star-effect filter by cutting a piece of aluminum screening to fit snugly inside the filter ring. You may want to paste clear plastic strapping tape to the surface of a skylight filter. The fiberglass threads in the tape should act to break up the image in unusual ways. Acquire some clear plastic from an art supply store and cut some circles to fit inside a skylight filter. Next, paint color on the plastic with colored magic markers in random patterns, multicolor swirls, stripes, or an overall hue. Split-color filters can also be made in this manner. Creating homemade filters is only limited by the breadth of your ingenuity.

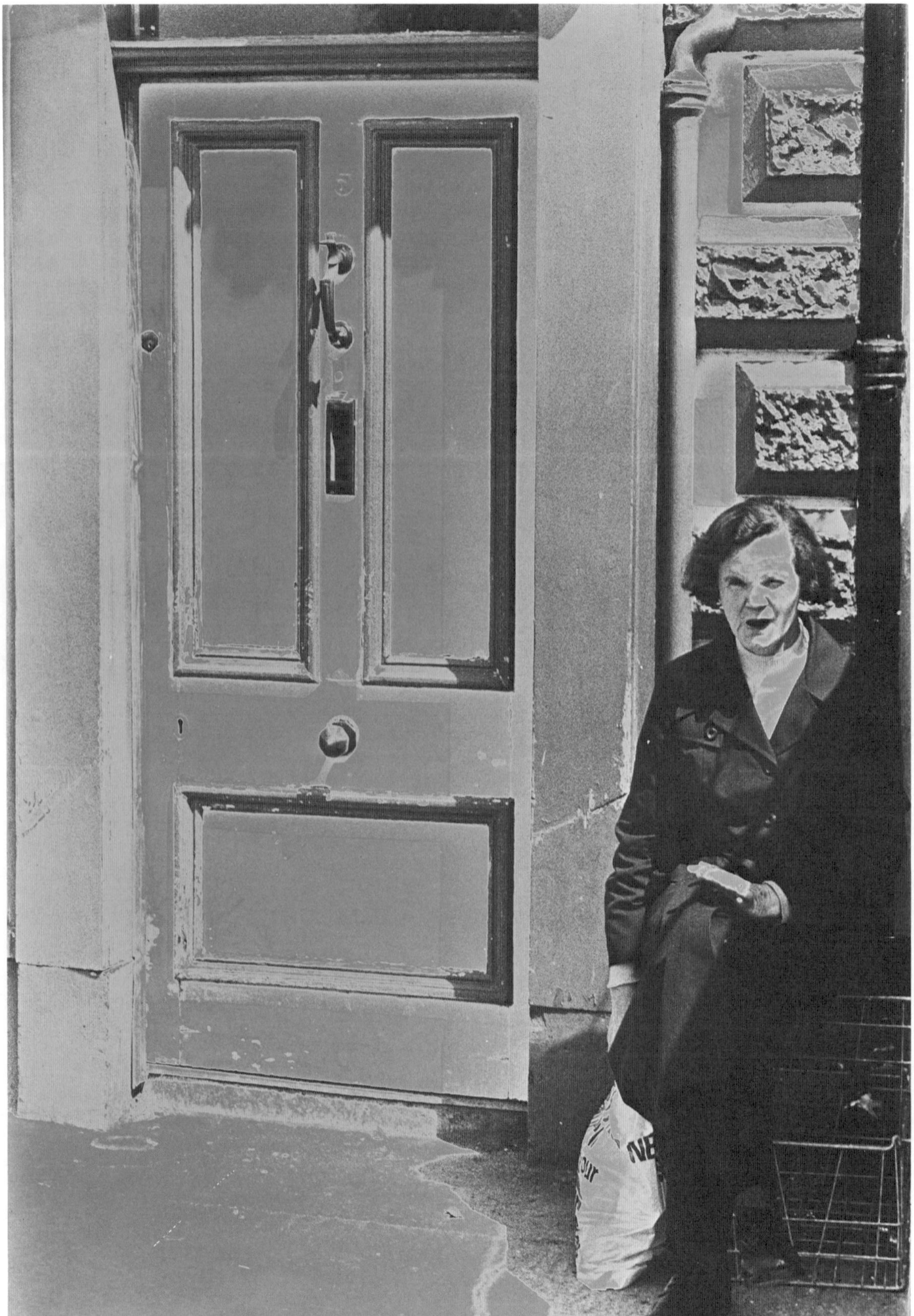

Chapter 5

Integrating Film into the Creative Process

Once upon a time, the best advice I could give fledgling photographers was to identify one film, learn it inside and out, and use it almost exclusively. I gave this advice back in an era when there was a large qualitative difference between films offered by major manufacturers. Now that gap has all but disappeared. No longer are there great differences in grain structure between slow, medium, and fast films. To be sure, differences still exist, but they are far less profound than they previously were. Today, instead of two or three high-quality films to choose from, we have somewhat of a burden of riches, with literally dozens of very high quality emulsions from which to select. Further, we no longer need to limit our selection to slow, fine-grained films. Almost all medium-speed and even several high-speed films are so good that they serve the needs of all but the most technically fussy.

In addition to having a much wider choice regarding film speed, photographers can choose from among the available films on the basis of each's particular performance strengths. You'll need to become highly selective if you hope to improve your photographic skills. Selectivity involves exerting control over every aspect of picture-taking. Some would say selectivity is everything. To be selective implies a certain amount of subjectivity. We consciously choose a particular fractional slice of time to freeze and then ask a single image to represent a range of visual and other sensory experiences over several seconds, minutes, or even longer. If we really think about it, that is asking one still photograph to do a great deal. In order for us to be the master, not the servant, it is imperative we understand the numerous differences between human and photographic (camera/lens/film) vision. Although both systems "record" visual impressions carried by photons of electromagnetic energy, one records everything objectively while the other system continuously filters, interprets as it recognizes, and retains some image components and rejects others. To make intelligent choices from among the vast number of tools at our disposal—cameras, lenses, lighting, filters, and the subject of this book, film—it seems to make sense to have a reasonable understanding of the similarities, differences, and limits of these two separate but interdependent vision systems.

Human and Camera Vision

Figure 5.1 Marc Levey. Kodak Technical Pan Film ISO 25, solarized.

As stated earlier, the two systems under study in this chapter do not form images in the same way. Still-camera systems, for all their electronic, mechanical, and chemical sophistication, function as nothing more than visual tape recorders. In and of themselves they make no judgment other than yes/no. Even the addition of the onboard computing ability that governs decision-making only takes us to the "maybe" level. Sensory input still remains one-dimensional in that it is still all visual. On the other hand human vision consists of an interface between visual inputs and a central processing unit (CPU), the brain, which not only records and evaluates visual signals, but also (and this is a big also), combines new visual information with that already stored, and then adds it to other associated sensory inputs, such as sound, smell, touch, and so forth. All this material is digested by the brain and the result is usually an enhanced image, existing in what we perceive to be real time, which then can be discarded, acknowledged, or stored, in total or in discrete pieces. Obviously no camera/lens/film triad yet devised can even approach this wondrous level of perfection. However, what manmade visual systems do, they do very well.

As earlier stated, cameras and film do not, in a number of significant ways, "see" what our eyes and brains do. Lens and film record some aspects of an image that the human vision system cannot physically perceive but, in comparison with the human system, camera/lens/film systems have very limited capacity to detect and reject images. It is important to understand that human vision is very different from what a camera lens can transmit to film. The fact remains that what you see is not always what the film records.

In some ways, human vision shares more in common with motion picture or video than it does with still photography, for we see things continuously over a time span and time divides what we see, sometimes into logical sequences, sometimes not. Our eyes transmit a continuous series of images to the brain and these "pictures" are analyzed and are then assigned meaning at a variety of levels. This process usually takes place in fractions of a second, appears to be instantaneous, and results in automatic response, at times not on the conscious level. The brain receives a tremendous volume of visual and other stimuli, and for even the simplest events it relies on memory, a highly efficient decoder, to make sense of these often-contradictory messages.

The situation is quite different for the photographer using a still camera. The eyes (camera and lens) and their memory bank (film) can only take in and record events or objects as single instants of human experience frozen in a specific and very short time-frame: typically from 30 seconds to 1/12000 second. Nevertheless, we assume what we saw at the time we took a picture is captured on film exactly as we experienced it. Unfortunately, this is not always so. Even multiple exposures add only a few more slices of experience on the original frame of film. The outcome? What we intend as the response to a scene we photograph does not always conjure up the same feeling in the viewer's mind when she looks at the finished end product. The trick is to train your eye and mind to think in two stages: first, how will the mechanical/chemical system modify the image that I perceive now; and second, what can I do, knowing something about the differences between human and photographic vision, to evoke a desired response from the viewer?

The human eye records and transmits objective reality. What the brain perceives—therefore, what we "see"—is subjective. A camera, on the other hand, "sees" (within the system's capacity) what is physically present and records what it "sees" without judgment or interpretation. There are several other significant differences between human and camera vision systems.

Speed

The human eye operates at a relatively constant speed (roughly 1/60 second). However, many cameras have shutters operating over a range of several seconds to 1/8000 or faster. Remember, the image we see at 1/60 second is not the same as what a camera lens transmits to film at 1/8000 second. At 1/8000 second, just about all motion is totally arrested and at 1/60 second it is not. The image in the photograph, therefore, will appear in a way we could never see unaided.

Focus

Our eyes constantly scan a scene within their field of vision, correcting focus automatically. Messages travel back and forth from the eyes to visual centers in the brain. The visual messages stimulate instantaneous corrections of focus. Even the best autofocus systems must contend with some time lag between initiation of an autofocus sequence and the actual focusing. This time lag is sometimes as much as one full second. The iris functions instantaneously, responding to very subtle changes in the volume of light present. Camera lenses have an automatic diaphragm controlling the amount of light striking the film; they do so only at a predetermined and fixed time for each exposure. Human vision constantly monitors and adjusts focus automatically. The most recent photo technology allows for automatic focus by user command, and for the most part, only for limited periods of time. Many cameras must still be focused manually and their diaphragm set by or at the command of the user, again for a specific time, usually only for the duration of the actual exposure.

Angle of View

Although our visual ability to discern fine detail is limited to about 20 degrees (10 degrees on either side of the optical axis), we are visually aware of a great deal beyond this 20-degree sweep. Peripheral vision permits us to "see" almost 180 degrees. While peripheral vision is not acute, that is, we cannot identify objects over the entire peripheral field, we do sense their presence. Camera lenses, however, have fixed angular coverage determined by their focal length and camera format. A 50mm lens used on a 35mm camera encompasses about 46 degrees, a 25mm lens on the same camera about 84 degrees, and so on. A zoom lens permits variable coverage but the controls are still manual rather than the automation of the eye.

Image Sharpness

With typical human vision, everything from about eight inches outward can be focused upon sharply. Farther away, every object within the range of vision will appear sharp. In other words, our eyes have virtually unlimited depth of field. Cameras are strictly limited in their ability to provide a wide range of sharply focused images at any one time. Each lens, depending on its focal length, aperture used, and distance to the subject, produces specific and limited depth of field. If we use the normal lens (50mm) in the 35mm format as a reference once again, set its aperture at a middle value of f-8, and focus on an object ten feet away, only the object itself and what is approximately one foot in front of and behind it will *appear* sharply focused in a resulting photograph. While at first this might seem to be a very great disadvantage, in practice this apparent limitation offers us the option of focusing selectively on portions of a subject and rendering them sharply while throwing others areas out of focus. To the viewer, sharp portions of the image advance and stand apart from the blurred ones.

HUMAN VISION SYSTEM

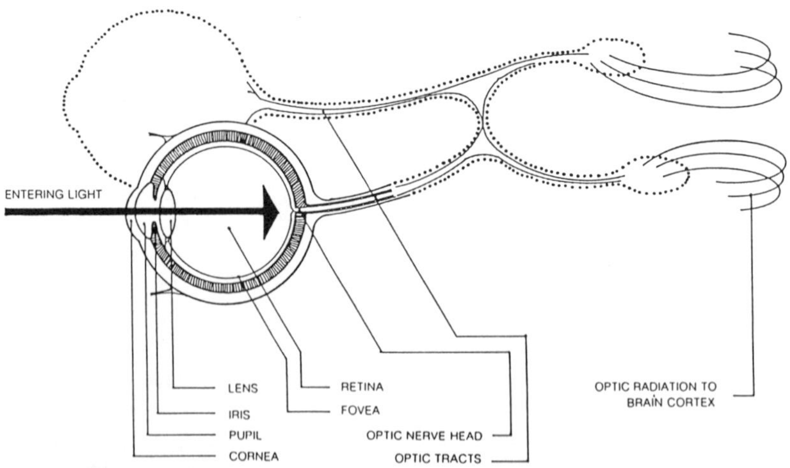

Figure 5.2 Human Vision System.

Brightness Range

We are capable of seeing a far greater range of tones (from dark to light) than cameras are able to record on film. The eye discerns detail over a contrast range of 1000:1 or more. Compared to this, the discernible contrast range of color negative or transparency film is less than 100:1. We must assume, then, that all the tonal values our eyes discern cannot possibly be recorded on film. Again, what you see is not what you get. Our vision perceives objects in both shadows and highlights that reproduce as either solid black or featureless white masses in a photograph or slide. The problem becomes particularly difficult for the color film user because color films accept and record detail within an even narrower tonal range than do black-and-white. One is forced to sacrifice detail in shadows in order to retain important detail in highlights or vice versa. It is important to be aware of modern films' limitations to record all the tonal information we are able to see. There are numerous occasions when the brightness range of a scene far exceeds *any* film's recording capacity.

Color Rendition

Human vision provides the brain with a series of messages that when translated become our working definitions of the range of colors visible to us. About 3 percent of the male population and a much smaller percentage of women are color-blind to the color qualities of red and green and sees these colors as shades of gray. These people see only five distinct colors: yellow, black, blue, white, and gray. For the rest of us, the seven visible colors are seen in an almost infinite variety of shades and tints.

Not so for cameras and film. First, black-and-white film transforms all visible colors into shades of gray; we only know what color an object is from memory. Second, *all color films exhibit a color bias.* They all tend to emphasize an overall color rendition. Warm films favor yellow, orange, and red, and cool films predominate in blues, greens, and violets.

Another factor influencing color rendition is that many lenses, because of the spectral qualities of their glass, impart a color bias of their own. Using lenses from different makers may lead to detectable differences in color rendition from lens to lens. At times we all have wondered why color images did not appear as we remembered them. The colors do not ring true! Possibly the lens, the film, or both imparted a color bias of their own to the scene.

The Appearance of Color

Human vision, with its great decoder the brain acting as arbiter of what is correct color, attempts to color-correct automatically in most lighting situations. Skin tones, whites, and blacks, seem to appear as we expect them to, even in extreme lighting. It is only when we consciously examine color rendition under nondaylight lighting that we come to realize that many colors are not what we have come to expect. Yet until we actually take the time to examine color consciously, our mind color-corrects enough to make most colors acceptable.

Film has no such magical power. It records subject colors as best it can given the limitations of the film itself. To some degree it imparts a color defect termed *color contamination* so that a scene may reproduce overly yellow using one film and too blue with another.

Dimension

Next, consider that the eyes together provide the brain with stereoscopic signals that are translated into true three-dimensional images of almost unlimited depth. But in a photograph, depth is no more than an illusion. In most photographic situations, the illusion of depth is a most important consideration. The third dimension missing in photographs is implied through manipulation of depth of field, through which we create figure-ground (subject-background contrast) differences. Other depth-implication techniques include emphasis on physical or color contrasts in the scene, aerial perspective of landscapes, and size or shape differences between significant visual elements in a scene.

Unique Capability of Photographic Films

Several specialized films record scenes we literally cannot see. Slow, fine-grained film combined with a macro focusing lens opens up a close-up world inaccessible to the unaided eye. Black-and-white or color infrared film offers possibilities by letting us use IR light that is outside the visible spectrum to image the unseen world. Using fast lenses and ultra-fast film allows us to freeze action that we see as just a blur. Techniques such as panning and intentional blurring also result in interesting images. With today's films you are free to select from among many contrast levels, color renditions, or sharpness levels.

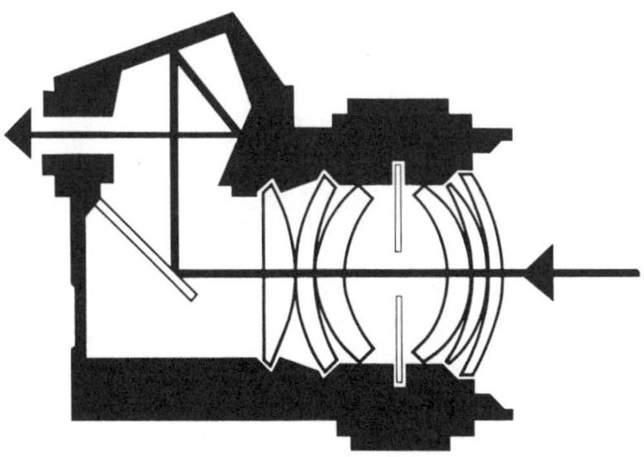

Figure 5.3 Typical Camera Optical System.

Angle of Acceptance

Through a camera's lens, the angle of acceptance is fixed and is determined by the len's focal length; a 50mm lens angle is about 46 degrees, a 28mm lens is 74 degrees, a 200mm lens is 12 degrees, and so on. Several lenses are designed to exceed the limits of human vision. Extreme wide angle lenses and long telephoto lenses take in far more than we can unaided. Fisheye lenses can actually see behind our back up to 220 degrees. Super telephotos of 200mm yield 40× magnifications so an object 100 yards away appears to be only 7 or 8 feet distant through the viewfinder.

The Image/Quality Tradeoff

Several aspects of technical image quality are related to the sensitivity or speed of film. How much fine detail film resolves is a product of the film's granularity, acutance, overall contrast, and, in the case of color emulsions, color saturation. In addition to these technical qualities of the film itself, scene contrast has much to do with the appearance of sharpness—high-contrast scenes reveal the greatest apparent sharpness, and low-contrast subject matter markedly less.

Most viewers define a quality image as one containing high levels of sharpness, plenty of detail and contrast, plus high color saturation. This definition pertains to all subjects, with the possible exception of some types of portraits. I suggest you think of quality in a different light. Quality is, after all, relative. To be tied to old definitions of image quality, which are based on technical criteria, is too limiting. For example, soft may be more desirable than sharp. So, great numbers of lines per millimeter may not always be a significant indicator of image quality. Resolving every wrinkle or facial flaw on your eighty-year-old grandmother's face may not necessarily serve your photographic purpose. Further, the muted palette offered by several of the current color print films as well as by one or two slide emulsions might be a most appropriate choice to photograph a field of wildflowers on an early spring morning. The list of exceptions to the commonly held technical definition of image quality is endless. I believe definitions of image quality are much more than yes/no responses to a series of technical questions. The definition also involves more than the measurement of lines per millimeter and percentages of contrast. While these and other objective criteria give us valuable information about film performance or print quality, they themselves do not tell the entire story. It does not always suffice to generate technically correct grainless negatives or slides: first, because lack of grain and minutely resolved image detail might not support the intent of the image-maker, and second, because an overconcentration on technical minutia can very well obscure what photographs are all about—to tell the world what we think and how we feel about the world around us. If we see the world as a gray, brooding, indistinct place, then our images should show that. They probably should not be bright, colorful, and sharply etched, for that presents a contradiction to the viewer and probably defeats their purpose.

After having said all this, I rush to admit a well-executed picture is most often preferable to one of obvious poor quality. It is also true that people generally prefer sharp to fuzzy, high to low contrast, and lots of color to muted tones. An image that is sharp, detailed, and well exposed is more visually assessable than one of lesser quality.

For many photographers, the pursuit of technical perfection overpowers all other image-making concerns. Technical imperfections such as visible grain are cause to immediately and forever condemn an image, regardless of its content. It is this single-minded pursuit of some technical nirvana that I find so counterproductive. When you have a choice, opt for traditional pictorial quality, but do so only within the context of your overall photographic purpose. Ask questions. Do I really want detail? Would a softer gradation enhance my image? What decisions under my control should I have made differently?

Everything photographic involves tradeoffs. If you choose to shoot in low light without supplementary lighting, you will need a fast lens, fast film, or both. Any of these decisions might compromise technical quality, with the result being less sharpness, more visible grain, loss of image contrast, and so on. If the picture is worth taking, these compromises are probably well worth the cost.

Film Selection Criteria

The usual advice for selecting film, black-and-white or color, is to choose the slowest emulsion that will allow you to accomplish your photo purpose. Thus, if the subject is stationary, as with still-life and landscape subjects, then by mounting your camera on a tripod you are free to use almost any film regardless of its ISO rating, provided the light is sufficient for you to select a desired f-stop/shutter-speed combination. However, film choice becomes more limited and therefore critical when subject or lighting conditions change. Consider indoor sporting events. Slow films are out of the question if you expect sharp, action-freezing images. If your goal is to deliberately blur, streak, or otherwise distort an image to imply motion, then and only then will slow films suffice. Usually, you will want sharp subject matter. Since you can't control light intensity, your only options revolve around selection of lenses and film. Most modern sports arenas will require film rated at ISO 400 to 1600, even when using fast telephoto lenses with maximum apertures of f-2.0 or 2.8. The problem you face is twofold. To isolate the action you will need a moderate-to-long telephoto lens. Shooting with the lens wide open affords barely enough light to expose at 1/125 second with ISO 1000–1600 film. Most sports photographers address this problem by using a form of supplementary support, usually a monopod. Used correctly, they can buy the equivalent of one or perhaps two extra shutter speeds of sharpness. Pictures shot with a monopod at 1/125 second look much sharper than their hand-held counterparts. The preceding is the time-honored approach to film selection, and in most instances it is a valid one. I do believe, however, that the above approach does not go far enough.

There are numerous times when going against the usual selection method makes esthetic sense. If, for example, you find a subject with an unusual textured surface and you want to emphasize it, then a faster and therefore grainier film will give you the visual bite needed to highlight this aspect of the subject. This is only one example of many. Each demonstrates that film selection should be made within the larger context of photographic intent. Ask yourself questions. Why am I taking this picture? What do I want it to say? Answering these and similar questions will go a long way toward determining what film, filter and technique to use.

Subjective Evaluation of Film by Selected Qualities

As a photo writer and teacher, I come in contact with all manner of photographers: neophytes, occasional amateurs, the seriously committed, and professionals. I have learned a great deal about film from all of them. I use this space to share their experience, observations, and (most important) considered judgments.

Remember, reading test reports is less useful than the judgments of knowledgeable practitioners. We have attempted to compile a consensus evaluation of selected films, identifying them by brand, name, and manufacturers' designation. In some instances, significant judgments are offered about some films that don't represent a true consensus. If I felt a comment was of sufficient importance, I included it, noting that it represents a minority opinion. Films in all three general categories were considered for inclusion. Not all the individual qualities we examined were the same for all films. Plainly, color qualities don't pertain to black-and-white and issues surrounding tonality are more important considerations to black-and-white film users. I chose to concentrate on the color qualities of transparency films principally because color rendition of print film is easily modified in the printing stage. Last, before moving into more specifics, I caution the reader that the following judgments must be considered subjective and I make no claims that they are in any way scientifically verifiable. They are nonetheless valid as far as the experiences of the test panel and I make them.

Qualities of Black-and-White Film

Ilford XP-1/XP-2

Surprisingly, from those individuals who used Ilford XP-1 (now replaced by XP-2), and that was almost 40 percent of the testers, universal praise is evident for the film. Flexibility in use was cited as its most useful attribute. Testers reported favorably on its ability to do more things well than any other film, black-and-white or color. It delivers high-quality images under a wider range of light conditions than any other film. Added to this is film processing that is all but glitch-free.

Kodak T-Max 100

Now in its second generation form, this film employs the now famous T-Grain technology to achieve its high level of performance. The film drew comments from all but three of the several dozen surveyed photographers who routinely work in black-and-white. The emulsion was praised for its micro grain, particularly at high enlarging magnifications. Also singled out were the film's wide tonal range and generally excellent image quality. Many appreciated the film's extra speed in relation to its grain size and high acutance. Several T-Max 100 users found processing the film not as predictable as they wished it to be.

Kodak Technical Pan Film

Quite a few black-and-white film users singled out this emulsion for comment. Most stated that they would choose it when they wished to produce very large (larger than 11 × 14) grain-free black-and-white prints. Users pointed out Tech Pan's ability to deliver full tonal range from a properly exposed and processed negative. They also liked the film because it shows no visible grain from a 35mm negative up to a 20" × 30" enlargement. While extolling its strong points, users acknowledged they needed to be very careful about exposure and camera technique in view of the film's low exposure index of ISO 16–25. Undoubtedly, this is not the film of choice for candid street or sports photographers but within its limits it provides exceptional quality.

Kodak T-Max P3200

Presently the fastest black-and-white film available, the emulsion is valued for its applicability to hand-held low-light picture taking. Those who use it regularly noted it delivers negatives with grain patterns more like ISO 400 Tri-X. A few users pointed out that the film is capable of yielding usable negatives at ISOs as great as 25,000.

Kodak High-Speed Infrared Film

Interestingly, more than 50 percent of the testers who use black-and-white film frequently have worked with black-and-white infrared, at least occasionally. They report everything from sheer delight to utter frustration with the film. The reason so many people are frustrated with this emulsion can be found in its lack of predictability. This is amply demonstrated in the mixed reviews the film receives. What does come out is a strongly held conclusion that for best results one needs to use a red 25A filter with this or its color equivalent, Ektachrome Infrared. Users suggest a starting exposure of f-11 at 1/125 second for outdoor subjects in daylight. This exposure works well for users of fully automatic program exposure cameras that lack provision for ISO or exposure manipulation. Most of these cameras will set ISO 100 for films not containing DX coding on their cassettes. Adherents enjoy its unpredictable results and surrealistic appearance. They point out that after one gains experience with the film, including careful observation of infrared renditions of several subjects, the photographer will begin to get a sense of what to expect on processed negatives.

Qualities of Color Negative Films

Fujicolor Realia

This relatively new film is experiencing widespread adoption because of its unique ability to deliver pleasingly natural color, especially in skin tones and under varied lighting, including electronic flash, tungsten, and even the old nemesis of color photographer, fluorescent. Realia users report the film to be extremely well behaved with regard to contrast, supplying open shadows and detailed highlights simultaneously. Its low-to-moderate scene contrast does not affect color contrast in the least. In fact, testers found the film to be among the most color-saturated of all color negative emulsions. In sum, this film exhibits medium but useful speed, very fine grain, excellent color behavior under difficult lighting, controllable contrast, and abundant color saturation. Users of color negative films consistently report Realia to have the fewest minuses and most pluses in the category. As an editorial comment, I add the observation that color negative film is only as good as the printing regimen or printer used to produce finished prints.

Kodak Ektar 25

The Kodachrome 25 of color print materials, this film seems to have found a niche among those who are interested in striving for the highest technical perfection and are willing to trade off in-use flexibility to achieve it. Photographers who use this film report it does indeed render richly color saturated, virtually grainless negatives with above-average contrast. These same people also report the film's slow ISO-25 speed rating does at times limit its usefulness. Many find themselves limited to static subjects in good light. Aside from this limitation, the film is the one most often chosen if very large ultra high quality prints are demanded from a color negative. A few observers point out that its low speed can be largely compensated for with studio floods or electronic flash.

Kodak Ektar 1000

Ektar 1000 is a very fast (ISO 1000) color negative emulsion that drew favorable reviews for its usability under adverse lighting or climatic conditions. And while Ektar-1000 users are fully aware of its distinct grain pattern compared to slower films, they offer the opinion that its pattern is tight and sharp-edged and not objectionable in prints of moderate magnification. Grain only becomes problematic in subjects containing major areas of featureless midtones as is found in a cloudless sky, with high magnification enlargements beyond 8 × 10.

Konica Super SR 100 and 200

Neither of these films is well known outside professional circles but both are nonetheless widely distributed as house brands such as Fotomat. Those who are familiar with these films testify to their excellent balance of pictorial qualities and performance. They incorporate the latest technological advances and reflect barely discernible grain, excellent neutral color, and superior contrast. The film is reportedly available at lower prices than more heavily advertised and promoted brands. What Konica offers is quality at a bargain price.

Scotchcolor 400

Like Konica, Scotch films are more likely to be found under other names, like Focal or Polaroid, than their own. Scotchcolor 400 is one of these. The film's parent, the giant 3M Company, has managed to design an excellent emulsion at a lower price than the competition. The film seems to be very compatible with the auto programs in processing machines usually set for Kodak products. Those who have shot with Scotchcolor report it capable of delivering prints indistinguishable from Kodacolor or Fujicolor.

Agfacolor Portrait 125

After I asked a number of photographers to try this film, they came back very impressed. Several liked its lower contrast combined with superb color characteristics. Beyond this, they were enthusiastic about the skin tones it delivered. As its name suggests, it is indeed an outstanding portrait film. The only problem with this film is its availability. I could not find a local source. It can't be found in mass-marketing outlets like K-mart and the like. As of this writing it should be plentiful in photo specialty stores in metropolitan areas.

Fujicolor HG 400 Professional

Introduced about two years ago, HG 400 is designed to signal a quantum leap for Fuji print products in the crowded ISO-400 color negative film market. Those of us who have used it can largely confirm its many virtues. It exhibits vastly reduced grain structure compared to other conventional ISO-400 color print emulsions. It behaves more like an ISO-100 film in this regard. Couple reduced grain with its very wide exposure latitude and you have a film some of my professional associates regard as a near-perfect color Tri-X. Weddings, natural light portraits, and candid street shooting are perfect applications for this versatile material.

Qualities of Color Transparency Films

Fujichrome Velvia

Almost everyone who uses Velvia eventually falls into one of two categories: those who love it and those who detest it. Why such divergent opinions? It's probably because the film's unique performance is unlike any other transparency film made today. The film is unique on several accounts, not just one. First is its color quality. It is intense, to say the least. Informed people I've talked with about the film claim it makes heavily overcast appear to be fully lit. All colors are bright and deeply color-saturated, even in the heaviest shadows or overcast. If color is an important component of your overall design strategy, this film will provide you color to spare—more, perhaps, than many need or want. Bear in mind that the color is not particularly accurate, especially when the film is underexposed even slightly. The color is very, very different from anything else.

Not better or worse, just different. The second unique quality about this film is its projectability. Of all the slide films I have used, this one projects with more drama and sparkle than any other. As with other Fujichromes it renders greens especially well on the screen, a vibrancy unmatched by any other film, Fuji or otherwise. Velvia's uniqueness extends to the way it handles contrast. Recall that there are two distinct types of contrast: color contrast and scene contrast. Although they sometimes go together, they don't always need to coincide. Velvia's color saturation is exceptionally high, but overall scene contrast has been kept within a very printable or reproducible range with detail apparent in both shadow and highlight areas over an impressively long range for a transparency film. Add to this a micro-fine grain pattern rivaling Kodachrome 25, and you have a really unique creative tool.

Kodak Ektachrome 100 Plus Professional and 100 HC

These two medium-speed siblings are examples of highly refined ISO-100 color transparency films whose performance is almost identical. Photographers who use these emulsions on a regular basis report them to be an almost optimum blend of fine grain, moderate contrast, neutral to slightly warm color rendition, and excellent color saturation. Each is capable of producing outstanding slides. Because they are E-6 films, they can be turned around very rapidly, even at home. The moderate contrast they exhibit is a bonus to those planning to make prints from their slides, and helps preserve detail in shadows and/or highlights of the enlargement. It is interesting to note that many of the comments aimed at the two Ektachromes were also made about other medium-speed color transparency films, such as Fujichrome RD 100 and Agfachrome CT 100i/RS 100+ Professional. It seems film-makers have gotten it right with the medium-speed color emulsions.

Scotchchrome 400

The latest generation of this medium/high-speed emulsion has a number of good things going for it. The film displays very good grain structure and color balance. These qualities hold up well when the film is exposed at higher exposure indices. The film has been designed to be push-processed one or two full f-stops, with minimal change in major color or grain qualities. Testers who used the film reported its performance at ISO 400 and 800 almost identical. At ISO 1600, viewers noticed slight contrast buildup and some moderate color shifting. As is the case with other Scotch film products, Scotchchrome 400 is sold at a lower price than comparable ISO-400 films.

Agfachrome RS 1000 Professional

I asked testers to shoot this high-speed film under bright but cloudy conditions, full daylight, and marginal light, normally the forte of the film. Results were rather interesting and somewhat unexpected. As I had found in years of shooting with this emulsion, it delivered pictorially unusual results under overcast conditions but it also delivered very interesting images in bright overcast and full light, provided it was underexposed by about ¼ to ⅓ f-stop (EI of 1250). Grain is prominent but tight. This in turn produced a pointilistic look that abstracted subjects while at the same time providing good image detail. Several photographers who used this film for the first time discovered they could generate a painterly version of subjects normally depicted realistically. Almost every tester agreed Agfachrome RS 1000 provided them with the capacity to take a fresh look at commonly shot subjects.

Fujichrome P1600 II Professional

As the world's fastest color transparency film, P1600 II permits photographers access to subjects formerly unphotographable. If a bit contrasty for some, the film does provide an opportunity to photograph subjects in a novel way. Now, image-makers can take advantage of super high speed shutters up to 1/8000 second but at apertures of f 5.6 or 8 rather than wide open. Shooting at these moderate f-stops and action-stopping shutter speeds is worth the quality tradeoff of using a relatively grainy high-speed film such as P1600 II.

Increasing Film Speed

Uses

Higher effective ISOs allow photographers to expose images at reasonable shutter speeds and moderate f-stops. Push-processing is one technique. The practical limit is one or two f-stops to minimize loss of quality. You can push more if you are willing to accept greater losses. On the minus side, there is bound to be some grain build-up. Whether it is objectionable or not is a matter of judgment. Also, increases in contrast in shadow areas leads to significant detail loss.

Increasing effective ISOs calls for processing compensation. With slide film, first developer times must be increased; for black-and-white film, the basic developing times need to be increased. Table 5.1 gives some guidelines for general-purpose developers such as D-76, used in conjunction with general-purpose black-and-white films.

Pulling Black-and-White Film

Reducing effective ISO has two effects. First, excessive contrast can be reduced, especially in the vital shadow areas. This is done by lowering the shooting ISO and decreasing the developing time by specific amounts. A further effect of this technique is to reduce apparent grain well below what is expected. For decades black-and-white photojournalists rated Tri-X be-

Table 5.1 Exposure Modifications When Using General-Purpose Developers

Change in Rated ISO	*Exposure Change Desired*	*Change in Development Time*
Black-and-White Film		
2×	–1 f-stop	+25%
4×	–2 f-stops	+75%
8×	–3 f-stops	+150%
1/2×	+1 f-stop	–30%
1/2×	+2 f-stops	–30% And Dilute by 50%
Color Negative Film		
2×	–1 f-stop	+30%
1/2×	+1 f-stop	–30%
Color Transparency Film		
2×	–1 f-stop	+25 %
4×	–2 f-stops	+75 %
1/2×	+1 f-stop	–25%
1/2×	+2 f-stops	–50%

tween ISO 250 and 320 instead of its standard ISO 400. When the derated film was given about 20 percent less exposure, visible grain decreased. One should be careful not to overuse pull-processing in situations where the subject exhibits normal or low contrast, for flat, dull negatives usually result.

Pushing and Pulling Color Negative Film

For the most part, either pushing or pulling most color negative films produces distinctly inferior negatives. Push-processing print film results in unsatisfactory color reproduction with an imbalance of colors in different portions of the negative. Trying to bring the colors back into balance using color printing filters in the darkroom is almost always a frustrating experience. As soon as you balance one portion of a pushed negative, another goes out of balance. Additionally, image contrast is adversely affected.

Pulling is even less desirable. The worst effect is a drastic reduction in negative contrast. If you use general-purpose color negative film, you don't really need either push or pull. Negative films possess impressive exposure latitude, sometimes exceeding 6 f-stops. Taking this into account, it is plain to see that it is questionable to employ either technique.

Color Transparency Films

All color slide films, including Kodachrome, are amenable to some extent to pull- and push-processing. The push/pull range is from about +2 f-stops to –1 f-stop. Increasing effective ISO by increasing first developer times almost always results in grain and contrast buildup. If the push is moderate, grain and contrast can be kept under control. Color will begin to shift usually toward red almost immediately even with ½ f-stop push-processing.

Some films, such as Fujichrome 100 and Agfachrome RS 100+, are particularly well suited for pushing. You will incur a minimum performance penalty if pushing is limited to no more than one f-stop. It is only when you choose to go beyond one f-stop that things get a bit dicey. However, if you need extra speed and pushing the film is the only option open to you, then the compromise in film performance may be acceptable. When you pull slide film, the result is an overall lowering of image contrast, distorted color, and marginally finer grain. Expect an overall color shift toward blue. Pulling slide film might be the only viable technique to tame extremely high contrast lighting where important visual information is contained in both shadows and highlights. Pulling is only practical if high color saturation is *not* a primary goal. Pull-processing invariably leads to muted, distorted colors.

Deliberate Under- and Overexposure without Processing Compensation

Overexposure of black-and-white or color print film leads to black-and-white negatives that are more contrasty and color negatives more color-saturated than normally exposed films. Some manufacturers purposely underrate their print films ½ f-stop slower than they really are. This works well for point-and-shoot camera users whose meters are notoriously inaccurate. Visually, prints made from overexposed color print negatives seem more color-saturated and snappier. The contrastier prints appear sharper. Straight underexposure of print film is not a recommended procedure unless shadow detail is absolutely essential and highlights can be sacrificed. In any event, limit underexposure of color print film to no more than 1 f-stop. Underexposure of black-and-white print film without attendant processing adjustments *will* increase film speed, but at a price. Both contrast and grain grow in proportion to the degree of underexposure. If the subject is flatly lit, showing little contrast, and contains lots of detailed mid tones, uncompensated underexposure can generate "snap" in the image through increased contrast, while at the same time maintaining reasonable grain. Underexposing color print film by one or two f-stops still leads to acceptable prints because of the wide exposure latitude of the material. Beyond two f-stops, underexposure color begins to wash out perceptibly.

Deliberate, uncompensated underexposure is most effective with color slide films. Used judiciously, underexposure increases effective ISO, color intensity, and contrast. Underexposures of ½ to 1 full f-stop can be tolerated by most slow- and medium-speed slide films. Underexposing higher contrast slide film like Fujichrome 50, Velvia, or Ektachrome 50 HC does indeed yield dramatic increases in color, but it does lead to decreased shadow detail to a point where important information may be totally lost. However, if you seek graphic, intense color, and a litho look, just underexpose an ultra high contrast slide film like Polaroid Presentation Chrome. Color should be as intense as it can be and contrast should compress considerably, adding to the drama.

The chart that follows summarizes the judgment of more than eighty experienced color photographers. Remember, judgments are relative and differences between films may be slight or pronounced and depend on more than the film itself. Factors such as camera calibration, lighting conditions, subject matter, picture-taking technique, or darkroom processes used contribute to the appearance of the finished slide or print.

Color Bias Chart of Selected Films

Cool (blue)	*Neutral*	*Warm (red)*
Ektachrome 64 Professional	Kodachrome 25	Ektachrome 64X Professional
Ektachrome 100 Professional	Ektachrome 50, 100 HC	Ektachrome 100X Professional
Ektachrome 200	Ektachrome 100 Plus	Ektachrome 400X Professional
Scotchchrome 100	Fujichrome 400	Ektachrome 400 HC
Scotchcolor 100, 200	Fujicolor HG 400 Professional	Fujichrome 50
Scotchcolor 160	Fujicolor Realia	Fujichrome Velvia
	Fujicolor 160S Professional	Agfachrome RS 100 Plus
	Konicacolor 160S	All amateur Fujicolor
	Kodak Vericolor 111 Professional	All amateur Kodacolor Gold Plus
	Agfacolor Portrait 160	
	Agfachrome RS 200	
	Kodak Ektar 25, 100	
	All Kodak Ektapress	

Table 5.2 summarizes the judgments of more than three dozen photographers who routinely make enlargements from their negatives or slides. Each was asked to examine same-size prints made from different-speed same-type films. All films were exposed and printed under uniform but not strictly controlled conditions in order to simulate what one might experience in the real world. Panel members were asked to answer one question: How much can films from each ISO category be enlarged and still produce sufficient detail, contrast, and/or color, maintaining reasonable grain to be judged as high-quality images? Even though the entire proceeding was almost completely subjective, as when judging the quality of anything, results were surprisingly consistent with 9 out of 10 agreeing where high quality ended. Judgments varied a great deal only when panelists were asked to consider the virtues of high-speed films. Overall, only twice did a judgment vary by more than one standard enlargement size (for example, by more than from 5 × 7 to 8 × 10).

Table 5.2 Maximum Enlargement* Film Selector

Film ISO	*Moderate to High-Contrast Subjects*	*Low-Contrast Subjects*
Black-and-White Film		
25–32	20 x 24+	16 x 20
100–125	16 x 20	11 x 14
400	8 x 10–11 x 14	8 x 10
1000	8 x 10	5 x 7+
1600–3200	Up to 8 x 10	5 x 7
Color Negative Film		
25	20 x 24+	16 x 20
50	Up to 20 x 24	16 x 20
100–125	16 x 20+	11 x 14+
200	11 x 14+	8 x 10
400	11 x 14	8 x 10
1000–1600	8 x 10+	8 x 10
3200	Up to 8 x 10	5 x 7
Color Transparency Film		
25–50	20 x 24+	16 x 20
64	16 x 20+	11 x 14+
100	16 x 20	11 x 14
200	Up to 11 x 14	8 x 10
400	8 x 10+	8 x 10
1000	8 x 10	5 x 7 +
3200	Up to 8 x 10	Up to 5 x 7

* In inches

Other Creative Techniques

Preflashing

Preflashing is a means of slightly fogging film or enlarging paper to reduce overall contrast. This procedure is usually confined to the darkroom but it can also be employed in-camera. If successful, preflashing will give you control over contrast of individual frames of film. For preflashing to work, you must own a camera that permits multiple exposures. The technique itself is not complicated, although you will need to purchase an 18 percent reflectance gray card.

Expose the subject as you normally would. Next, position a gray card so that the same light falls on it as on the subject. You are now ready to preflash an individual frame of film. To do so, completely fill the viewfinder with the gray card and defocus the lens. Set exposure for 1/100 of that indicated to properly expose the gray card. Remember to follow multiple-exposure procedures so that the frame of film does not advance. To achieve a moderate preflash effect, exposure should be reduced by 6 to 6½ f-stops. With ISO-50 film in the camera, a typical outdoor day exposure under bright but cloudy sunlight is around f-11 at 1/60 second. A preflash exposure would work out to be around f-22 at 1/1000. If the calculated preflash exposure is beyond the range of available f-stops and shutter speeds, use a polarizer or neutral density filter to reduce the light uniformly by one or more f-stops. The effects of preflashing vary greatly from film to film so it's a good idea to bracket exposures several f-stops on either side of the calculated exposure. Use your favorite film and run some tests at different exposures.

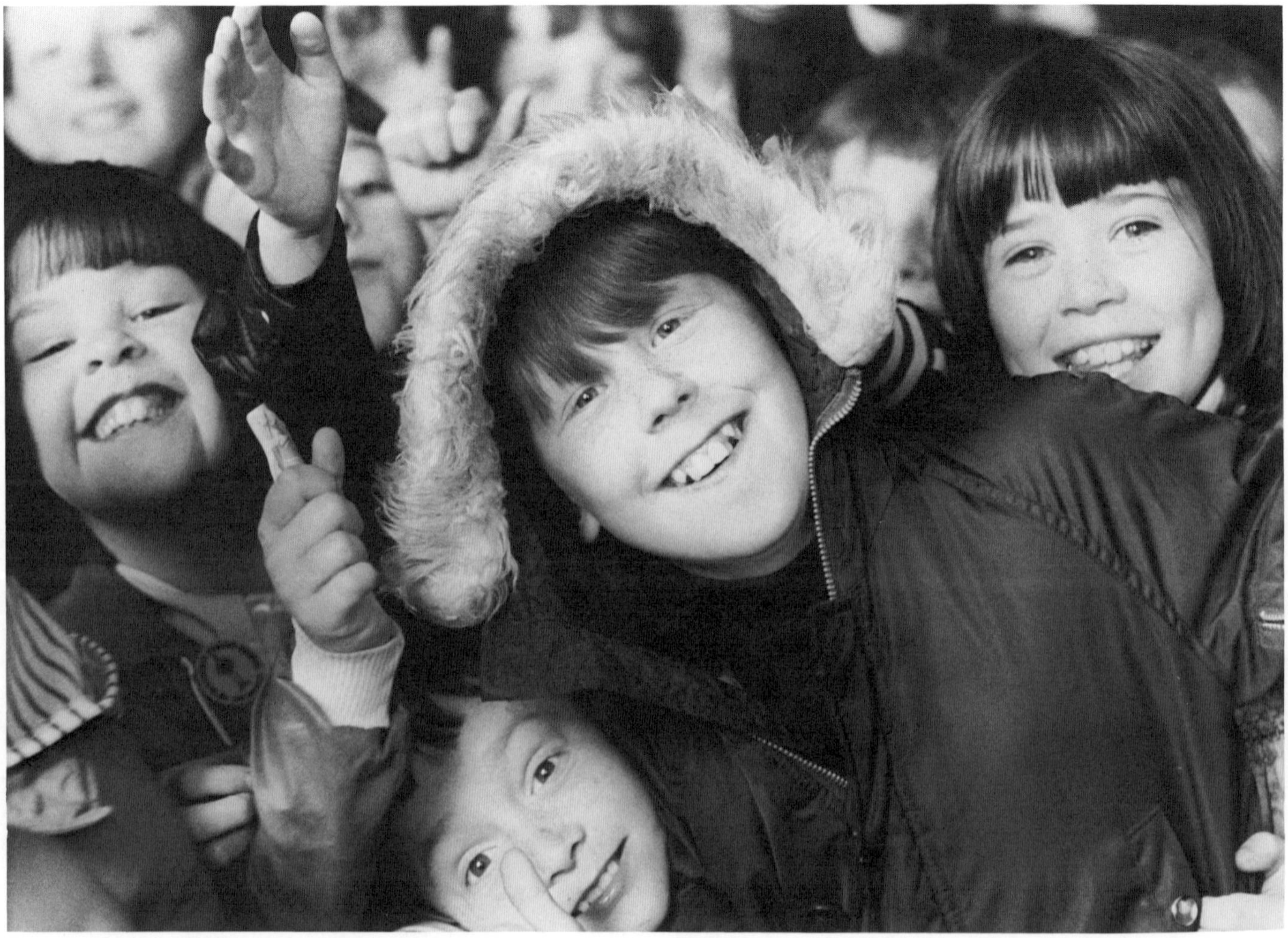

Figure 5.4 Marc Levey. Agfa Pan APX 100 at ISO 80.

Exaggerated Grain

Aside from using high-speed film whose grain is normally pronounced, there are a number of grain-increasing techniques at your disposal. One such technique involves modifying exposure and, at the same time, modifying development (see over- and underexposure in this chapter). Another way to produce increased grain is via a grain screen, which can be purchased or can be made by photographing various grain patterns on fine-grain black-and-white film and then sandwiching them with a slide or negative. They may even be mounted inside the guide rails to cover the film opening inside the camera itself. Commercially made micro-grain screens may be used over the lens much like an ordinary filter.

Slides from Black-and-White or Color Negatives

Excellent quality slides are possible from black-and-white or color negatives using a special slide film made by Kodak, designated Vericolor Slide Film 5072/SO 279. This tungsten-balanced emulsion is a low-contrast material. Very little if any resolution, color saturation, or contrast is lost in the production of second-generation slides in this manner. With this film, experienced darkroom workers can closely match the color of the original. The emulsion is processed in ordinary C-41 color negative film chemistry.

Negatives, Slides, and Prints from a Single Film

For several years, a few specialized color processing labs have offered a service whereby negatives, slides, and prints could be had from one film—Eastman Type 5247 35mm motion-picture film. Recently, other films have been added from Kodak and Fuji. When first introduced, the end products, although convenient, were below the quality of prints for negative film or slides made directly from transparency film. Frequently, contrast was low and color less than perfect. Today, the labs offering the all-in-one service deliver a much higher level of quality, as seen in finished prints and slides.

Labs formerly limited to C-22 process motion-picture films now accommodate all color negative films processed in the up-to-date C-41 chemistry. The ability to utilize any of the many superb color print films on the market plus several motion-picture films allows processors to supply original prints of near-professional caliber. At the same time, they are able to produce slides of high color saturation and sharpness. These slides are often equal or superior to in-camera originals. Unquestionably, there is widespread disagreement as to which color film, slide, or print is preferred. The ability to use negative film with its expanded exposure latitude and still get excellent slides is compelling. (See the next section for a summary of arguments for and against the use of color slide or negative film.) Several individuals I know use this service regularly and are quite satisfied. A few others report some disappointment with slides in particular. My experience is limited, but the best results from slides and prints from one film have been equal to any of my usual lab's work with conventional films and processing.

Do not expect to drop your film off at the local drugstore. You will need to mail exposed color negative film to one of three specialty labs currently offering the service. You have the option of several combinations of services: negatives and slides; negatives and prints; and all three—negatives, prints, *and* slides. For all three, prices range from $14 to $16.50. This compares favorably with individual services. The labs and their addresses are included in Appendix B. You may write for an ordering packet, which includes mailers and price lists for this and other services.

Using the Wrong Film at the Right Time

Purposely loading tungsten film into your camera for use under daylight conditions would *seem* to be self-defeating, but not always. Sometimes knowing the color cast a particular film produces helps make possible strong, monochromatic images emphasizing a certain mood. For example, tungsten film exposed in daylight will display a strong blue cast. Under strong overcast—the kind that already exhibits a cool, bluish hue—tungsten film will exaggerate and strengthen the coolness of the scene. Underexposing this combination produces a moonlight effect. Conversely, employing daylight color film under tungsten lights produces color that is mostly described as an overly warm yellow-orange. In special circumstances, such as mixed daylight and incandescent light, the effect of using daylight film can be visually arresting. Figure studies, some types of portraits, and certain indoor scenes benefit from an overemphasis on the warm portion of the color spectrum. The rule with either is to not use the wrong film without thinking through the probable consequences.

Color Slides Processed in Negative Chemistry

Sometimes it is possible to talk the local color lab into running a roll of E-6 color slide film through standard C-41 color negative chemistry. You may have to work hard to convince them that such processing *will not* adversely affect other print films being processed at the same time.

What can you expect? First you will see vastly increased contrast. Next, colors will be reversed, with complementaries flip-flopped. Green grass shifts toward its complement, magenta. Blue sky become yellow, whites go black, and caucasian skin tone a ghoulish yellow/green. If you wish to limit contrast buildup, underexpose or preflash the film.

To Use Color Slides Versus Color Negatives

On the color print side:

1. Color print film has much greater exposure latitude, amounting to at least 2 f-stops under- and as much as 4 stops overexposure. Slide films allow only about ½ f-stop over- and 1 f-stop underexposure before showing adverse effects.
2. Print film gives photographers better contrast control with more open shadows and highlights.
3. Variations in contrast and color balance in the film can be adjusted in the printing stage, offering more control over these variables.
4. When making enlargements, first-generation negatives offer sharper and less grainy results than using a second-generation internegative from an original slide.
5. Color negatives facilitate printing quality black-and-white prints on panchromatic paper. Slides don't.
6. Color negatives, good slides, and prints can be made from print film all at the same time and at a competitive price.
7. Color print film produces more pleasing color balance under tungsten, fluorescent, and mixed light.
8. Most people prefer to view prints over slides. Prints are far more accessible.

The case for slide film:

1. Similar-speed color slide film is both finer grained and of a higher resolution than print film.
2. There are many more choices of slide films. Slide films differ more from one another than print films do. Slide films have distinct "personalities."
3. Almost all publications require color transparencies.
4. Because of slide film's limited exposure latitude, photographers must expose more precisely than when shooting print film.
5. Finished slides are more economical than color prints. Further, slides may be used to choose images to be printed. Negatives must first be printed before judging the final results.
6. Projected slides create a more dramatic and dynamic visual experience than viewing a print by reflected light.

To Use Color Versus Black-and-White

Along with the negative-versus-slide controversy, there exists another age-old debate—the question of the merit of black-and-white and color as vehicles for expression. The following discussion does not attempt to answer the question, but merely to put forth the major arguments on both sides.

The question of whether to photograph a subject in color or black-and-white has no ready answer. Photographers and critics have argued for decades for the use of each. Still, no one has discovered a pat formula that seems to satisfy the photo world. To explore this issue, I present a case for both black-and-white and color. As you will see, both are compelling.

The case for black-and-white usually takes two forms:

1. Arguments extolling the virtues of black-and-white photography.
2. Criticism of color photography.

Adherents argue that black-and-white is by its very nature an abstraction, and therefore ideal as a vehicle for interpretive image-making. Enthusiasts point to the inherent pictorial qualities of black-and-white that allow photographers to emphasize pure form, line, texture, and tone. They state that when these qualities are of prime importance, black-and-white is the clear choice. Beyond this, they suggest black-and-white had been the only photographic material traditionally taken seriously as "art."

Those who disparage color see it as a gimmick. They imply color often dominates a photograph at the expense of content—in other words, color obscures meaning. They argue further that color makes photography too explicit; photographs become mere replications of reality rather than personal interpretations. In addition, many supporters of black-and-white believe it spurs creativity and innovation expressly because there is no color on which to lean. In other words, they suggest it is more difficult to produce an outstanding photograph in black-and-white.

Color enthusiasts, on the other hand, believe that in fact color brings photography alive. They believe color photographic images have the power to reach out and touch us in a very intimate way. Of all the visual communications media, color photography is the most real. Black-and-white, they claim, is sterile, analytical, and aloof. Color does deal with tone, line form, and texture, and does so better because of the added dimension of color. Last, adherents point to the billions of color photographs taken each year that represent over 90 percent of all photos taken worldwide and ask, Can this many people be wrong?

In my view, both positions contain grains of truth, although each is perhaps overstated and misrepresentative, at least to some degree. The first point to consider is that certain photographs are great because they live over time. Their appeal lies in the sense of wonder, excitement, anger, or surprise they evoke. Simply because a photo is rendered in black-and-white or color is not the issue. Neither can possibly have a monopoly on truth. And truth is the quality that allows a photograph to live. I believe (and this I ask you to test) that black-and-white and color have very different expressive potentials. The properties unique to each demand their use under different circumstances. In other words, each has its place as an expressive tool. It is not whether black-and-white is better than color but how well each can be used to communicate the photographer's intent. Your task is to decide when the expressive properties of each can be used to best advantage.

Since color has become a pervasive force in modern still photography it might be well to consider the effects of color on viewer perception. The following attempts to give the reader some background in this important area.

The Psychological Effect of Color

Individual colors project quite specific meanings. However, the overall color emphasis a photograph projects can do far more to influence viewer perception than any one color. Thus, color can give an image a feel much the same as the tonal relationship does in a black-and-white image.

As we saw in the preceding discussion, black-and-white purists reject color as a key ingredient in the creative mix, viewing it only as a crutch often masking inferior images. Because color is a dominant force, it seems logical to understand its potential as a creative tool although we may choose not to use it. There is overwhelming evidence that color is a potent persuader. Examining the work of contemporary advertising and fashion photographers is very revealing in this regard. These photographers use color to manipulate our perceptions, creating an emotional climate conducive to our acceptance of the client's product or idea. For them, color is a critical component in what is essentially a rhetorical situation. Advertisers have learned that subtle changes in overall hue, tint, or shade often result in disproportionate shifts in image emphasis and meaning. For a fuller understanding of color, we also need to bear in mind constantly that color as we see it is different from the color recorded on film (this is a subject covered in some depth in another section of this book). The qualities of color we see (chroma, tint, shade) define the relative importance of a color perceptually. Color may be subdued and quiet, alive and energetic, or anywhere in between. The task the photographer has is twofold: to understand how color is recorded on film and how color alters our perceptions of the photographic subject. For a better understanding of the action of color as it relates to perception find a copy of Josef Albers's classic *Interaction of Color.* I recommend it highly.

35mm Film Recommendations by Subject

Candid Portraiture

Several industry surveys demonstrate conclusively that pleasing skin tone is the most important ingredient when people determine the success of any portrait. Films that render skin tones naturally and with soft color palette are generally preferred over more color-saturated color films.

Lenses: 28–85mm, fast 85–90mm, shoot close to wide open.

Films: ISO 400 black-and-white or color; Ilford XP-2 at ISO 800; Fujicolor HG 400; Scotchchrome 400 at ISO 800; Scotchchrome 640-T; Ektar 1000; Agfachrome RS 1000; T-Max P3200.

Outdoor (General)

Lenses: 28–85mm, 70–210mm zooms, 300mm 2.8–4.0

Film: In bright daylight: T-Max 100; XP-2 at ISO 200–400; almost any ISO-100 transparency film; Ektar 100; Realia; Konica SR 100–200; Scotchcolor 100; Agfa Portrait 125. In marginal light (overcast, rain, snow, and fog): XP-2; T-Max 1000; Ektachrome 400X, 400 HC; Fujichrome; Fujicolor 400; Konica Super SR 400; Kodacolor Gold 400 Plus; Scotchchrome Color 400.

Formal Portraiture

Lenses: 85–135mm, 28–105mm short telezooms. 28–35mm wide angles for groups.

Film: T-Max 100; Agfa Pan APX 25; Fujicolor HG 400 Professional; Ilford FP4 Plus; Vericolor III Professional; Realia; Agfacolor Portrait 125; Ektachrome 100 Professional; Agfachrome CT-100i, Agfachrome RS 100 Plus.

Children

Children must be stalked like animals in the wild. Fast film allows you to capture fleeting expressions, rapid movement. Shoot wide open if you can.

Lenses: fast 85–200mm telephotos, zooms.

Film: In good light: medium-speed films in the ISO 100–200 range; T-Max 100; XP-2; FP-4 Plus; Ektachrome 64X, 100X; Fujichrome 100RD; Ektachrome 100 HC, 100 Plus; Fujicolor Realia; Kodak Ektar 100; Vericolor III; Fujicolor 160S; Konica Super SR 200. In marginal light: XP-2; Fuji Neopan 400; HP-5 Plus; Kodachrome 200; Fujichrome 400; Fujicolor HG 400 Professional; Kodacolor Gold 400 Plus; Konica Super SR 400; Scotchcolor 400.

Sports

Most sporting events move faster than slow fine-grain film can adequately handle. The primary consideration in most sports photography is to freeze action at its peak. To do this consistently, you'll need three things. First, and most important, is an understanding of the nuances of the sport itself. Second is the use of a relatively fast long lens. Finally, you will need to employ medium to fast film, ISO 100–400. There will be occasions where lighting will call for much faster film up to ISO 3200.

Lenses: Fast telephotos for stadium sports such as football, baseball, soccer: 200–300mm f-2.8, 400–600mm f-4.0. Basketball and other close-range sports such as tennis, gymnastics, ice hockey, track: 80–200 f-2.8 zoom. 24–28mm wide angle.

Film: Black-and-white; Ilford XP-2; Kodak T-Max 400, T-Max 3200; Fuji Neopan 1600 color; Ektar 1000; Ektachrome 400X HC; Kodachrome 200; Fujichrome 400; P-1600 II Konica SR 3200. If lighting is very bright, film recommended for portraiture can be used.

Animals in the Wild

Much of what is true of sports is also true of wildlife photography. Be aware that you will need to contend with contrast-reducing atmospheric haze when shooting distant subjects.

Lenses: Fast, high-quality high-contrast telephotos are recommended. Shooting wildlife calls for lenses of at least 300mm focal length. If you are willing to give up some sharpness and fine grain, use faster film and a slower 300–500mm lens in the f-4–8 range.

Film: Accomplished nature photographers who use fast lenses tend to load slower, fine grain slide films in their SLRs. Kodachrome 64, Fujichrome 50, or Velvia are some of the favorites. In Europe, Agfachrome RS 50 and 100 Plus have attracted a devoted following. I have found the extra 1/3 to 1 f-stop conferred by the current crop of fine-grain medium-speed films offers advantages of faster shutter speeds that far outweigh the small amount of grain increase their use produces.

Film: Black-and-white: Ilford XP-2, T-Max 100. Color: Kodachrome 64; Fujichrome Velvia; Fujichrome 50; Fujichrome 100 RD; Ektachrome 100 Plus; Kodachrome 200; Ektar 100.

Landscapes

These subjects are encountered in all types of light. It is therefore difficult to recommend specific lenses or film for all situations but here are some general observations.

Lenses: Landscape subjects respond well to the use of wide-angle lenses from 20mm to 28mm. The great inherent depth of field conferred by W/A lenses enhances front-to-back visual organization guiding the viewer's eye through the picture. Also, many landscape photographers employ polarizing filters regularly to improve color rendition and to compress scene-contrast range, thus making shadows and highlights more printable.

Film: Black-and-white: Agfapan 25; Forte 100; Technical Pan; Ilford Pan F; T-Max 100. Color: Fujichrome 50; Velvia; Agfachrome RS 50 Plus; Ektachrome 64X, 100X; Kodachrome 25, 64; Ektar 25; Agfa Ultima 50; Konica Impresa 50; Fujicolor Realia. Consult chart in this chapter for film recommendations on cool and warm subjects.

Architecture

Architecture demands attention to detail and lenses and film that record detail faithfully are recommended.

Lenses: Perspective control or tilt-and-shift lenses are most useful. These are usually moderate wide-angle optics in the 28mm–35mm range. Short telephoto macro lenses offer the highest levels of definition and contrast for this demanding work. Urban architectural subjects such as bridges and skyscrapers sometimes require longer telephotos. For sharpest results, use a tripod whenever possible.

Film: Black-and-white: Technical Pan; Ilford XP-2 at ISO 100; T-Max 100. Color: Kodachrome 25; Agfachrome RS 50 Plus; Fujichrome Velvia; Fujichrome 100 RD; Fujicolor Realia; Ektar 25; Agfacolor Optima; Konica Impresa 50.

Close-up/Macro

Close-up work of any subject demands precise composition, exposure, and attention to depth of field. You will be working at very small f-stops and slow shutter speeds so careful work is a must. As a general rule, some form of specialized close-up equipment is necessary. These can be as simple as screw-in close-up filters to a true flat-field macro lens.

Lenses: Supplementary close-up lenses in several strengths, fully corrected macro lenses in 50, 100, or 200mm focal lengths. There is also at least one special 1–3× zoom macro available. If you desire, you can convert your normal lens into a macro of reasonable performance via a lens reversing adaptor. Reversing a normal lens makes its macro performance much better but you lose all automation, including autofocus and automatic metering.

Film: Mounting the camera/macro combination on a tripod permits use of slow (ISO 25–50) high-definition black-and-white or color film. There are occasions when ISO 64–100 film is a good choice in that it allows smaller f-stops with reasonably fast shutter speeds. As with most film choices, you should choose the slowest film that will satisfy depth of field and shutter speed requirements for the particular subject. There are exceptions, of course, but it's a good guideline to follow for macro work. Black-and-white: Technical Pan Film; Ilford Pan F; T-Max 100; Ilford XP-2 at ISO 50–100. Color: Kodachrome 25, 64; Fujichrome Velvia, 50D; Agfacolor Ultima 50; Agfachrome RS 50 Plus; Ektar 25; Fujicolor Realia; Konica Impresa 50.

Underwater

Underwater photography poses some special problems. Down to about sixteen feet (five meters), light in clear water is sufficient to take pictures without artificial lighting. Below that depth, waterproof underwater electronic flashes will be necessary.

As depth increases, water becomes more blue with red and yellow being absorbed in greater and greater amounts. In shallow water, a CC 030 or 050R filter corrects much of this problem.

Lenses: Specialized UV lenses corrected for underwater work are desirable for critical applications. Several camera-makers offer waterproof point-and-shoot cameras for use down to about 4.5 meters depth.

Film: Black-and-white: Fast (ISO 400 or greater) are recommended. Color: Films with warm color rendition are highly recommended. Using electronic flash, Ektachrome 100X is a good choice, as is Fujichrome 100 RDP, Agfachrome RS 100 Plus, or Kodachrome 200. Fujicolor Super HG and Kodacolor 400 HC and Kodachrome Gold 400 Plus provide speed and warm, saturated color.

Nighttime

The night provides us with a wide variety of subjects. At one time, ISO-400 film was the most sensitive film available and night photography was limited to comparatively well-lit subjects that were almost static. With the advent of ultra-fast film, we now have more flexibility in both subject and film. Note: Moonlight is reflected sunlight and as such is well matched to the spectral response of daylight color film. Table 5.3 gives exposure recommendations for common night scenes.

Lenses: For candid, hand-held photography, the faster the better. Lenses such as 85mm f-1.4–1.7, 200mm 2.8, 35mm f-1.4 or normal 50mm and sensitive film give the shooter options slower combinations don't.

Film: For stills on a tripod, any medium- or high-speed black-and-white or color film of your choice will do. For hand-held work at night: black-and-white: T-Max P3200; Fuji Neopan 1600 or Ilford XP-2 rated at ISO 1600. Color: Fujichrome 400 or Scotchchrome 400, both push-processed to ISO 800–1600, for lighting rich in tungsten, Scotchchrome 640-T. All other situations calling for color: Fujichrome P1600 II Professional at ISO 1600; Scotchchrome 1000 or Agfachrome RS 1000; Kodak Ektapress 1600; Ektar 1000.

The Sun

When the sun is the primary element in a photograph, great care must be taken not to damage your eyes by looking directly at the sun through the viewfinder. Focus by placing a piece of white cardboard an inch or two from the viewfinder eyepiece and move the camera until the sun's disc appears on the card. Trial and error will give you some idea of the sun's position in the frame.

Sun exposures can be made over a very wide range. By altering exposure, many different effects are possible, from a pastel noon look through to a deeply colored disc. Viewable images are possible over a 5 to 6 f-stop spread. With selected filters such as green, blue, and especially 25A red, surrealistic silhouettes are attainable. Several photographers I know have

Table 5.3 Exposure Recommendations for Common Night Scenes

Scene	*Film ISO*	*Exposure*
Fireworks	100	f-8 @ 20 sec.
	400	f-11 @ 10 sec.
	1600	f-16 @ 5 sec.
Moonlight (full, clean sky)	100	f-8 @ 20 sec.
	400	f-11 @ 10 sec.
	1600	f-16 @ 5 sec.
Neon marquees, billboards	100	f-8 @ 4 sec.
	400	f-11 @ 2 sec.
	1600	f-16 @ 1 sec.
Portraits under streetlight	100	f-5.6 @ 8 sec.
	400	f-4.0 @ 1 sec.
	1600	f-2.8 @ 1/8 sec.
	3200	f-2.0 @ 1/60 sec.
Candlelight (one or two candles)	100	f-5.6–8 @ 4 sec.
	400	f-5.6 @ 1 sec.
	1600	f-2.8 @ 1/15 sec.
Candids on street	100	f-5.6 @ 10 sec.
	400	f-4.0 @ 1 sec.
	1600	f-2.0 @ 1/15 sec.
	3200	f-2.0 @ 1/60 sec.
Night interiors (average illumination)	100	f-5.6 @ 1/8 sec.
	400	f-5.6 @ 1/30 sec.
	1600	f-5.6 @ 1/125 sec.

In all instances, it is wise to bracket exposures + one or two f-stops.

shot dozens of different sun shots and used them to sandwich with slides or negatives of other scenes. This can be done with the moon as well.

Figure Studies

The human figure is not an easy subject to photograph well. Careful study and sensitivity to shape, form, texture, line, and tonality are essential to success. Traditionally executed in black-and-white, nude figure studies are increasingly being done in color. Also, more and more male figure studies are being exhibited in well-known galleries. Nonetheless, the female form still dominates photographic figure study.

Lenses: Short telephoto lenses in the 80–135mm range are the preferred optics for this type of photo work. They produce natural-size relationships front to back and do not distort the relative size of any one portion of the body over another.

Film: Black-and-white: Slow- to medium-speed film that yields wide tonal range and good tonal separations highly desirable. T-Max 100, Ilford FP-4 Plus, and Neopan 100, are effective choices. Color: Soft, lower contrast films often produce the qualities that enhance the human form. The following films are lower in contrast and/or color saturation than other color emulsions but the differences are not profound. The same soft pastel rendition is achievable through the use of special soft-focus filters such as Tiffen's Soft Net series or a variety of other makers of soft-focus or diffusion filters. Ektachrome 100 Professional; Agfachrome CT-200; Fujicolor Realia; Fujicolor HG 400 Professional; Fujicolor 160S; Scotchchrome 100; Vericolor III Professional; Agfacolor Portrait 160 Professional; Konica SRG 160.

For the male figure, all of the above films work well. If you seek a more contrasty and strongly color saturated look, then the following films should do nicely: Fujichrome Velvia; Kodak Ektachrome 50 HC, 100 Plus; Ektachrome 64X, 100X; Agfacolor Ultra 50; or any Kodacolor Gold Plus, Fujicolor Super HG, Konica print or Scotchcolor print films. Additionally, you may wish to experiment with Scotchchrome 1000 or Agfachrome RS 1000 rated at ISO 2000 for a pointilistic, super high grain effect.

Theater, Indoor Performance

Check the lighting type to achieve proper color match between it and the color film. Consult the lighting color balance chart elsewhere in this chapter for starting filtration. Lighting in theaters generally requires use of fast lens/film combinations.

Lenses: 80–200mm or 300mm f-2.8; 85–100mm fixed focal length f-1.4–2.0 lenses when working close to the stage.

Film: Experience dictates ISO-400 film is about the slowest emulsion most lighting set-ups in theaters permit. Black-and-white: Fuji Neopan 400; Kodak T-Max 400; T-Max P 3200; Fuji Neopan 1600; Ilford XP-2 at ISO 400–800. Color: Fujichrome 400; Scotchchrome 400 at ISO 800; Fujichrome P- 1600 II Professional; Agfachrome RS 1000 Professional; Scotchchrome 640T for tungsten illumination without the use of filters; Ektar 1000; Ektapress 1600 Professional; Konicacolor SR 3200.

Film Travel Kits

The recommendations made here are for several itineraries, from a day trip to travel of two or more weeks.

Basic guidelines include:

1. Carry more film than you think you will need. Experienced travel photographers carry at least 50 percent more film than they calculate needing.

2. Mix film types with an eye to probable subject matter and lighting and photographic purpose. Always have a few rolls of fast (ISO 400) or ultra-fast (ISO 1000+) film on hand. Also carry a few tungsten-balanced rolls.

Day Trip

one roll fast black-and-white or color film, ISO 400 or faster
one roll tungsten balanced color film
three to four rolls medium-speed film, color or black-and-white, ISO 100–125.

Note: if you take more than one camera, add 50 percent more film to the above recommendations.

Weekend Trip

Essentially double the day-trip numbers. Mix the same. If you are traveling to a metropolitan area and have the time, consider purchasing your film in a high-volume photo discount store. You can save money.

two rolls fast film
two rolls tungsten-balanced film
six to eight rolls medium-speed film

One-Week Vacation Trip

If you will be doing a good deal of sightseeing, you will probably need more film than if you are going to one spot (like a beach house) and staying put.

two to three rolls fast film
one roll ultra-fast film (ISO 1000+)
one roll slow, fine-grained film (ISO 25–50)
eight to ten rolls medium-speed film
one to two rolls tungsten-balanced film

Two or More Weeks Trip

Planning an extended trip involving photography as a major activity requires you consider factors such as the local availability of film, luggage space, and your film-usage habits. If traveling overseas, particularly to remote or third world areas, assume the film with which you begin the trip will be the only film available to you. When traveling to exotic locations, count on exposing more film than normal.

Buy 36-exposure rolls for medium-speed film and shorter lengths if possible for other types.

three to four rolls high-speed film
two rolls tungsten-balanced film
twelve to twenty-four rolls medium-speed film
two to three rolls ultra-fast film (ISO 1000+)

If you are going on a safari or other nature/wildlife expedition, you should carry a higher proportion of fast film (50 percent) and/or fast lenses.

I average over five rolls of 36-exposures film each shooting day on trips to locations such as East Africa or Europe. Keep in mind my film consumption is on the high side. And because I shoot mostly slides, I often expose two or more frames of important subject matter to give me multiple "original" dupes.

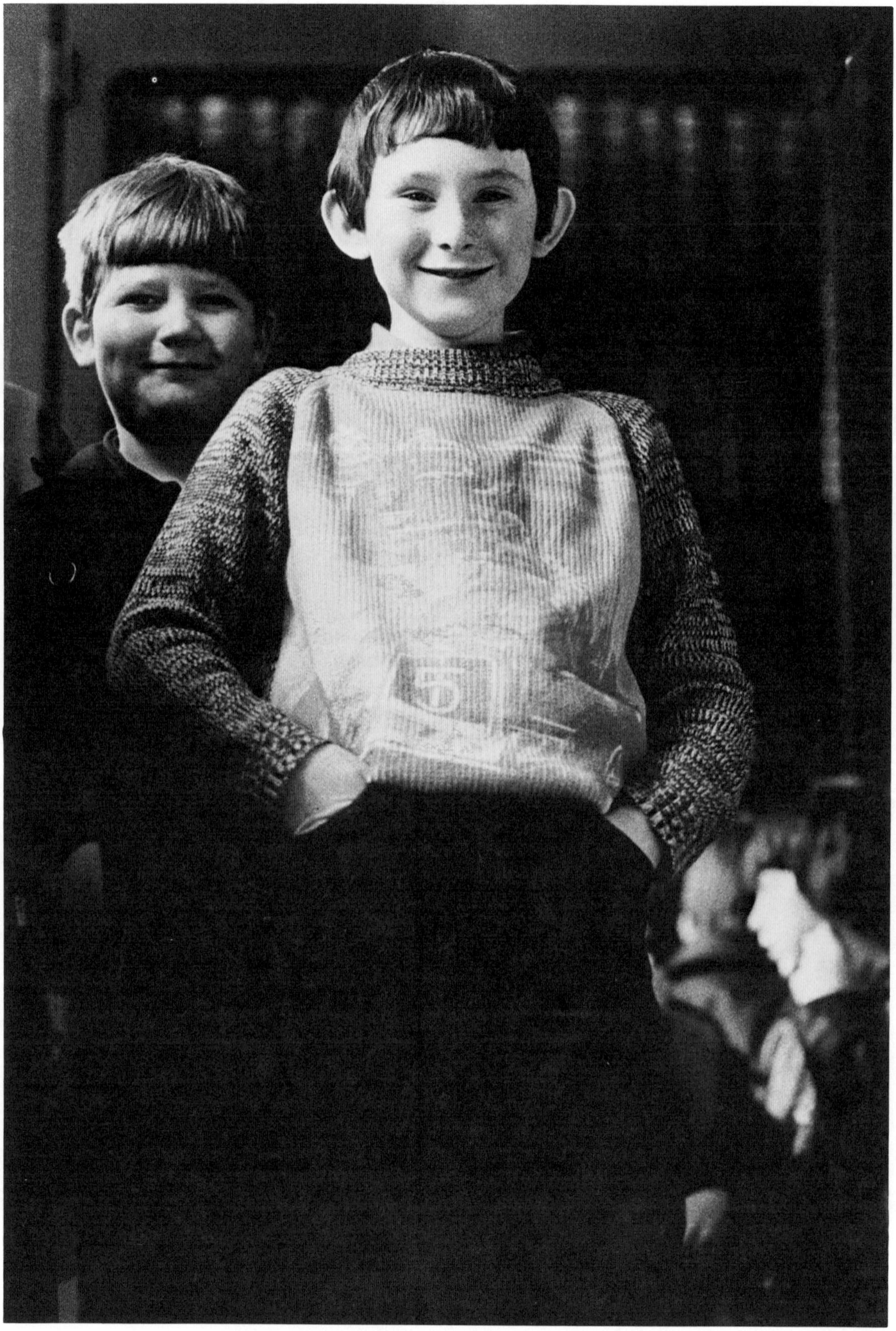
5

Chapter 6

Data and Evaluation of Film

Contained within chapter 6 are extensive data and evaluations of over 110 35mm films being offered as of the date of this book's publication. Each individual film type is categorized, that is, grouped within a general type such as color negative, color transparency, special-purpose, and so on.

A standardized format for presenting information was adopted to allow easy access to data and evaluations as well as to facilitate cross-referenced comparisons. A word of caution: comparisons can do only so much and in this instance are intended as subjective beginnings rather than absolute pronouncements. If they serve to whet the creative appetite and encourage experimentation, they serve their purpose. But, even though they are indeed opinion, they are critical and thoughtful opinion. Further, each represents the thinking and experience of more than one dozen seasoned professional and amateur film users.

With this as preface, examine the sample form and individual entry explanations that follow directly to understand more fully what we intend each data entry to represent.

Sample

Manufactured by: Film-maker, division, and country of origin
Film Type: Category of 35mm film, i.e., color, B&W, negative, transparency
Speed: Normal ISO rating *Push and/or Pull Feasible?* If uprating or downrating is useful, not just possible
Processing: Preferred processing chemistry
Size(s): 35mm *Roll Lengths:* Cartridge lengths
Long Rolls Available: Either metric or English
Other Sizes Available: Medium or large format
Diffuse RMS Granularity Value: Relative measure of grain within each film type
Resolving Power: A standard measure of film sharpness.
High Contrast 1000:1 =
Low Contrast 1.6:1 =
Exposure Latitude: Practical working limit
Contrast Characteristics: Subjective user judgments (in relation to film of similar type)
Film Structure: Special qualities built into film's structure
Color Balance: Lighting under which film is intended to be used
Color Bias: Subjective observation of film's color tendencies
Recommended Applications: Normal and typical

Figure 6.1 Marc Levey. Ilford HP5 Plus ISO 320.

Special Uses: Creative, unusual, or offbeat
Reciprocity (recommended shutter-speed range): Range within which film retains normal spectral response.
Exposure Correction Table
Exposure Time: In seconds
Correction: Adjustments in exposure and/or filtration needed to retain
Filtration (CC): Spectral response
Storage, Keeping Properties, and Special Handling: Self-explanatory
Availability: Where film can usually be found
In-Use Evaluation: Consensus evaluation of author and test panel of professional and experienced amateur photographers

Designation: **Kodachrome KM**
Manufactured by: Eastman Kodak Co., USA
Film Type: Extremely fine grain color transparency film
Speed: ISO 25 *Push and/or Pull Feasible?* Yes, limited
Processing: Special Process K-14, Kodak plus about 2 dozen custom labs
Size(s): 35mm *Roll Lengths:* 24, 36 exposures
Long Rolls Available? No
Other Sizes Available: 120
Diffuse RMS Granularity Value: 9
Resolving Power
High Contrast 1000:1 = 100 ll/mm
Low Contrast 1.6:1 = 63 ll/mm
Exposure Latitude: +1/2, –1 f-stop
Contrast Characteristics: Medium to high
Film Structure: Conventional
Color Balance: Daylight
Color Bias: Cyan-blue (slight)
Recommended Applications: Still-life, detailed close-ups, macro, landscapes
Special Uses: Preferred by many for high-magnification enlargements, publication, nature
Reciprocity (recommended shutter-speed range): 1/10000–1/10 second
Exposure Correction Table:

Exposure time:	1/10000	1/1000	1/100	1/10	1	10
Correction:	none	none	none	none	+1/2 f-stop	N/R
Filtration (CC):	none	none	none	none	none	N/R

Storage, Keeping Properties, and Special Handling: Professional film. Refrigerate until use.
Availability: General
In-Use Evaluation: This is the finest grained of all color films. It is also capable of resolving more detail than any other color emulsion. Color is a bit on the cool side but saturated. If quality is your main consideration and you can live with its slow speed, this film is a good choice for a good variety of subjects. Additionally, magazine editors, art directors, and ad people have shown a strong preference for Kodachromes.

Designation: **Kodachrome PKM**
Manufactured by: Eastman Kodak Co., USA
Film Type: Extremely fine grain color transparency film
Speed: ISO 25 *Push and/or Pull Feasible?* Yes, limited

Processing: Special Process K-14, Kodak plus about 2 dozen custom labs
Size(s): 35mm *Roll Lengths:* 24, 36 exposures
Long Rolls Available? No
Other Sizes Available: 120, 220
Diffuse RMS Granularity Value: 9
Resolving Power
High Contrast 000:1 = 100 ll/mm
Low Contrast 1.6:1 = 63 ll/mm
Exposure Latitude: +1 f-stop, –1 f-stop
Contrast Characteristics: Medium to high
Film Structure: Conventional
Color Balance: Daylight
Color Bias: Nearly neutral
Recommended Applications: Still-life, detailed close-ups, macro, landscapes
Special Uses: Preferred by many for high-magnification enlargements, publication, nature
Reciprocity (recommended shutter-speed range): Same as Kodachrome KR
Exposure Correction Table: Same as Kodachrome KR
Storage, Keeping Properties, and Special Handling: Professional film. Refrigerate until use.
Availability: Photo specialty stores
In-Use Evaluation: Many of the comments applied to KM also apply to PKM. Since PKM is "ripe," it should be exposed and processed as soon as it is received. PKM seems to be more neutral in color cast than its amateur version. The pro film has more tolerance for overexposure.

Designation: **Kodachrome KPA**
Manufactured by: Eastman Kodak Co., USA
Film Type: Extremely fine grain color transparency film
Speed: ISO 40 *Push and/or Pull Feasible?* Yes, limited
Processing: Special Process K-14, Kodak plus about 2 dozen custom labs
Size(s): 35mm *Roll Lengths:* 36 exposures
Long Rolls Available? No
Other Sizes Available: None
Diffuse RMS Granularity Value: 9
Resolving Power
High Contrast 1000:1 = 100 ll/mm
Low Contrast 1.6:1 = 63 ll/mm
Exposure Latitude: +1/2, –1 f-stop
Contrast Characteristics: Medium to high
Film Structure: Conventional
Color Balance: 3400K photofloods
Color Bias: Warm with slight magenta cast
Recommended Applications: Primary uses under studio lights for stills, portraits, or advertising. Also useful for color copy work.
Special Uses: Can be used with 3200K tungsten lamps (82A filter needed), or daylight with 85 series filter
Reciprocity (recommended shutter-speed range): 1/10000–1/10 second
Exposure Correction Table:

Exposure time:	1/10000	1/1000	1/100	1/10	1	10
Correction:	none	none	none	none	+ 1/2 f-stop	N/R
Filtration (CC):	none	none	none	none	none	N/R

Storage, Keeping Properties, and Special Handling: Professional film. Refrigerate until use.
Availability: Photo specialty stores
In-Use Evaluation: The extra contrast of this film is useful in studio situations to add an extra punch in flatly lit scenes. This film yields particularly well saturated reds and oranges. Even though it is almost a full f-stop faster than K-25, the films are all but identical in terms of grain and resolution.

Designation: **Kodachrome KR**
Manufactured by: Eastman Kodak Co., USA
Film Type: Very fine grain, color transparency film
Speed: ISO 64 *Push and/or Pull Feasible?* Yes, limited
Processing: Special Process K-14, Kodak plus about 2 dozen custom labs
Size(s): 35mm *Roll Lengths:* 24, 36 exposures
Long Rolls Available? No
Other Sizes Available: 120, 220
Diffuse RMS Granularity Value: 10
Resolving Power
High Contrast 1000:1 = 100 ll/mm
Low Contrast 1.6:1 = 63 ll/mm
Exposure Latitude: +1, –1 f-stop
Contrast Characteristics: Medium to high
Film Structure: Conventional
Color Balance: Daylight
Color Bias: Warm and a bit magenta especially in the whites
Recommended Applications: Still-life, detailed close-ups, macro, landscapes, general photography
Special Uses: Preferred by many for high-magnification enlargements, publication, nature
Reciprocity (recommended shutter-speed range): 1/1000–1/10 second
Exposure Correction Table: Same as KM except at 1 second, add 1 f-stop exposure and use a CC10R filter
Storage, Keeping Properties, and Special Handling: Normal
Availability: General, widespread
In-Use Evaluation: Kodachrome 64 is probably the most widely used color slide film in the world especially by professional nature landscape and still-life photographers. It combines an excellent mix of sharpness, very fine grain, excellent color saturation, and high contrast. While not as fine-grained as K-25, its speed makes it far more useful in hand-held or low-light situations. Also, K-64 offers better exposure latitude than its slower cousin.

Designation: **Kodachrome PKR**
Manufactured by: Eastman Kodak Co., USA
Film Type: Very fine grain color transparency film
Speed: ISO 64 *Push and/or Pull Feasible ?* Yes, limited
Processing: Special Process K-14, Kodak plus about 2 dozen custom labs
Size(s): 35mm *Roll Lengths:* 36 exposures
Long Rolls Available? No
Other Sizes Available: 120
Diffuse RMS Granularity Value: 10
Resolving Power
High Contrast 1000: 1 = 100 ll/mm
Low Contrast 1.6: 1 = 63 ll/mm

Exposure Latitude: +1, –1 f-stop
Contrast Characteristics: Medium to high
Film Structure: Conventional
Color Balance: Daylight
Color Bias: Warm, slightly magenta
Recommended Applications: Still-life, detailed close-ups, macro, landscapes, critical professional applications
Special Uses: Preferred by many for high-magnification enlargements, publication, nature, and critical professional applications
Reciprocity (recommended shutter speed range): 1/10000–1/10 second
Exposure Correction Table: Same as KR
Storage, Keeping Properties, and Special Handling: Professional film. Refrigerate until use.
Availability: Photo specialty stores
In-Use Evaluation: PKR film is very consistent within each batch number. Data sheets that come with the film state a rated speed that is tested and is accurate to within 1/6 f-stop. This film has a somewhat different color bias than its amateur version, being more yellow with reduced but still vibrant reds. Skin tones are more natural, if warm.

Designation: **Kodachrome KL**
Manufactured by: Eastman Kodak Co., USA
Film Type: Medium fine grain color transparency film
Speed: ISO 200 *Push and/or Pull Feasible?* Yes, limited
Processing: Special Process K-14, Kodak plus about 2 dozen custom labs
Size(s): 35mm *Roll Lengths:* 24, 36 exposures
Long Rolls Available? No
Other Sizes Available: 120
Diffuse RMS Granularity Value: 16
Resolving Power
High Contrast 1000:1 = 100ll/mm
Low Contrast 1.6:1 = 63 ll/mm
Exposure Latitude: +1/2, –1 f-stop
Contrast Characteristics: Medium
Film Structure: Conventional
Color Balance: Daylight
Color Bias: Slight magenta
Recommended Applications: Still-life, detailed close-ups, macro, landscapes, action, sports, and low-light general photography
Special Uses: High-quality available light situations
Reciprocity (recommended shutter-speed range): 1/10000–1/10 second
Exposure Correction Table: Same as KR
Storage, Keeping Properties, and Special Handling: Normal
Availability: Photo specialty stores and many discount outlets
In-Use Evaluation: The fastest Kodachrome, this film has the "Kodachrome look" but with substantially more grain and somewhat reduced sharpness. It is still the finest grained, sharpest, and most saturated 200-speed color slide film available. Photographers have grown to love its gritty high contrast combined with powerful colors. A terrific film if grain is not critical.

Designation: **Kodachrome PKL**
Manufactured by: Eastman Kodak Co., USA
Film Type: Medium fine grain color transparency film
Speed: ISO 200 *Push and/or Pull Feasible?* Yes, limited
Processing: Special Process K-14, Kodak plus about 2 dozen custom labs
Size(s): 35mm *Roll Lengths:* 36 exposures
Long Rolls Available? No
Other Sizes Available: 120
Diffuse RMS Granularity Value: 16
Resolving Power
High Contrast 1000:1 = 100 ll/mm
Low Contrast 1.6:1 = 63 ll/mm
Exposure Latitude: +1/2, –1 f-stop
Contrast Characteristics: Medium
Film Structure: Conventional
Color Balance: Daylight
Color Bias: Slight magenta but less than KL
Recommended Applications: Same as KL but more useful for situations where super-accurate speed calibration is necessary. Whites appear to be cleaner in this professional version.
Special Uses: Use when color and/or film speed is critical from cassette to cassette
Reciprocity (recommended shutter-speed range): 1/10000–1/10 second
Exposure Correction Table: Same as KL
Storage, Keeping Properties, and Special Handling: Professional film. Refrigerate until use.
Availability: Photo specialty stores
In-Use Evaluation: The professional version of K-200 shares the same major characteristics as the amateur version except that it is ripe and ready for use immediately. Colors seem to be cleaner and more neutral than K-200. I find the added grain useful for subjects such as candid street portraits and action photography like rodeos or motor sports.

Designation: **Agfachrome RS 50 Plus**
Manufactured by: Agfa Geveart AG, Germany
Film Type: 35mm color transparency
Speed: ISO 50 *Push and/or Pull Feasible?* Yes
Processing: Standard E-6
Size(s): 35mm *Roll Lengths:* 36 exposures
Long Rolls Available? No
Other Sizes Available: 120, sheets
Diffuse RMS Granularity Value: Less than 10
Resolving Power
High Contrast 1000:1 = 125 ll/mm
Low Contrast 1.6:1 = 70 ll/mm
Exposure Latitude: +1/2, –1 f-stop
Contrast Characteristics: Medium
Film Structure: Conventional, with new, cleaner dye formulation
Color Balance: Daylight
Color Bias: Neutral
Recommended Applications: Excellent studio film for still-life and fashion. Good choice for brightly lit outdoor subjects.
Special Uses: None

Reciprocity (recommended shutter-speed range): 1/1000–1 second
Exposure Correction Table:

Exposure time:	1/1000	1/100	1/10	1	10
Correction:	none	none	none	+1/2 f-stop	+1 f-stop
Filtration (CC):	none	none	none	050B	075B

Storage, Keeping Properties, and Special Handling: Professional film. Refrigerate until use.
Availability: Photo specialty stores
In-Use Evaluation: Brighter more saturated colors are a feature of these relatively new films identified with a "Plus." While other high color saturation films are also contrasty, RS 50 Plus retains its predecessors, moderate contrast and tonal neutrality. Overall, the film is improved in almost every performance aspect.

Designation: **Agfachrome RS 100 Plus Professional**
Manufactured by: Agfa Geveart AG, Germany
Film Type: 35mm color transparency
Speed: ISO 100 *Push and/or Pull Feasible?* Yes
Processing: Standard E-6
Size(s): 35mm *Roll Lengths:* 36 exposures
Long Rolls Available? No
Other Sizes Available: 120
Diffuse RMS Granularity Value: 10
Resolving Power
High Contrast 1000:1 = 100 ll/mm
Low Contrast 1.6:1 = 50 ll/mm
Exposure Latitude: +1/2, –1 f-stop
Contrast Characteristics: Medium
Film Structure: Conventional with improved color dyes
Color Balance: Daylight
Color Bias: Neutral to very slightly cool
Recommended Applications: All general purpose photography
Special Uses: None
Reciprocity (recommended shutter-speed range): 1/1000–1 second
Exposure Correction Table:

Exposure time:	1/1000	1/100	1/10	1	10
Correction:	none	none	none	+1/2 f-stop	+1 f-stop
Filtration (CC):	none	none	none	050B	075B

Storage, Keeping Properties, and Special Handling: Professional film. Refrigerate until use.
Availability: Photo specialty stores
In-Use Evaluation: Professional counterpart of Agfachrome CT 100i. Manufactured to ISO tolerance of ± 1/6 f-stop. Color balance a bit cooler than CT 100i, but saturation remains very high. The emulsion is an improved version of an already excellent all-purpose film.

Designation: **Agfachrome RS 200 Professional**
Manufactured by: Agfa Geveart AG, Germany
Film Type: Fine grain 35mm color transparency
Speed: ISO 200 *Push and/or Pull Feasible?* Yes
Processing: Standard E-6
Size(s): 35mm *Roll Lengths:* 36 exposures
Long Rolls Available? No
Other Sizes Available: 120

Diffuse RMS Granularity Value: 12
Resolving Power
High Contrast 1000:1 = 90 ll/mm
Low Contrast 1.6:1 = N/A
Exposure Latitude: +1/2, –1 f-stop
Contrast Characteristics: Medium to low
Film Structure: Conventional, with enhanced edge sharpness
Color Balance: Daylight
Color Bias: Neutral to slightly warm
Recommended Applications: General amateur and professional photography
Special Uses: Situations where color fidelity and speed are two primary concerns
Reciprocity (recommended shutter-speed range): 1/1000–1 second
Exposure Correction Table:

Exposure time:	1/1000	1/100	1/10	1	10
Correction:	none	none	none	+1/2 f-stop	+1 f-stop
Filtration (CC):	none	none	none	025	050B

Storage, Keeping Properties, and Special Handling: Professional film. Refrigerate until use.
Availability: Photo specialty stores
In-Use Evaluation: This medium-speed film exhibits many of the same color qualities as its slower brand mates. Being faster, it is noticeably more grainy, but it is still very sharp with excellent edge definition. Its lower contrast compared to the other pro Agfas or Kodachrome 200 Professional makes it suitable for delicate skin tones and many landscapes.

Designation: **Agfachrome RS 1000**
Manufactured by: Agfa Geveart AG, Germany
Film Type: Medium to coarse grain 35mm color transparency
Speed: ISO 1000 *Push and/or Pull Feasible?* Yes, one stop
Processing: Standard E-6
Size(s): 35mm *Roll Lengths:* 36 exposures
Long Rolls Available? No
Other Sizes Available: 120
Diffuse RMS Granularity Value: 26
Resolving Power
High Contrast 1000:1 = 70 ll/mm plus
Low Contrast 1.6:1 = N/A
Exposure Latitude: +1/2, –1 f-stop
Contrast Characteristics: Medium
Film Structure: Conventional
Color Balance: Daylight
Color Bias: Neutral to slightly warm
Recommended Applications: Low-available-light photography, sports action, street portraiture in bad weather
Special Uses: I use this film to create dramatic, richly textured candid portraits on location. Grain enhances texture.
Reciprocity (recommended shutter-speed range): 1/1000–1 second
Exposure Correction Table:

Exposure time:	1/1000	1/100	1/10	1	10
Correction:	none	none	none	+1/3 f-stop	+2/3 f-stop
Filtration (CC):	none	none	none	none	none

Storage, Keeping Properties, and Special Handling: Professional film. Refrigerate until use.
Availability: Photo specialty stores
In-Use Evaluation: 1000 RS displays outstanding sharpness for an ultra-fast color film. Its color and contrast do not differ from other Agfa films in any significant way. While this emulsion is quite grainy, the grain is tight and very even. Of all the ultra-fast color slide films, 1000RS possesses the highest D-Max or deepest blacks. This is an important quality consideration since blacks tend to go purple or blue-gray as film ISO increases. Push-processing one stop has no noticeable negative effects on the major qualitative aspects of this film.

Designation: **Agfachrome CT 100i**
Manufactured by: Agfa Geveart AG, Germany
Film Type: 35mm color transparency
Speed: ISO 100 *Push and/or Pull Feasible?* Yes
Processing: Standard E-6
Size(s): 35mm *Roll Lengths:* 24, 36 exposures
Long Rolls Available? No
Other Sizes Available: No
Diffuse RMS Granularity Value: 10
Resolving Power
High Contrast 1000:1 = 100 ll/mm
Low Contrast 1.6:1 = 50 ll/mm
Exposure Latitude: +1/2, –1 f-stop
Contrast Characteristics: Neutral
Film Structure: Conventional with boosted color saturation
Color Balance: Daylight
Color Bias: Neutral to slightly warm
Recommended Applications: All general purpose color photography
Special Uses: None
Reciprocity (recommended shutter-speed range): 1/1000–1 second
Exposure Correction Table:

Exposure time:	1/1000	1/100	1/10	1	10
Correction:	none	none	none	+1/2 f-stop	+1 f-stop
Filtration (CC):	none	none	none	05B	075B

Storage, Keeping Properties, and Special Handling: Professional film. Refrigerate until use.
Availability: Photo specialty stores
In-Use Evaluation: Amateur version of RS 100 Plus with its general performance characteristics. However, CT 100i seems a bit warmer than its professional sibling.

Designation: **Agfachrome CT-200**
Manufactured by: Agfa Geveart AG, Germany
Film Type: Fine grain 35mm color transparency
Speed: ISO 200 *Push and/or Pull Feasible?* Yes
Processing: Standard E-6
Size(s): 35mm *Roll Lengths:* 27, 36 exposures
Long Rolls Available? No
Other Sizes Available: No
Diffuse RMS Granularity Value: 12

Resolving Power
High Contrast 1000:1 = 70 ll/mm plus
Low Contrast 1.6:1 = N/A
Exposure Latitude: +1, –1 f-stop
Contrast Characteristics: Medium
Film Structure: Conventional
Color Balance: Daylight
Color Bias: Neutral to slightly warm
Recommended Applications: General photography
Special Uses: None
Reciprocity (recommended shutter-speed range): 1/1000–1 second
Exposure Correction Table:

Exposure time:	1/1000	1/100	1/10	1	10
Correction:	none	none	none	+1/2 f-stop	+1 f-stop
Filtration (CC):	none	none	none	025B	050B

Storage, Keeping Properties, and Special Handling: Normal
Availability: Photo specialty stores, some discount outlets
In-Use Evaluation: Similar to the old CT-100 with increased grain, lower sharpness. A good all-round performer generally available at a favorable price. Excellent color separation.

Designation: **Konica Chrome/Sakurachrome 100**
Manufactured by: Konishiroku Photo Co., Japan
Film Type: 35mm fine grain color transparency
Speed: ISO 100 *Push and/or Pull Feasible?* Yes
Processing: Standard E-6
Size(s): 35mm *Roll Lengths:* 24, 36 exposures
Long Rolls Available? No
Other Sizes Available: None
Diffuse RMS Granularity Value: 10
Resolving Power
High Contrast 1000:1 = 100 ll/mm
Low Contrast 1.6:1 = 50 ll/mm
Exposure Latitude: +3/4, –1-1/2 f-stops
Contrast Characteristics: Medium
Film Structure: Conventional
Color Balance: Daylight
Color Bias: Neutral to slightly cool
Recommended Applications: General-purpose photography
Special Uses: None
Reciprocity (recommended shutter-speed range): 1/1000–4 seconds
Exposure Correction Table:

Exposure time:	1/1000	1/100	1/10	5	10
Correction:	none	none	none	+1/2 f-stop	N/R
Filtration (CC):	none	none	none	none	N/R

Storage, Keeping Properties, and Special Handling: Normal
Availability: Presently only available in the U.S. as a private-label product such as Fotomat
In-Use Evaluation: A fully competitive medium-speed transparency emulsion that looks very much like Fujichrome 100 RD. In some respects it features superior performance compared to like films such as Fujichrome and Ektachrome in that it features somewhat greater exposure latitude on the overexposure side. Since it is essentially a

private-label product when it does get into the country and is not identified, it's almost impossible to know when you're using it.

Designation: **Ektachrome EPY**
Manufactured by: Eastman Kodak Co., USA
Film Type: 35mm fine grain color transparency
Speed: ISO 50 *Push and/or Pull Feasible?* Yes
Processing: Standard E-6
Size(s): 35mm Roll Length: 36 exposures
Long Rolls Available? Yes
Other Sizes Available: 120, sheets
Diffuse RMS Granularity Value: 13
Resolving Power
High Contrast 1000:1 = 125 ll/mm
Low Contrast 1.6:1 = 50 ll/mm
Exposure Latitude: +3/4, –1 f-stop
Contrast Characteristics: Normal skin tones otherwise, normal
Film Structure: Conventional
Color Balance: Tungsten 3200K
Color Bias: Neutral to cool
Recommended Applications: Studio portraits, still-life, advertising
Special Uses: Slide copying, when duplicating film is not available or when slightly increased contrast is desired
Reciprocity (recommended shutter-speed range): 1/100–1 second
Exposure Correction Table:

Exposure time:	1/1000	1/100	1/10	1	10
Correction:	N/R	none	none	none	N/R
Filtration (CC):	N/R	05C	none	05R	N/R

Storage, Keeping Properties, and Special Handling: Professional film. Refrigerate until use.
Availability: Photo specialty stores
In-Use Evaluation: An excellent E-6 tungsten-balanced film that is extremely fine grained with very saturated colors, particularly blues and greens. For a slide film, this emulsion exhibits an exceptionally broad tonal range from shadows to highlights.

Designation: **Ektachrome EPR**
Manufactured by: Eastman Kodak Co., USA
Film Type: Very fine grain 35mm color transparency
Speed: ISO 64 *Push and/or Pull Feasible?* Yes
Processing: Standard E-6
Size(s): 35mm *Roll Lengths:* 36 exposures
Long Rolls Available? Yes
Other Sizes Available: 120, 220, sheets
Diffuse RMS Granularity Value: 12
Resolving Power:
High Contrast 1000:1 = 125 ll/mm
Low Contrast 1.6:1 = 50 ll/mm
Exposure Latitude: +3/4, –1 f-stop
Contrast Characteristics: Low to medium
Film Structure: Conventional
Color Balance: Daylight
Color Bias: Neutral to cool

Recommended Applications: General photography
Special Uses: Useful when softer than normal rendition is required
Reciprocity (recommended shutter-speed range): 1/10000–1/10 second
Exposure Correction Table:

Exposure time:	1/10000	1/1000	1/100	1/10	1
Correction:	none	none	none	none	+1/2 f-stop
Filtration (CC):	none	none	none	none	CC05R

Storage, Keeping Properties, and Special Handling: Professional film. Refrigerate until use.
Availability: Photo specialty stores
In-Use Evaluation: Soft working, smooth-edged film with a distinctive non-Kodak look. Brilliant color rendition somewhat counteracted by reduced sharpness compared to other slow- to medium-speed transparency films.

Designation: **Ektachrome EPX Professional**
Manufactured by: Eastman Kodak Co., USA
Film Type: Very fine grain 35mm color transparency
Speed: ISO 64 *Push and/or Pull Feasible?* Yes
Processing: Standard E-6
Size(s): 35mm *Roll Lengths:* 24, 36 exposures
Long Rolls Available? Yes
Other Sizes Available: 120
Diffuse RMS Granularity Value: 11
Resolving Power
High Contrast 1000:1 = 125 ll/mm
Low Contrast 1.6:1 = 50 ll/mm
Exposure Latitude: +3/4, –1 f-stop
Contrast Characteristics: Medium
Film Structure: Conventional
Color Balance: Daylight
Color Bias: Warm, CC06 Y, CC02R added to enhance warm end response
Recommended Applications: General amateur and professional photography
Special Uses: Film's added yellow response makes it very useful for portraiture, electronic flash, under overcast or in shadows
Reciprocity (recommended shutter-speed range): 1/10000–10 seconds
Exposure Correction Table:

Exposure time:	1/10000	1/1000	1/100	10	1
Correction:	none	none	none	none	+1/2 f-stop
Filtration (CC):	none	none	none	none	CC05R

Storage, Keeping Properties, and Special Handling: Professional film. Refrigerate until use.
Availability: Photo specialty stores
In-Use Evaluation: Added warm response of this popular film makes its use under overcast or in shadows possible without filtration. Eliminates the so-called Ektachrome blues. Also this film should brighten up bad weather scenes. This should be popular film for location shooting.

Designation: **Ektachrome 50 HC EM**
Manufactured by: Eastman Kodak Co., USA
Film Type: Very fine grain 35mm color transparency
Speed: ISO 50 *Push and/or Pull Feasible?* Yes

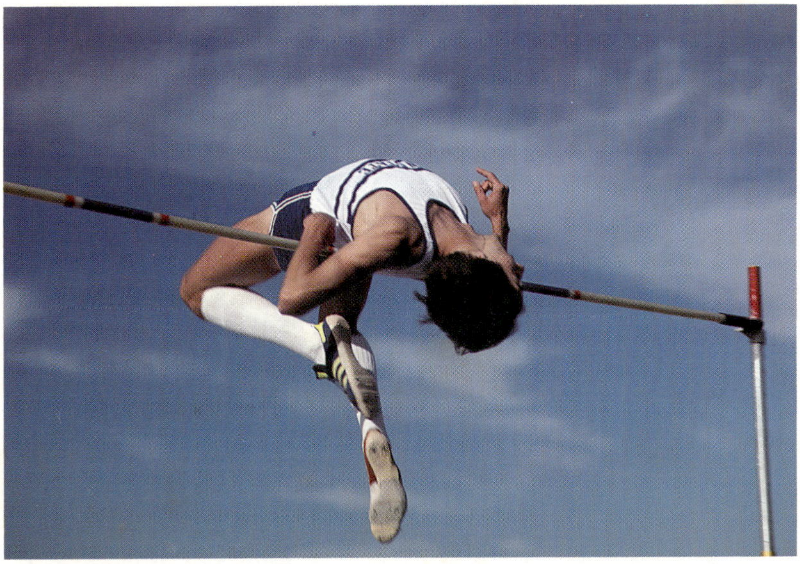

1. The ISO-200 color negative film used to take this picture yields a good balance of speed, grain, and sharpness. Agfacolor XRG 200 @ ISO 200.

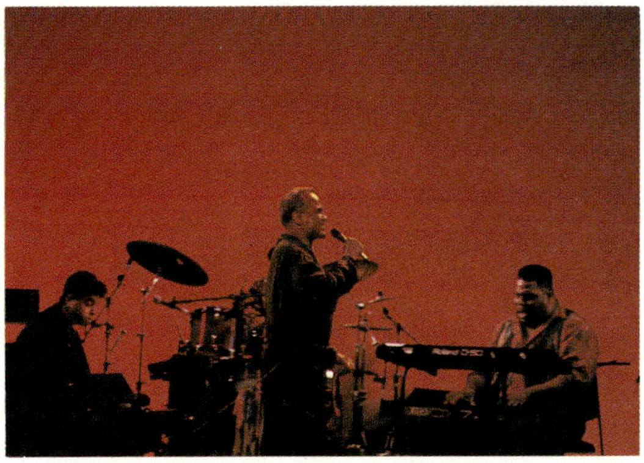

2. This hand-held image of performing-great Harry Belafonte was made possible by using ISO 1600 film. Overall red color cast due to color gel spotlights. Fujichrome P1600II @ ISO 1600.

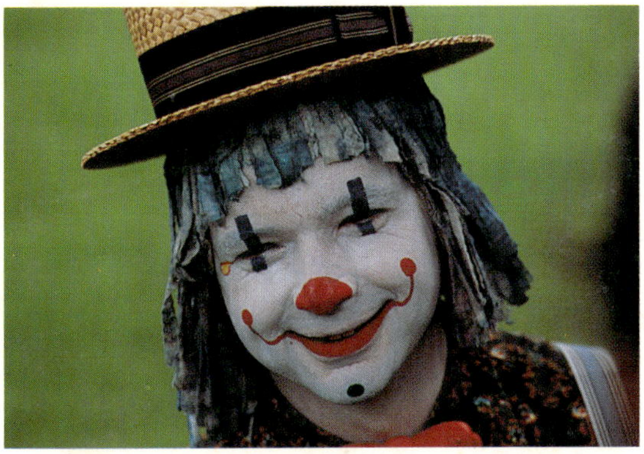

3. Slow, contrasty, and relatively high color saturation film results in very detailed, brilliant color imagery. Underexposure further heightens color saturation. Fujichrome 50 D @ ISO 64.

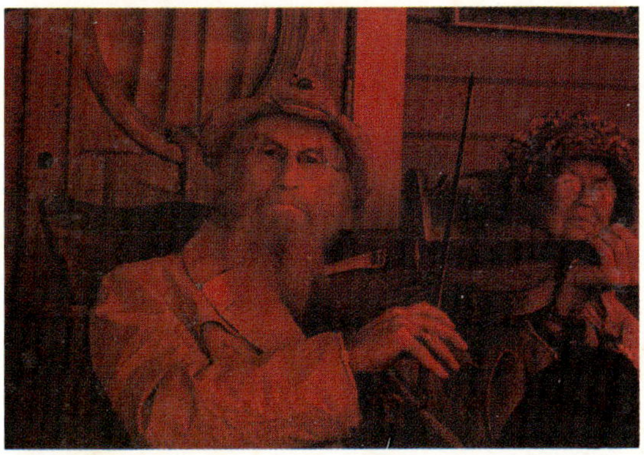

4. Typical of infrared color is the grainy texture and false color as seen in this image. Ektachrome Infrared, green filter. Initial rating ISO 125, then bracketed.

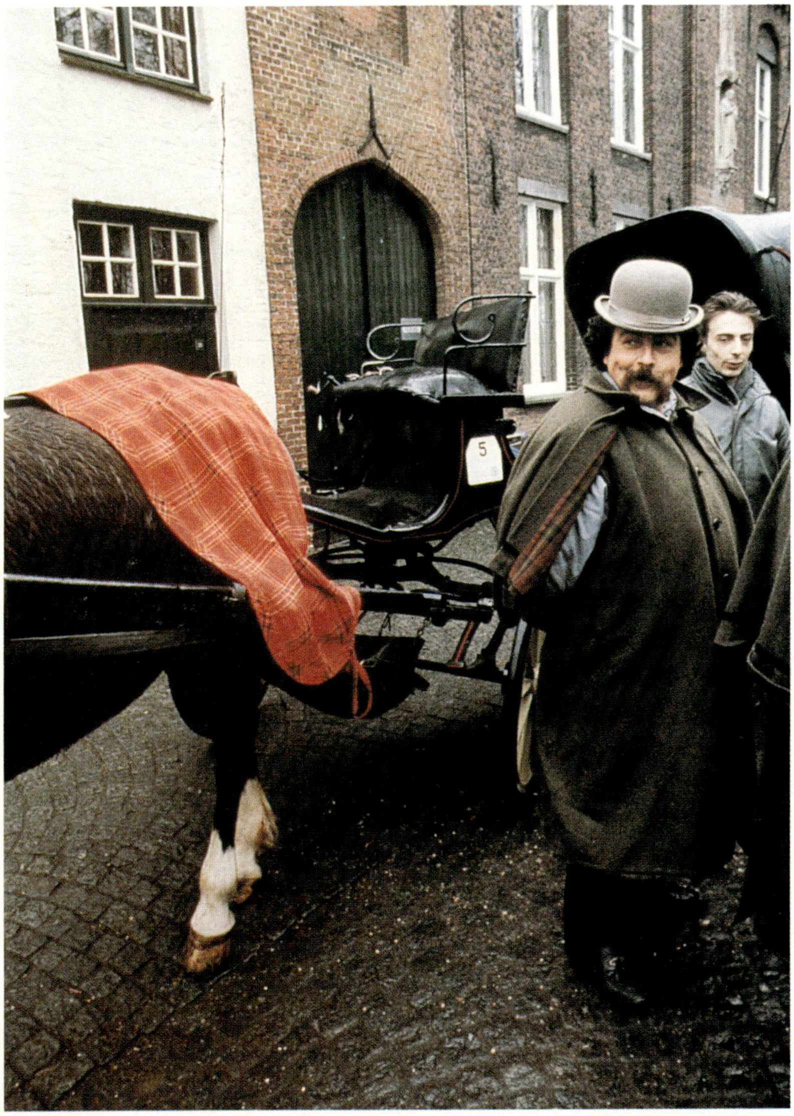

5. The extra speed and inherently rich color rendition helps this image significantly. Overcast light boosts apparent high color contrast. Kodachrome 200 @ ISO 250.

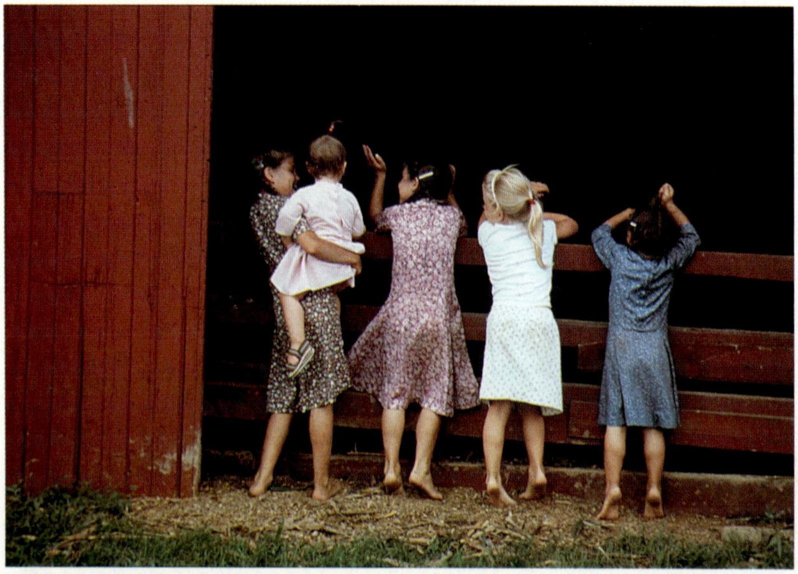

6. Up-to-date ISO-400 films employing latest grain and color technologies can satisfy many of the photographic needs of much slower, conventional films. Ektachrome 400X @ ISO 400.

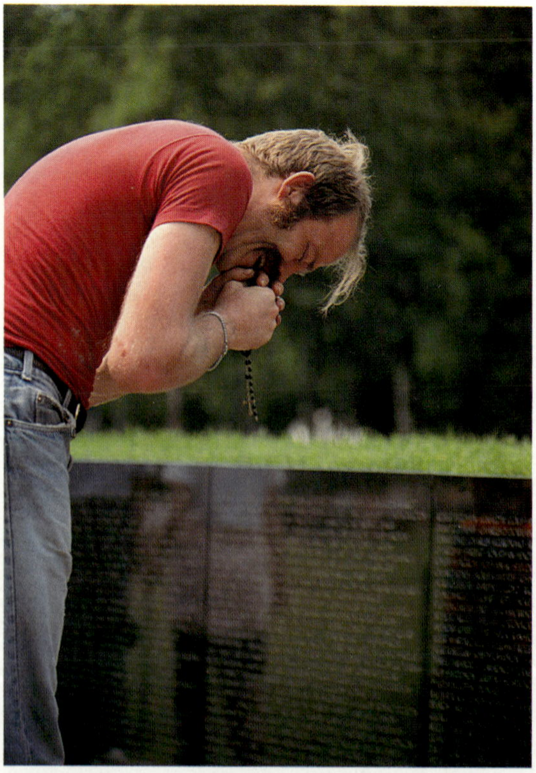

7. For location photography in changing light, trading off some sharpness and grain for more film speed makes good sense. Fujichrome 400 @ ISO 400.

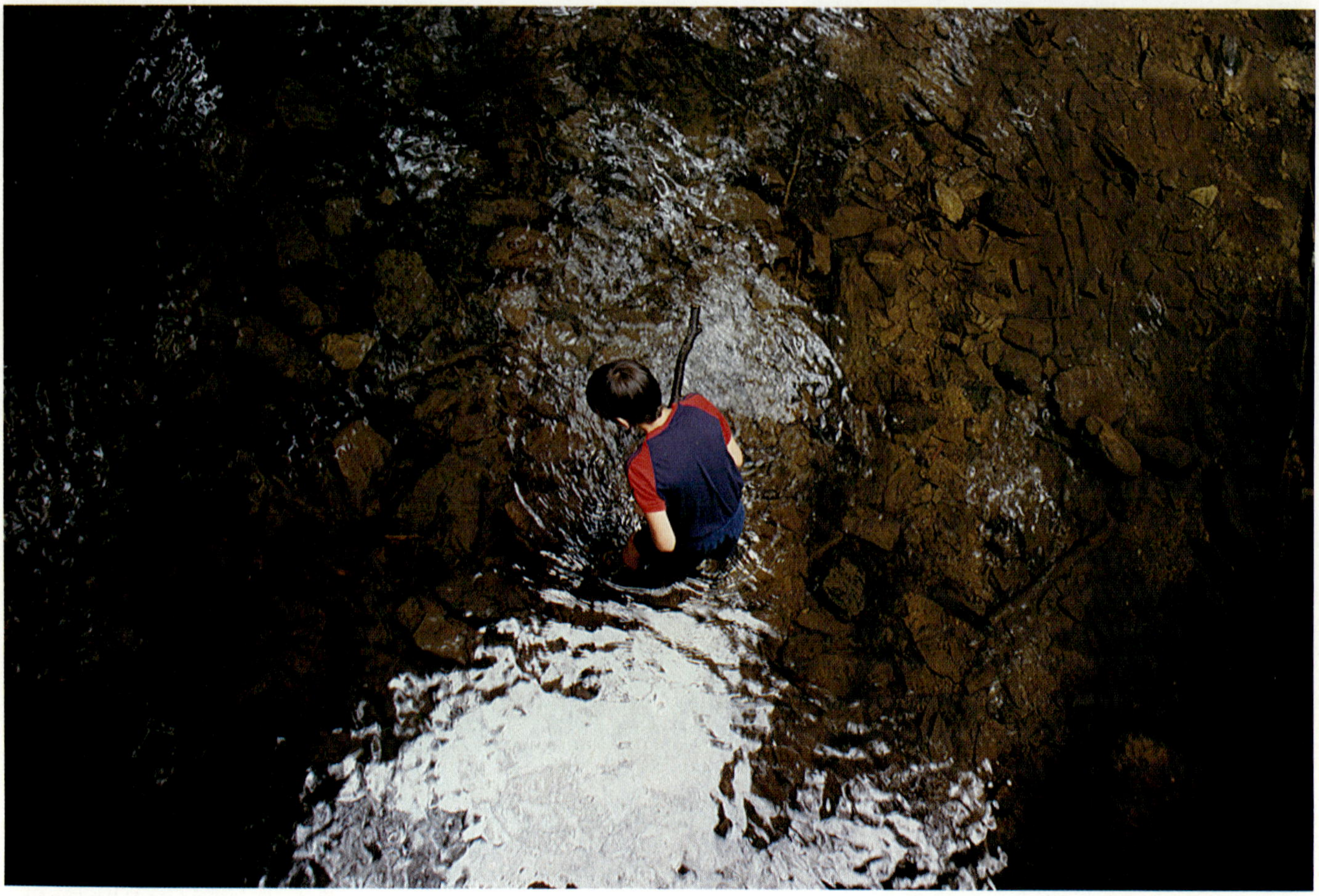

8. Mounting the camera on a tripod allowed the use of a slow, ultra fine grain film. Ektar 25 @ ISO 25.

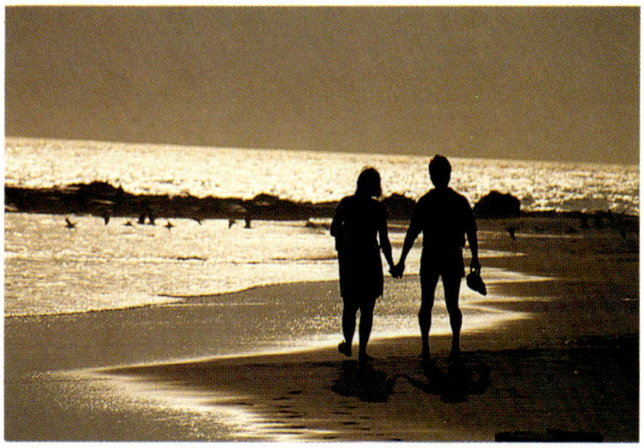

9. Underexposing about two f-stops and metering the sky produced this silhouette at sunset. Agfachrome RS 200 @ ISO 250.

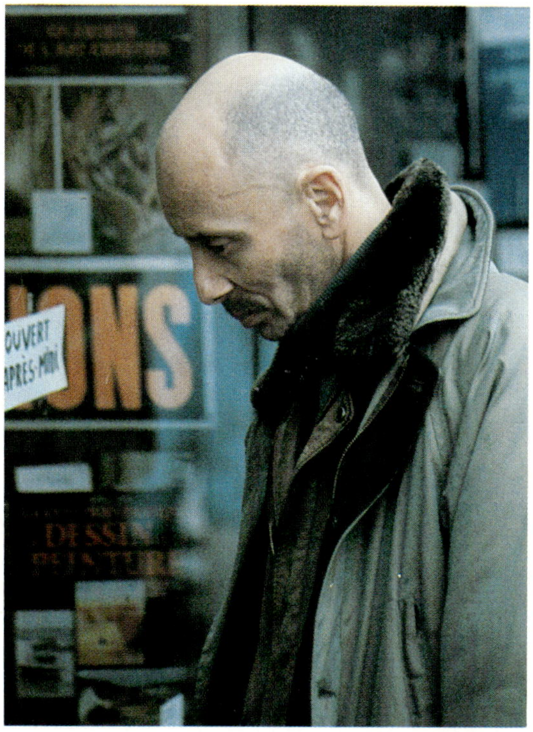

10. Employing ultra high speed color film in normal daylight makes several effects possible: first, it can create moody grain and texture; second, it allows the use of high shutter speeds coupled to reasonable f-stops; and third, this type of film usually generates subdued tones and soft color. Agfachrome RS 1000 @ ISO 1250.

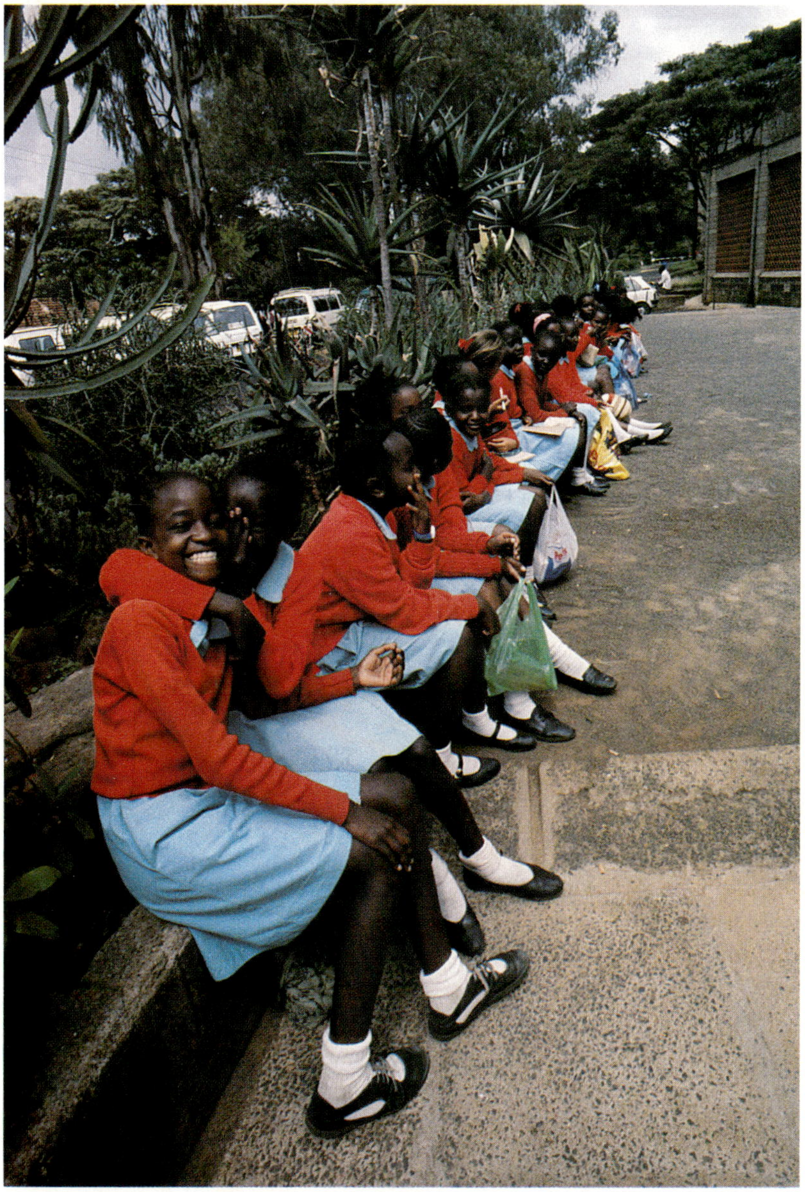

11. Compared to previous color slide films, today's medium-speed emulsions feature extremely deep and brilliant colors. Fujichrome 100D @ ISO 125.

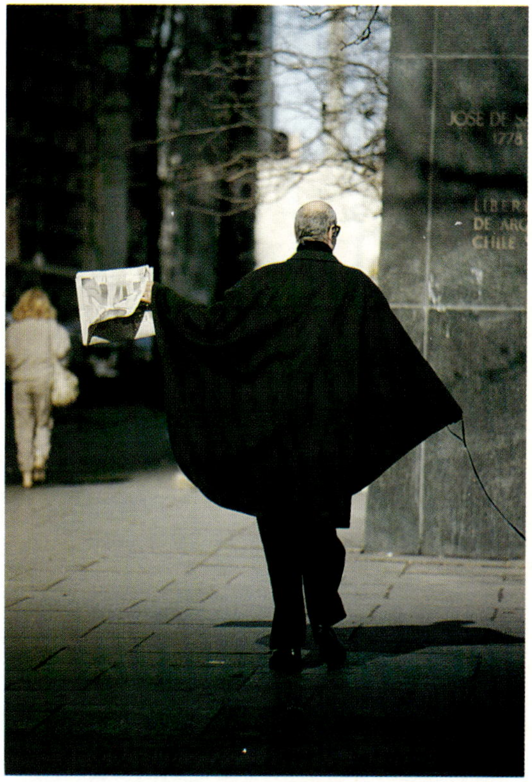

12. ISO-200 color print films are a natural for travel photography. They are fast enough to allow the use of motion-stopping shutter speeds and are particularly forgiving of exposure errors. Kodacolor Gold 200 Plus @ ISO 200.

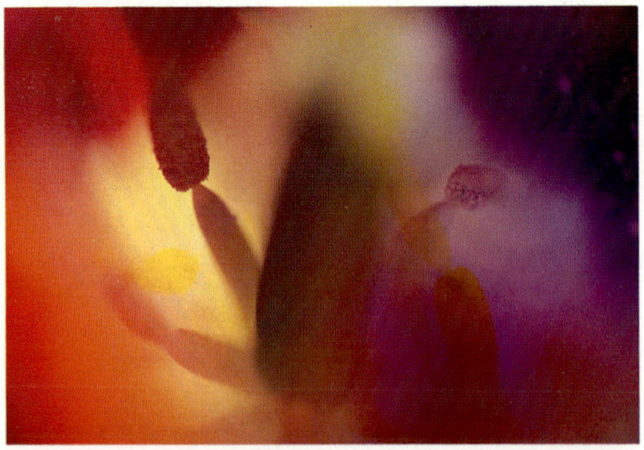

13. Ektachrome 100 Professional offers a softer color palette to most other super high color saturation medium speed slide emulsions. Ektachrome 100 Professional @ ISO 100.

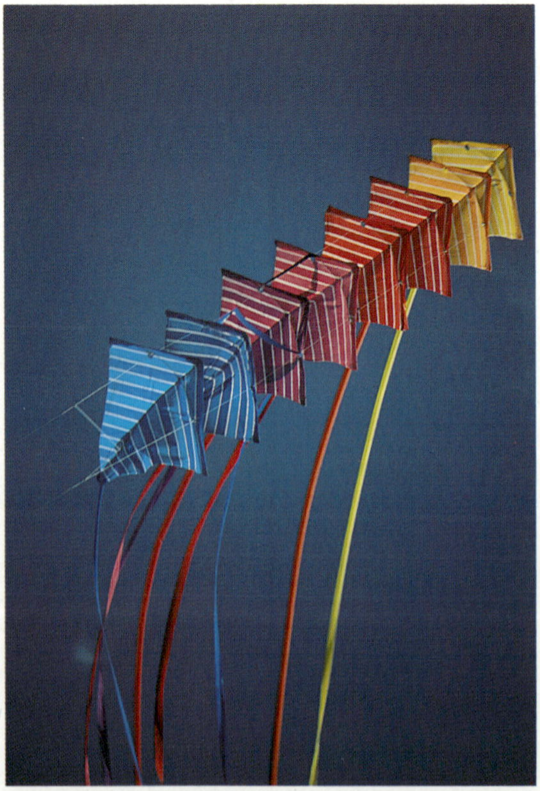

14. Following the kites with the camera mounted on a tripod retained image sharpness even with a slow, ultra fine grained film. Fujichrome Velvia @ ISO 64.

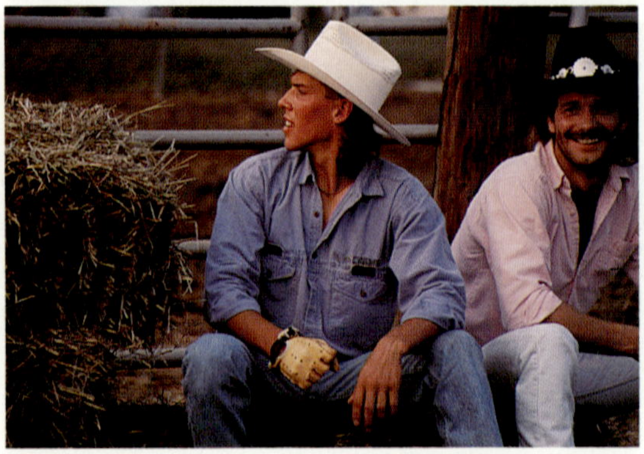

15. When I went to shoot this candid portrait, I realized I had relatively slow Kodachrome loaded. Bracing the camera against the side of a pole provided the needed stability to expose the picture at 1/30 second. Kodachrome 64 @ ISO 80.

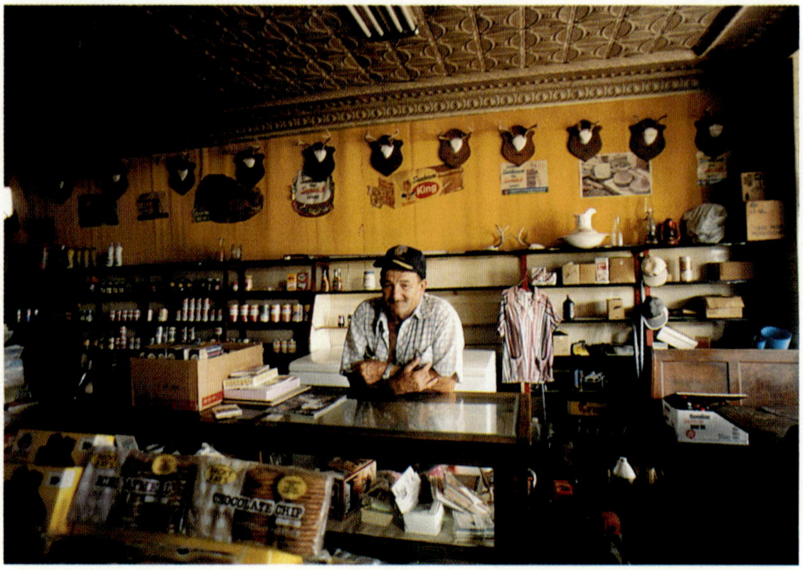

16. Ektachrome 100 Plus's excellent color fidelity and some window light helped keep skin tones from becoming objectionably warm. Ektachrome 100 Plus @ ISO 125.

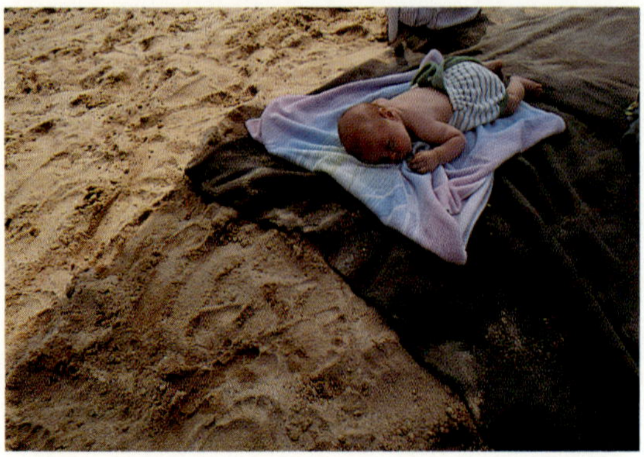

17. The high contrast of this scene is handled very well by Konica SRG 160 Professional print film at ISO 160.

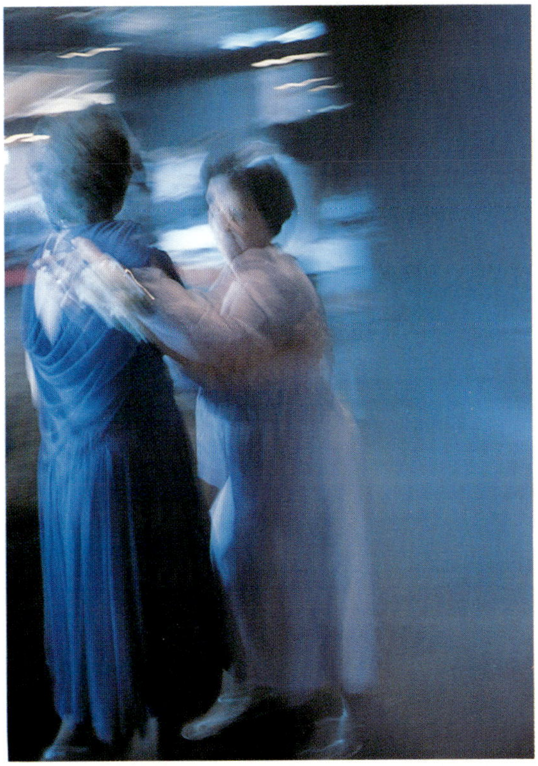

18. This image is an example of using a film mismatched to the prevailing light, but still producing an interesting image. Ektachrome 160 (Tungsten) @ ISO 160, exposed at 1/4 second to achieve a dreamlike appearance.

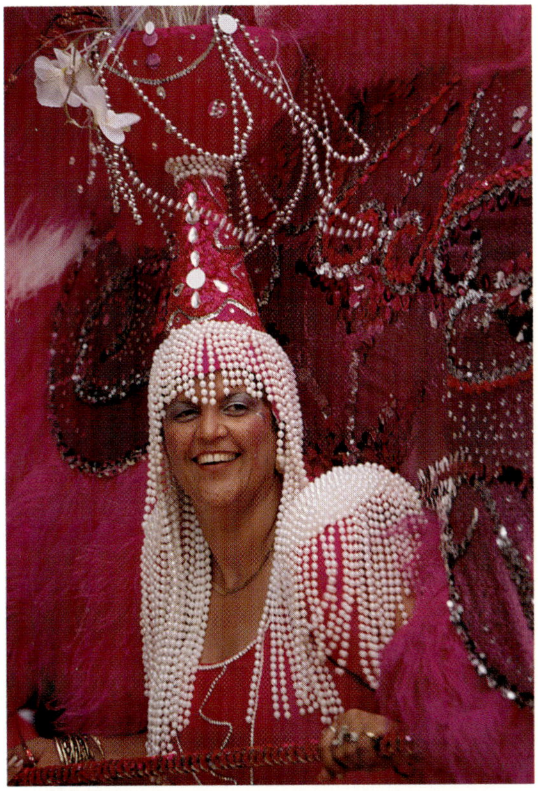

19. Modern color print film now rivals its slide siblings for color saturation and brilliance. Ektar 100 @ ISO 100.

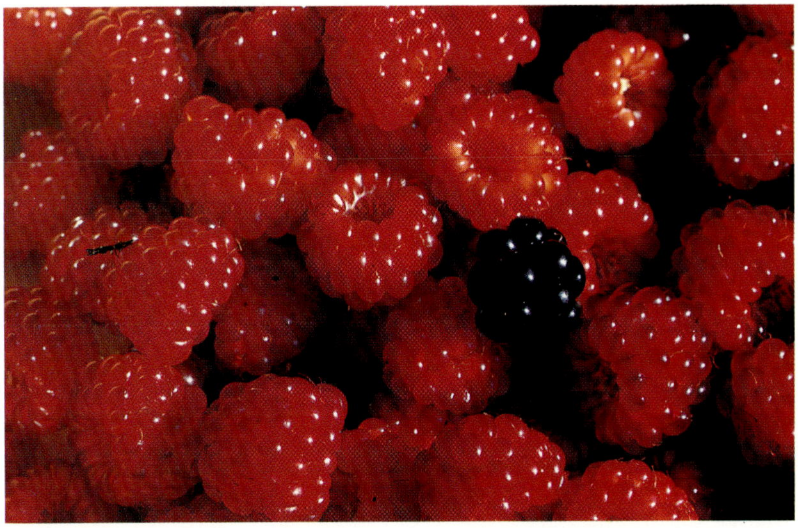

20. Close-up calling for extreme detail resolution demands the use of the sharpest, lowest grain color film available. This still means Kodachrome 25. Kodachrome 25 @ ISO 25.

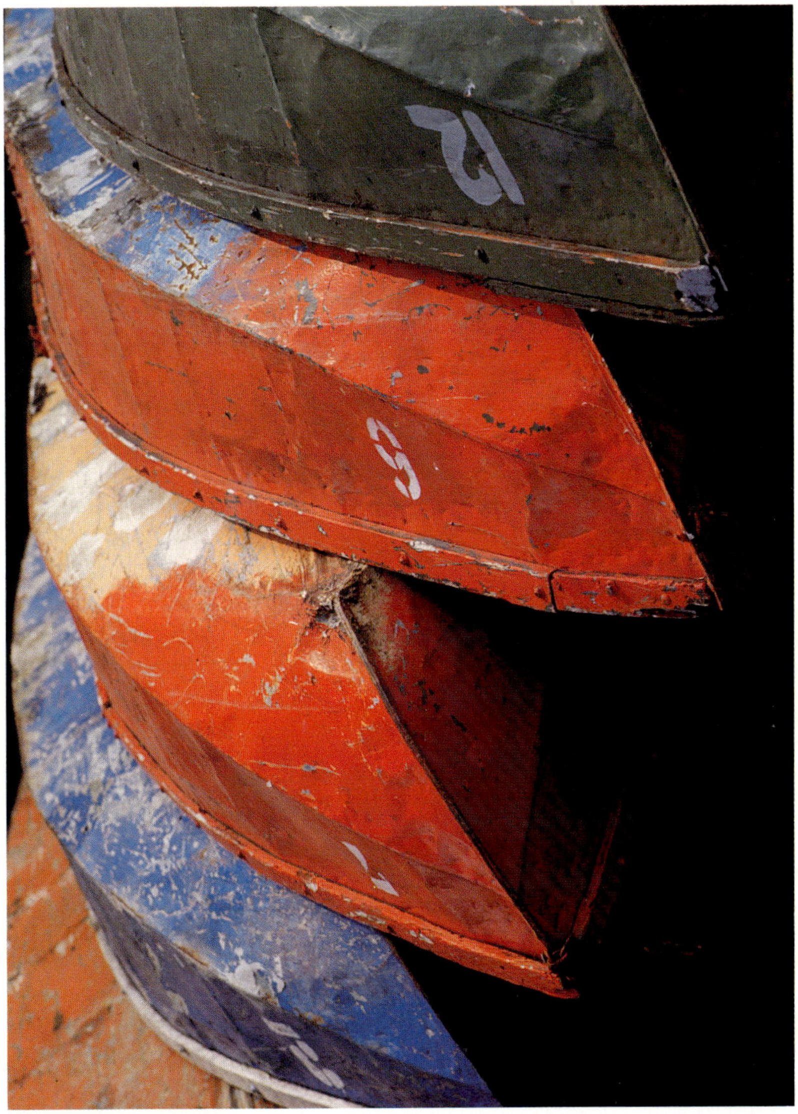

21. Virtually all medium-speed color print films emanating from major producers are capable performers offering as they do rich color, reduced visible grain, and excellent detail rendition. Scotchcolor HR 100 @ ISO 100.

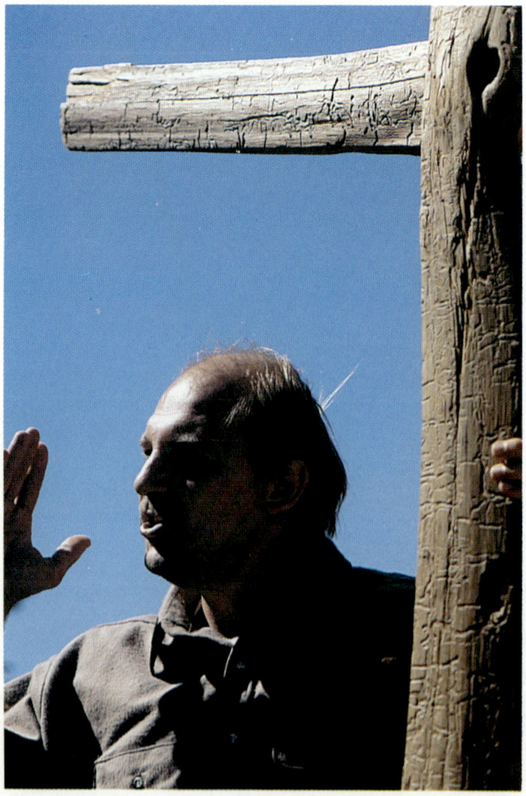

22. If you desire an available-light outdoor image much as the eye sees it, try using one of several color print films designed for professional studio work. Fujicolor 160 Professional S @ ISO 160.

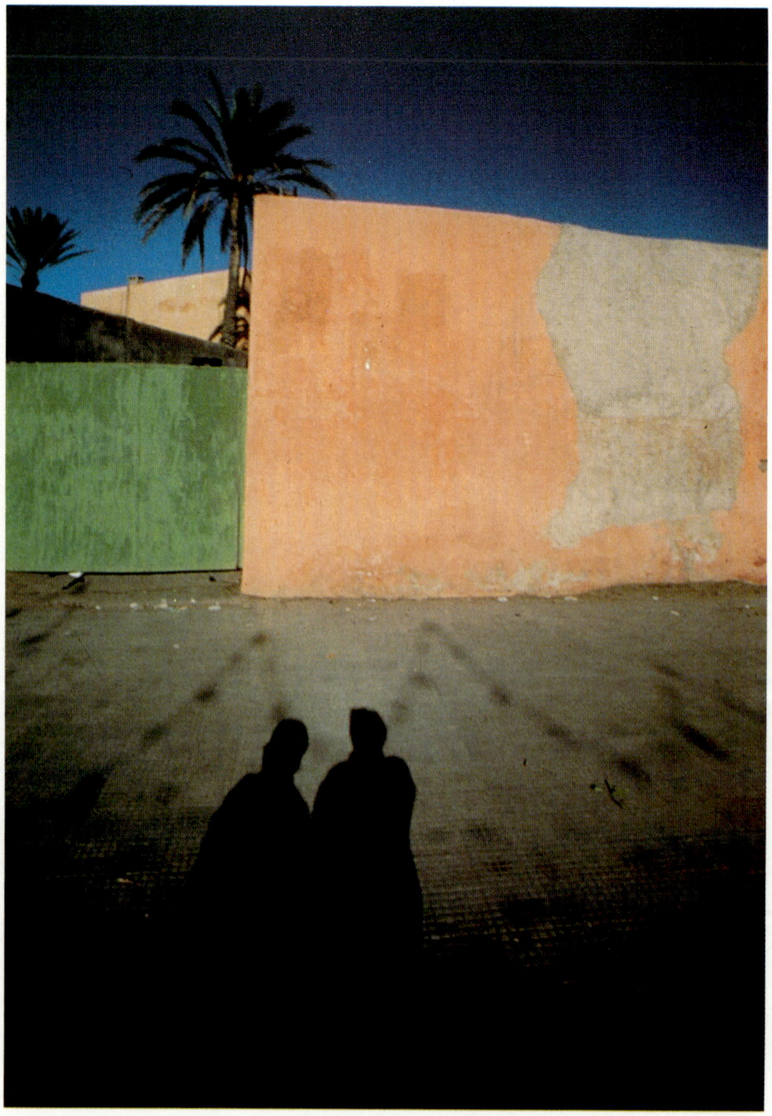

24. Even though grain is visible, ISO-400 color films deliver sharp, rich color that is comparable to much slower film of only a few years ago. With the full application of the technologies now available, greater quality advances will soon become apparent. Scotchchrome 400 @ ISO 400.

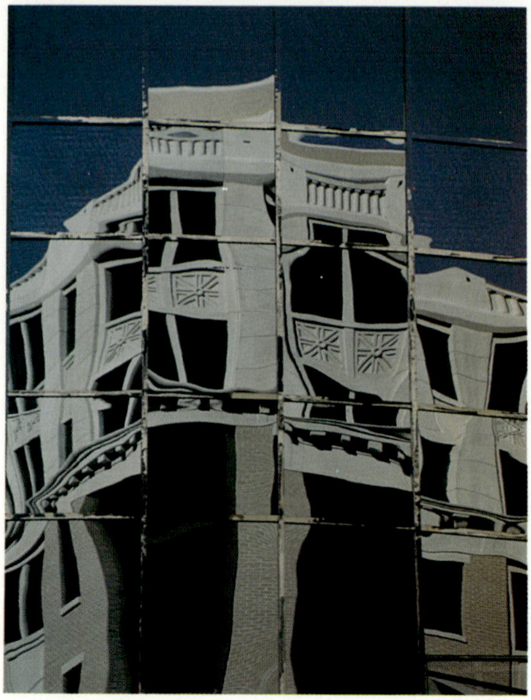

23. Among slide films, Ektachrome 100 and 200 offer more natural reduced contrast than most other emulsions of their type. These films prove very useful to control excessively contrasty scenes. In spite of lower-than-average contrast, they both exhibit very good color characteristics. Ektachrome 200 @ ISO 250.

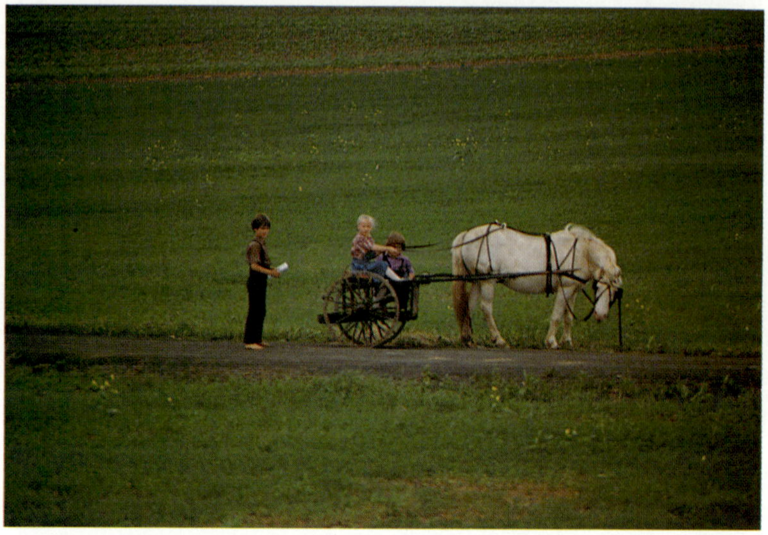

25. Fuji films are especially prized for their lush reproduction of greens. Fujichrome 50 D @ ISO 64.

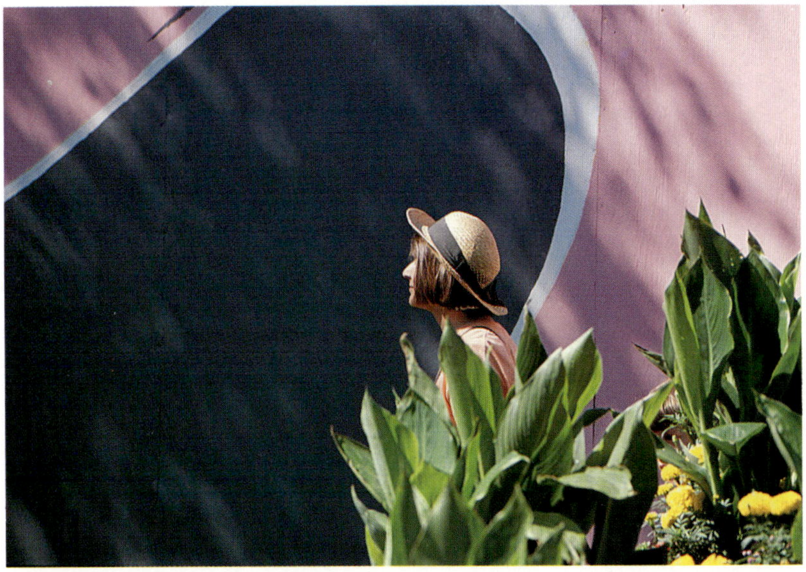

26. Virtually any medium-speed color print or slide film should be capable of producing excellent color in good light. Polaroid One Film @ ISO 200.

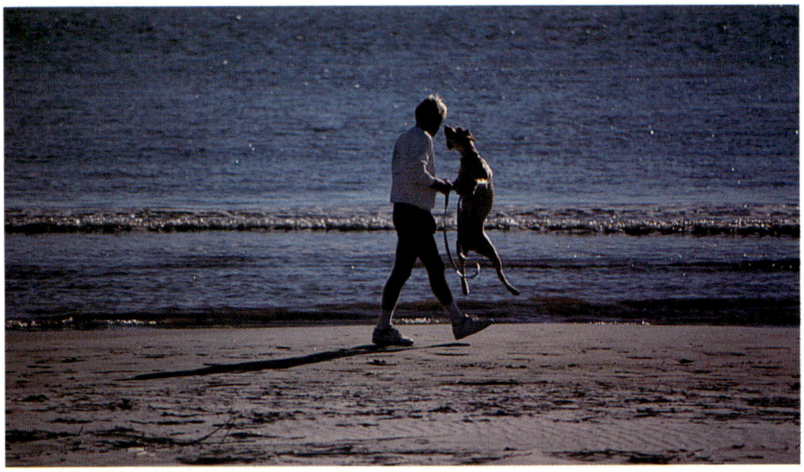

27. Agfachrome CT-100i is a new slide film displaying almost no vices. In addition, it is often available at a lower price than its more widely advertised competitors. Agfachrome CT-100i @ ISO 100 (Jeanne Miller).

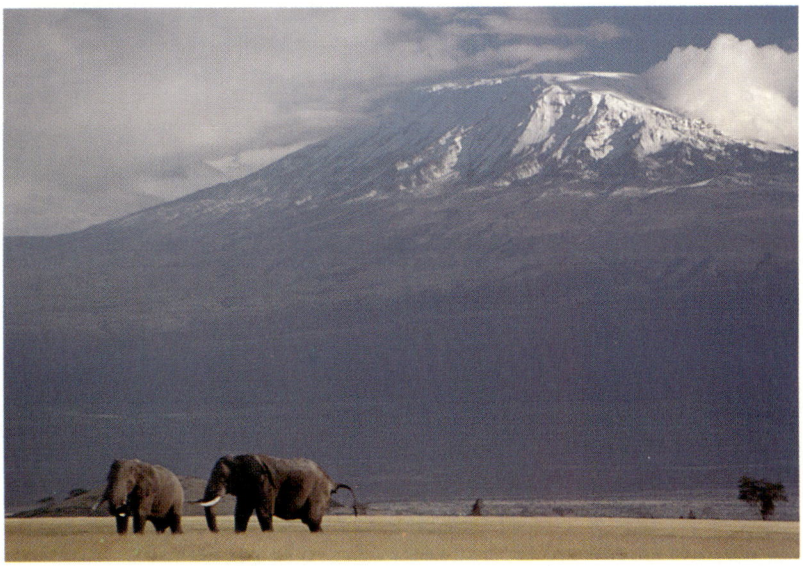

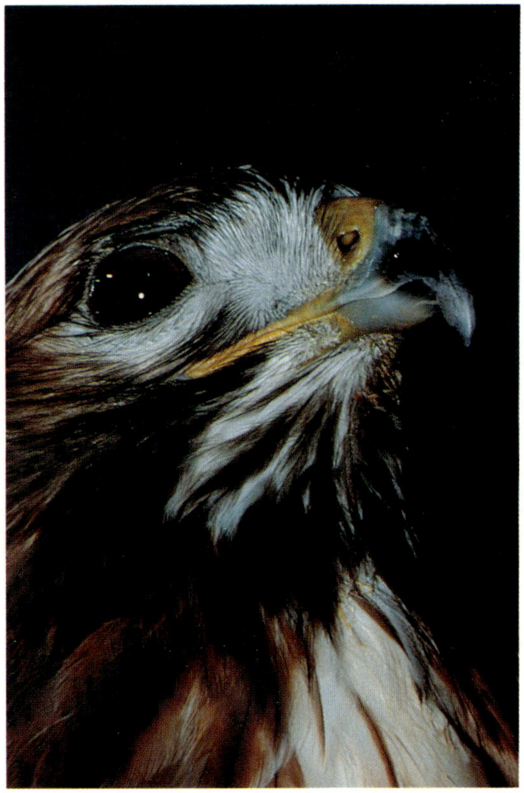

28. A Minolta ringlight close-up electronic flash unit provided the shadowless light to this "portrait." The cool blue of the electronic flash was compensated for by warmer-than-normal Ektachrome 64 X slide film. Ektachrome 64 X @ ISO 80.

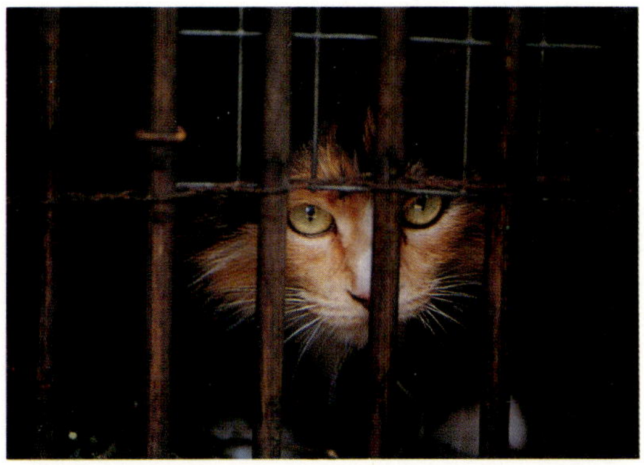

29. Tiffen 812 filter adds a bit of warmth to this striking study. Scotchchrome 100 @ ISO 160.

30. When shooting on the run, employing a faster film than usual can yield a higher percentage of acceptably sharp pictures. Fujichrome 400 @ its rated speed.

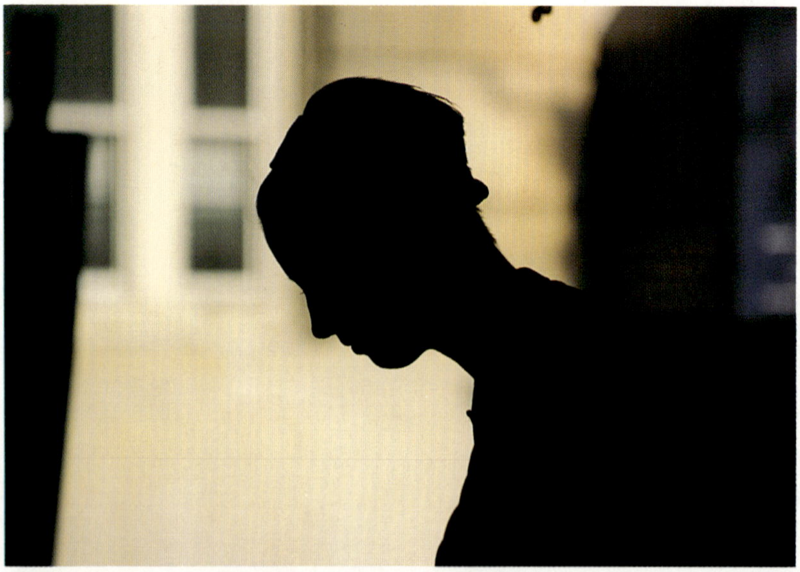

31. Exposing highlights with the spot metering option of a multimode SLR resulted in this image from high-speed color negative print film. Kodacolor Gold 400 Plus @ ISO 160.

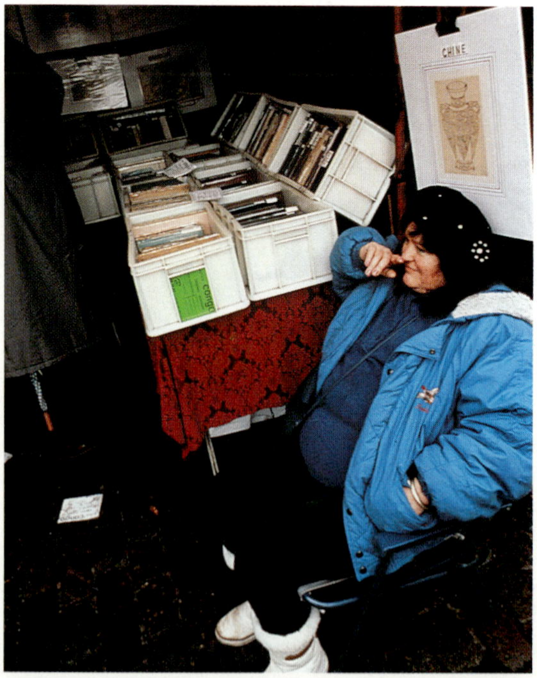

32. Overcast light combined with Kodachrome 200 to produce the brilliant color evident in this travel image made in Brussels. Kodachrome 200 @ ISO 250.

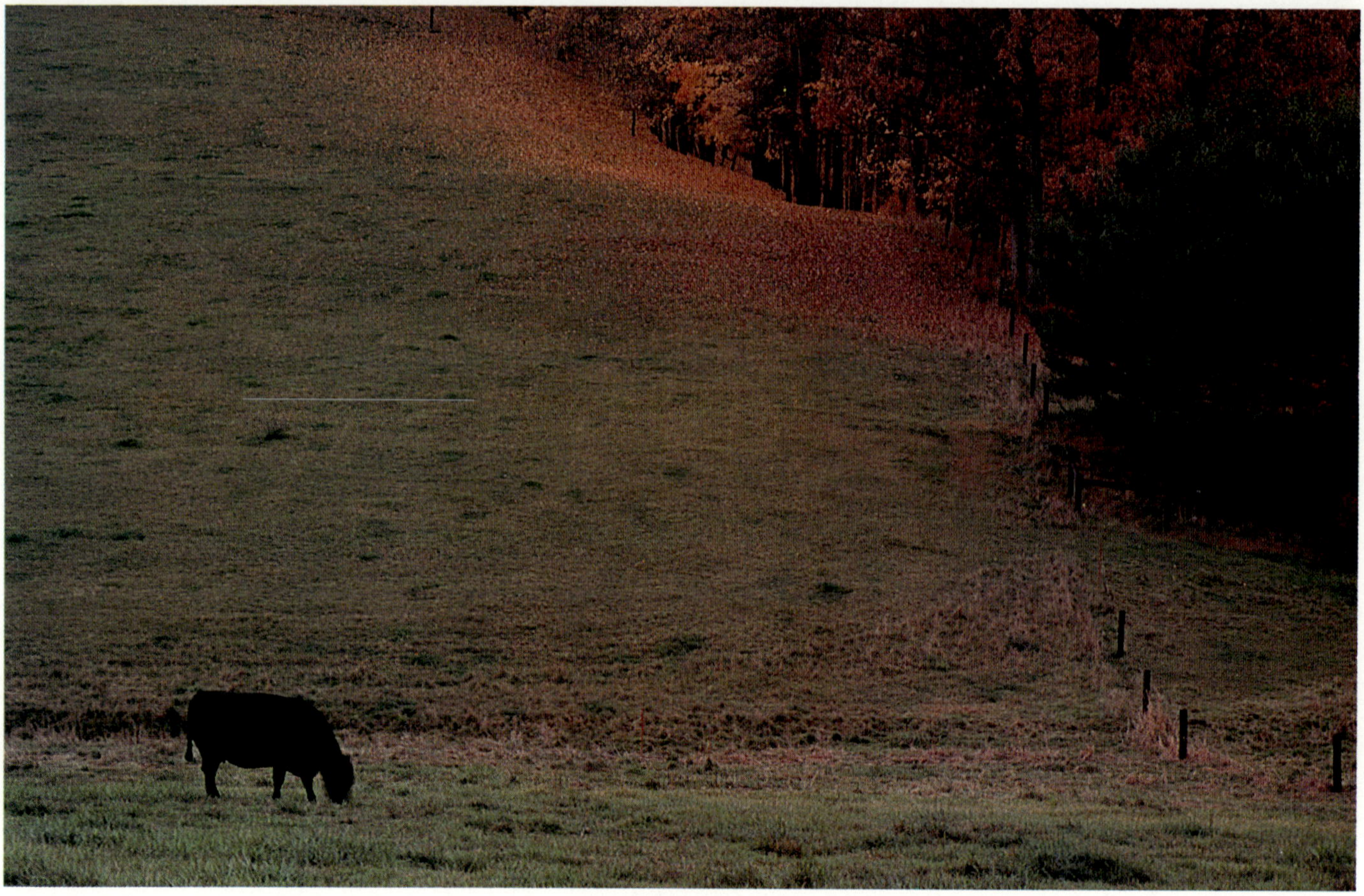

33. Tiffen's unique color enhancing filter boosts saturation of reds, oranges, and yellows while leaving the rest of the palette relatively unaffected. It is the only filter that acts in this manner. Ektachrome 50HC @ ISO 50.

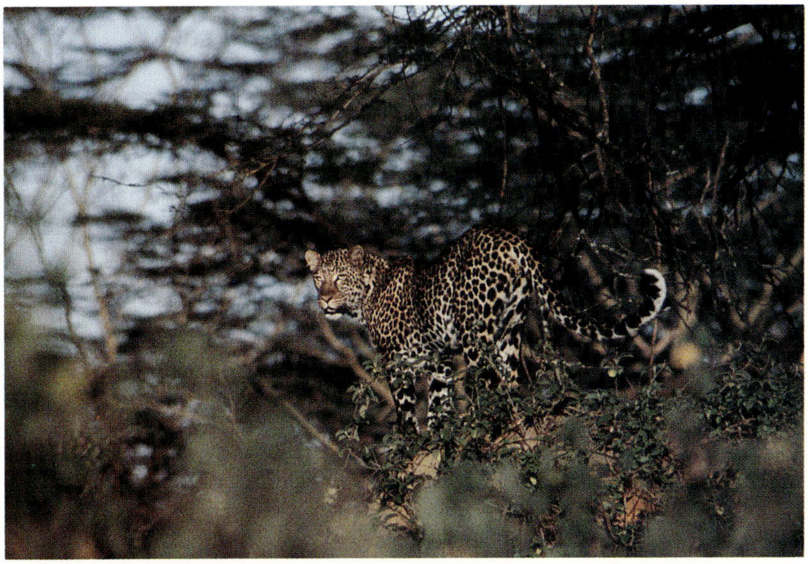

34. Ultra high-speed films have been improved to the point where they can deliver very good quality 8x10 prints, and usable color transparencies. Ektar 1000 @ ISO 1000.

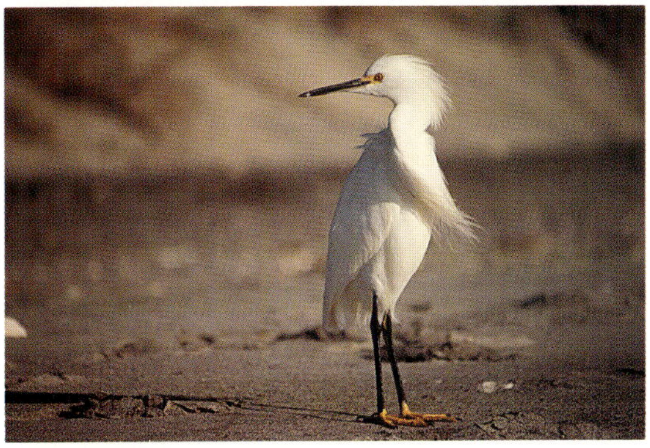

36. Metering the bird's feathers and overexposing by about one-half stop resulted in a much more accurate rendition of the white tones. Without these adjustments, light tones would probably appear a darker grayish color. Agfachrome CT-200 @ ISO 160 (Jeanne Miller).

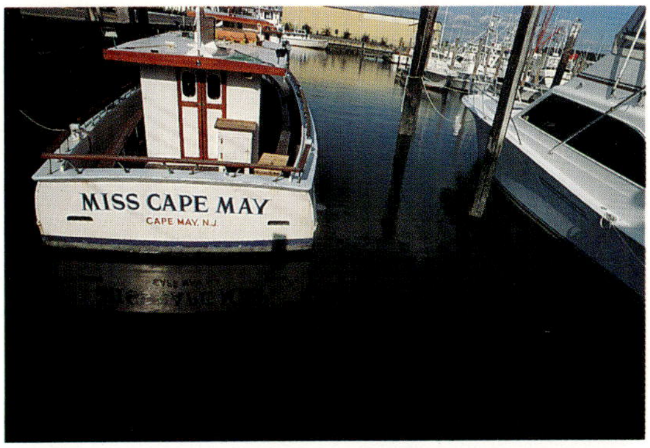

37. Rich, all-but-grainless, high-contrast image mark this photo as one taken on the distinctive Agfacolor Ultra 50 film. Agfacolor Ultra 50 @ ISO 50.

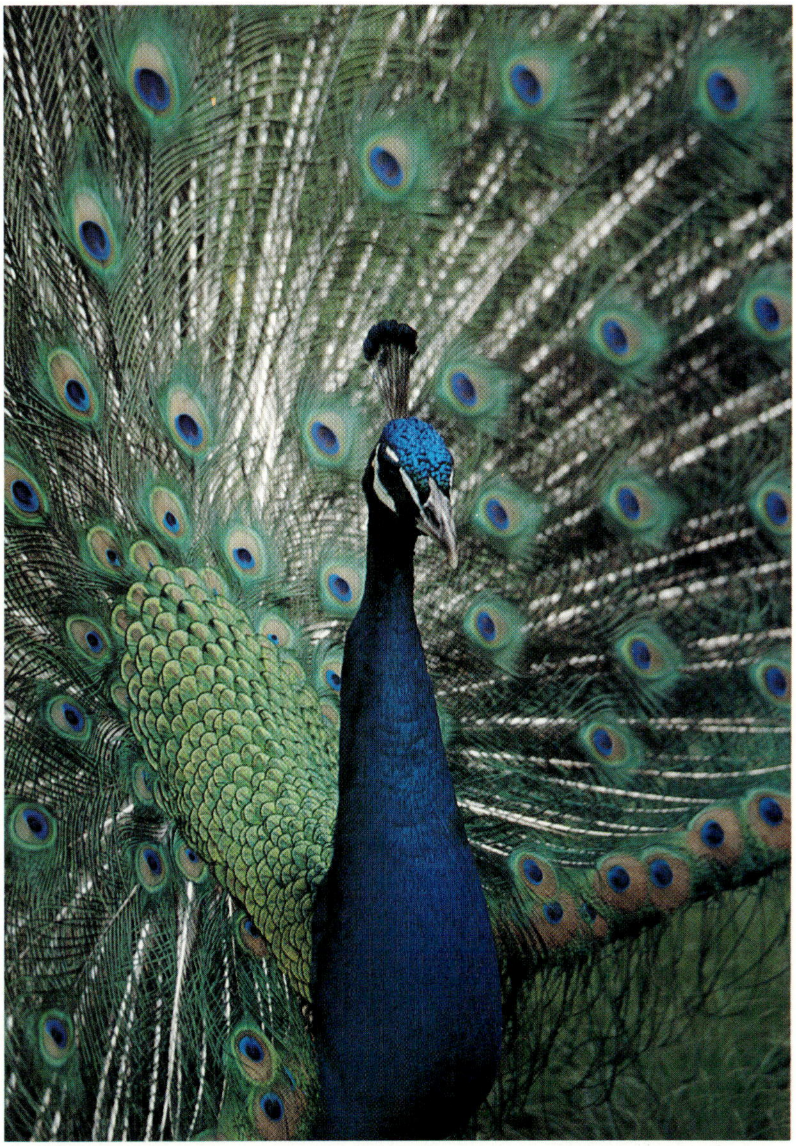

35. Extraordinary detail, super-saturated color, and E-6 processing convenience are qualities that have made Fujichrome Velvia an immediate success among working professionals. Velvia @ ISO 64.

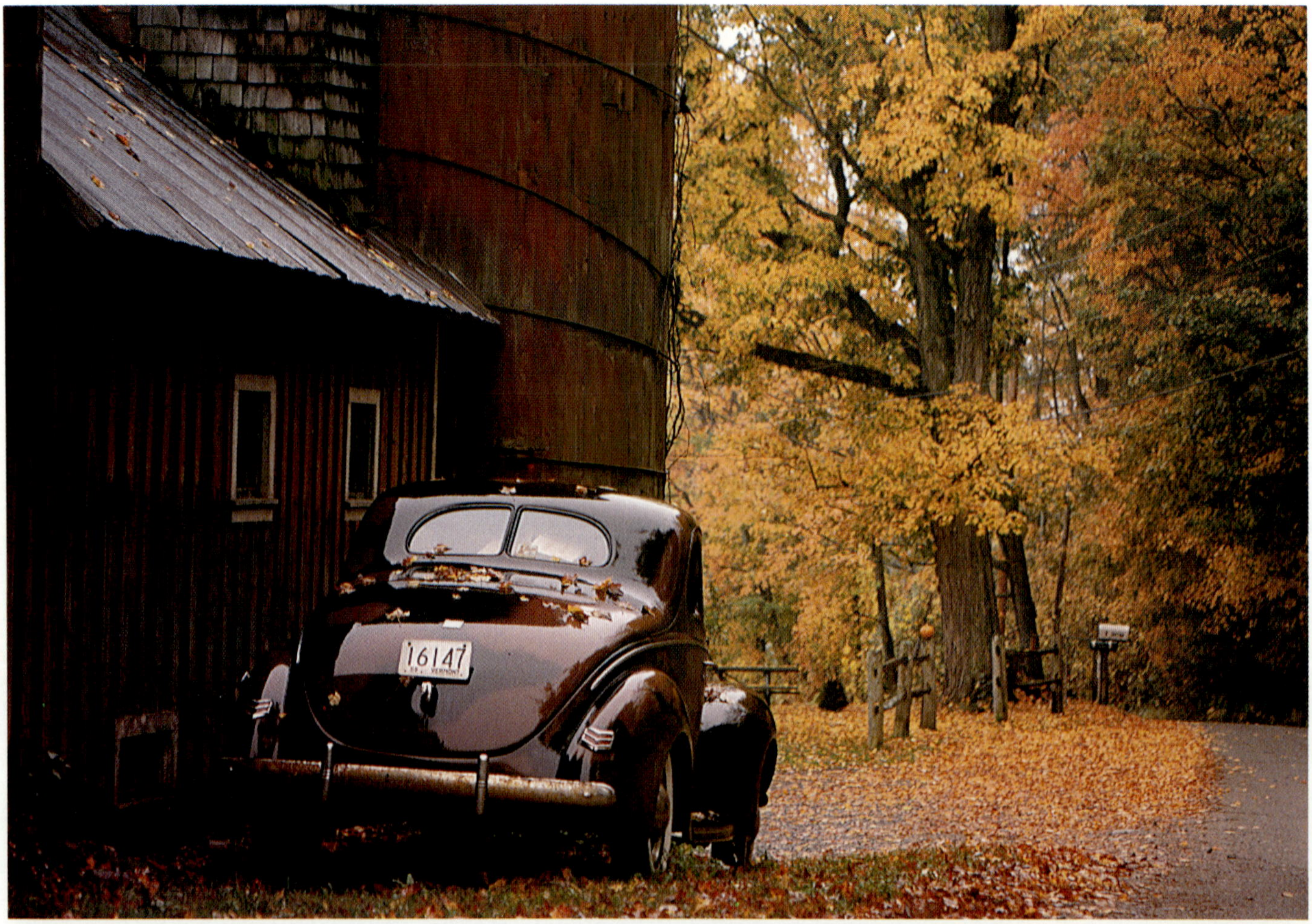

38. Full-bodied color and excellent wider-than-normal tonal separation are design goals of Agfacolor Optima 125. Film exposed @ ISO 125 (Jeanne Miller).

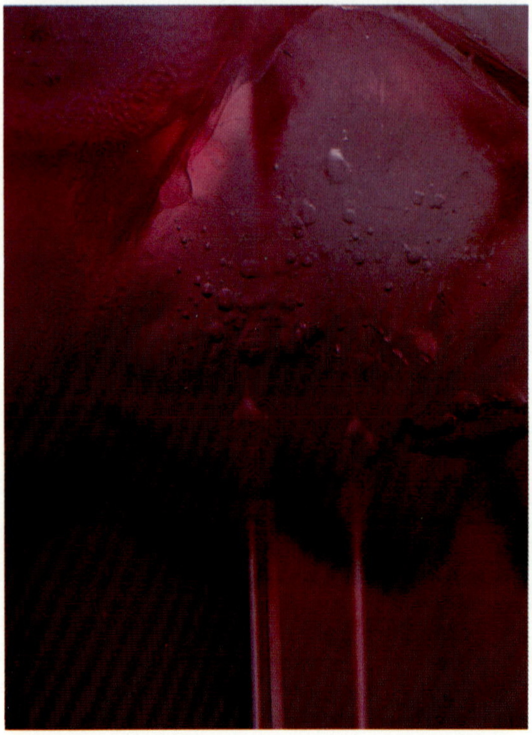

39. This still-life was lit with tungsten photofloods. Kodachrome 40 was selected for its sharpness and fine grain and because its spectral response is very close to that of the light source. Kodachrome 40 @ ISO 40.

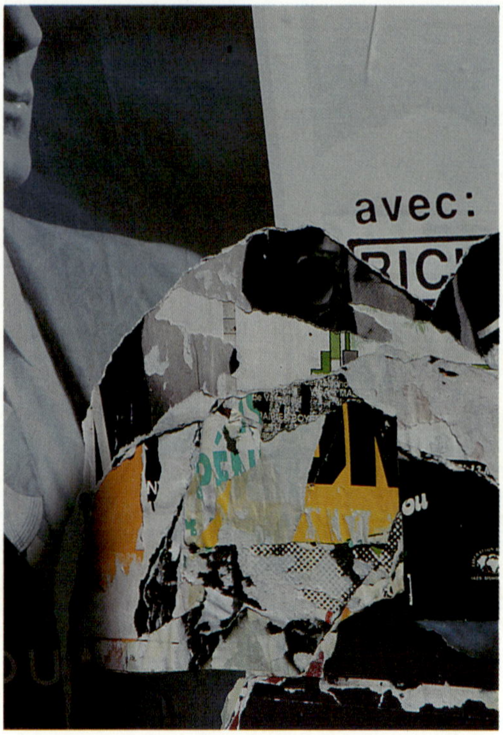

40. The camera was braced against streetlight in order to use long lens of 300mm focal length. Fujicolor Super HG 200 @ISO 200.

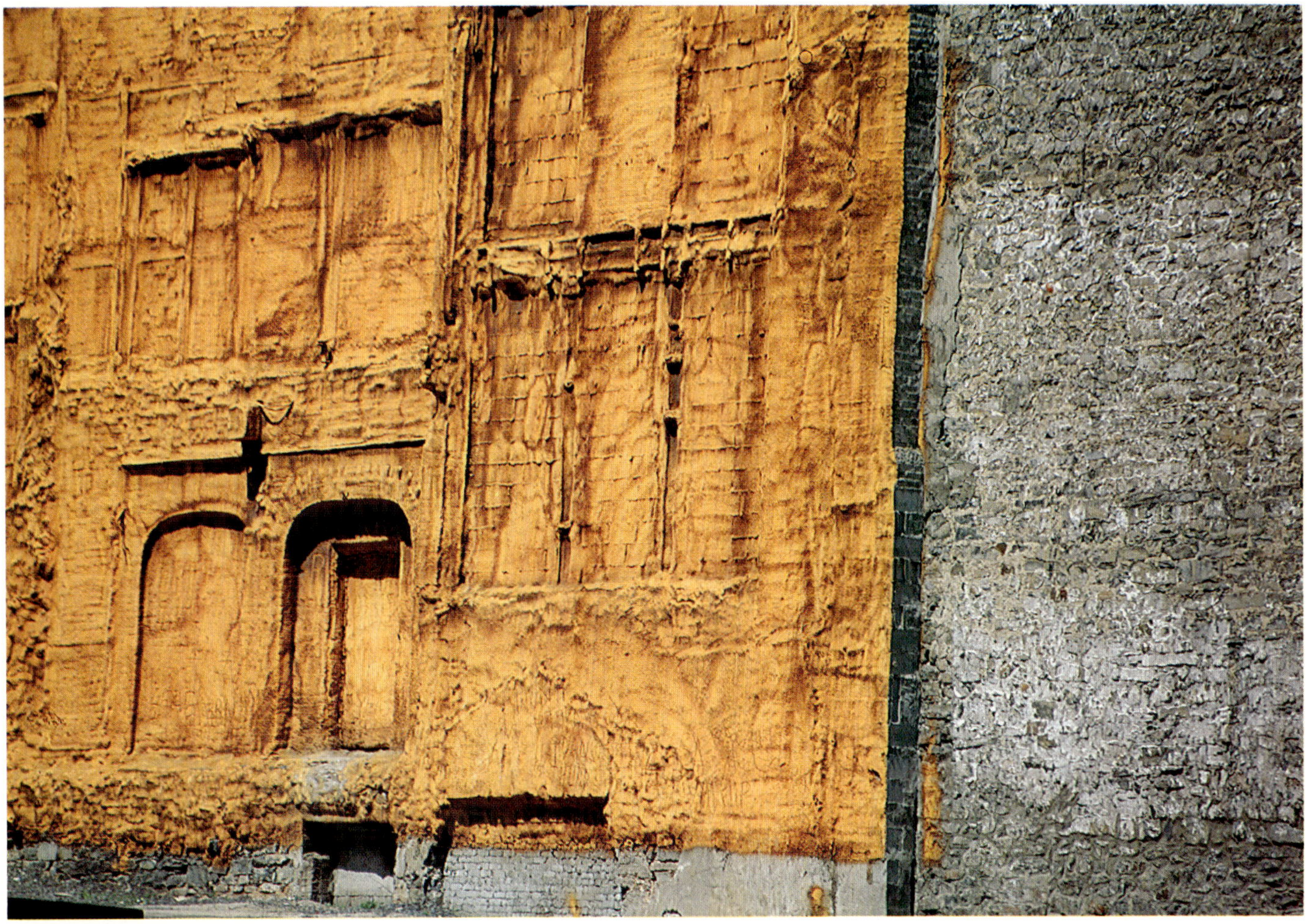

41. This scene benefited from the use of a polarizing filter to enhance color reproduction. Fujichrome 100 RDP @ ISO 125.

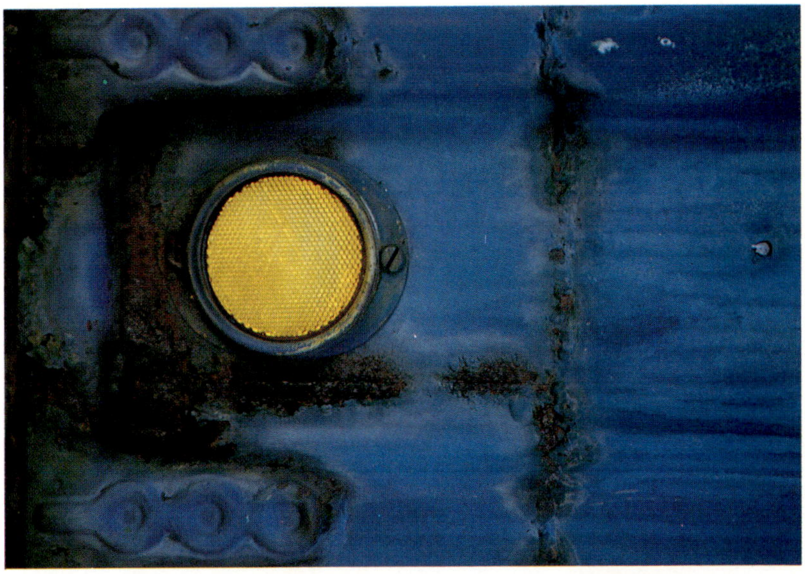

42. Macro image takes advantage of recently introduced Agfachrome RS 50 Plus with its fine grain and moderate, controllable contrast. Agfachrome RS 50 Plus @ ISO 64.

43. Scotchchrome 100 displays the ability to render both pastels and deep shades equally well in the same image. This is an excellent performance that some other similar-speed films would find difficult to match. Scotchchrome 100 @ ISO 100.

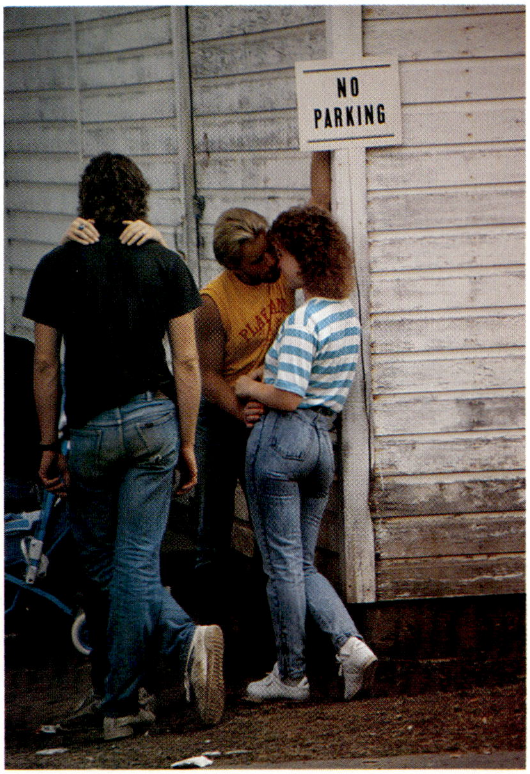

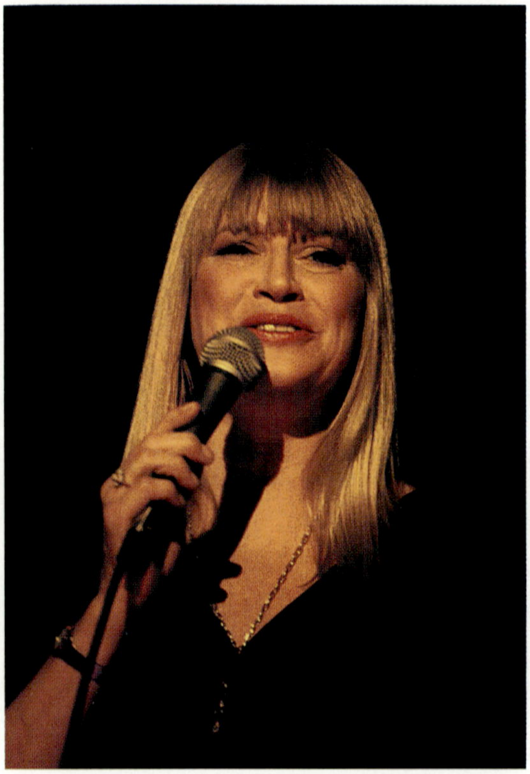

44. Warm daylight-balanced film used here merely added to the already golden glow provided by colored gel spotlights. Film pushed one stop by rating it at ISO 800. Ektachrome 400 X @ ISO 800.

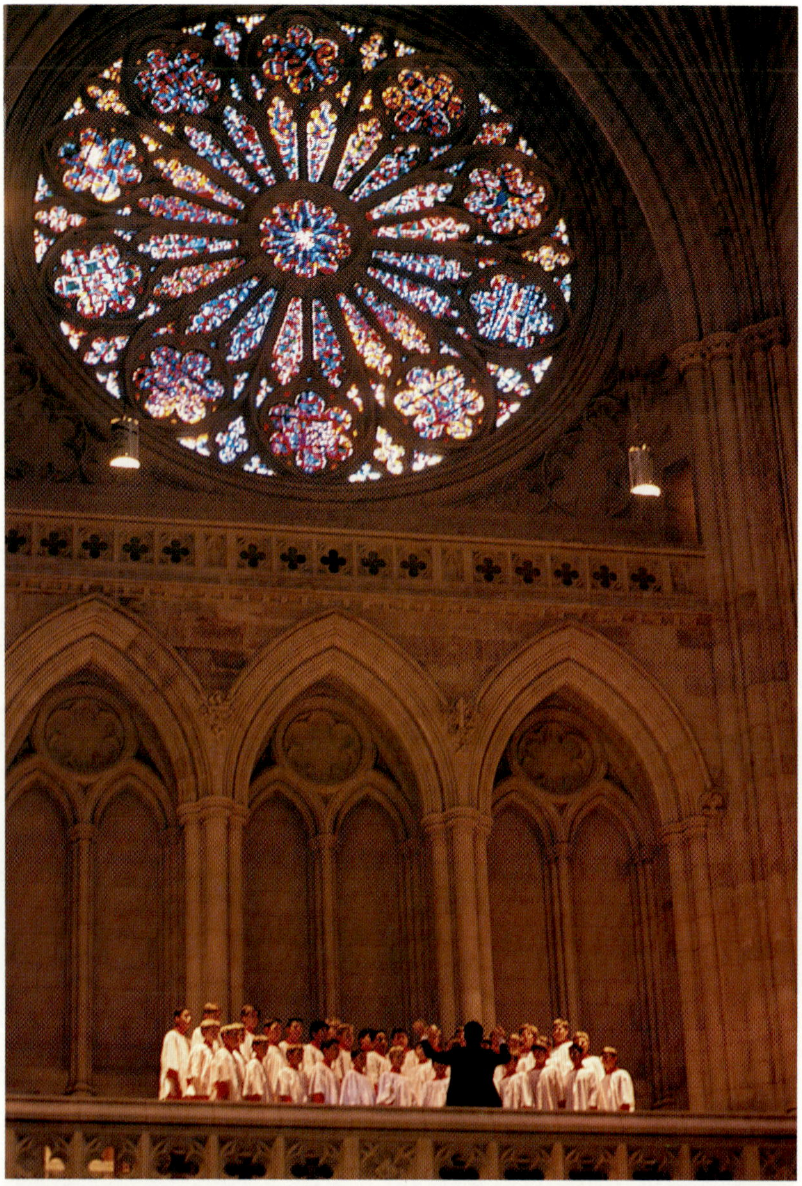

45. Resting the camera on the back of a pew made this picture possible with moderate-speed print film. Konica Super SR 160 S @ ISO 160.

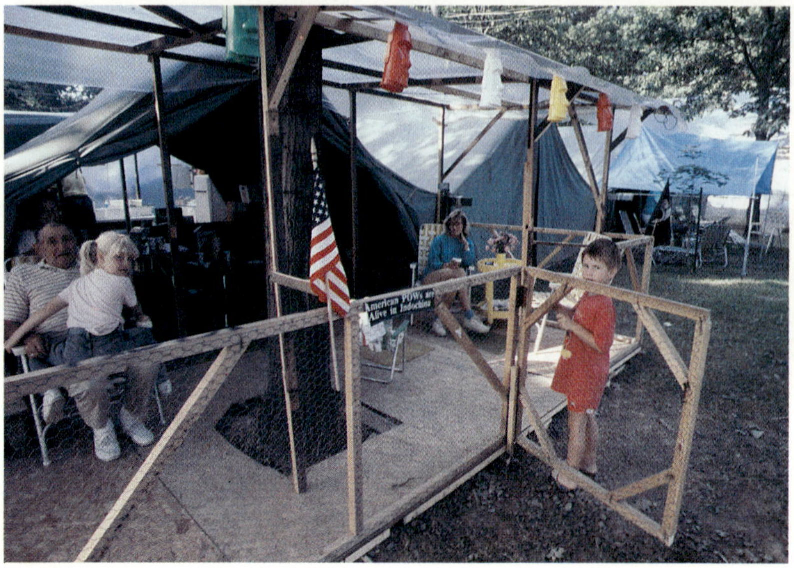

46. In order to get detail in both highlights and shadows, I selected a film noted for its ability to reproduce a wide range of tones but still deliver excellent color rendition and speed. Fujicolor 400 HG Professional @ ISO 400.

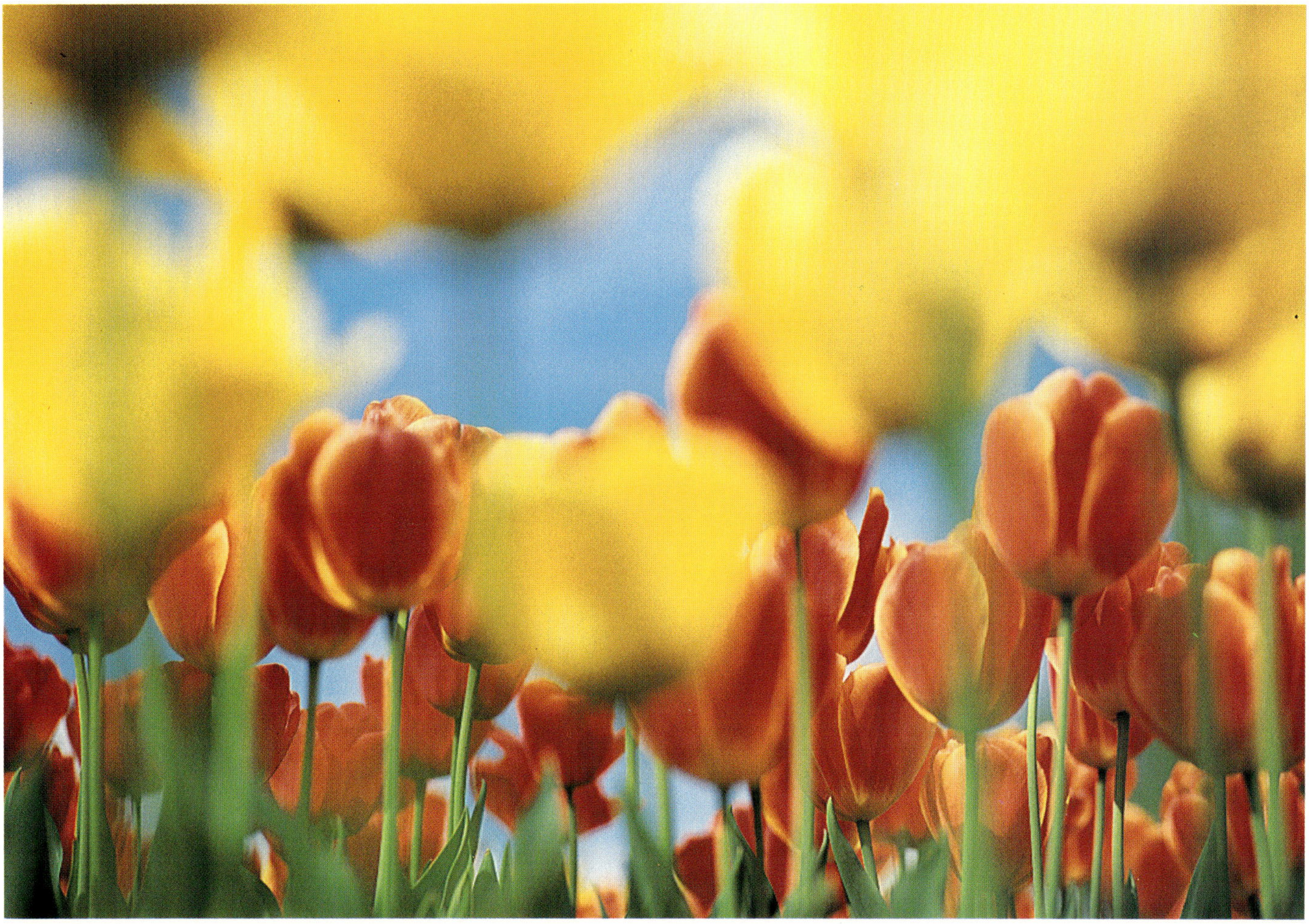

47. Athough Fujichrome Velvia has extremely high color contrast, its overall image contrast is moderate. These two qualities make brilliantly colored but finely detailed images possible. Fujichrome Velvia @ ISO 50.

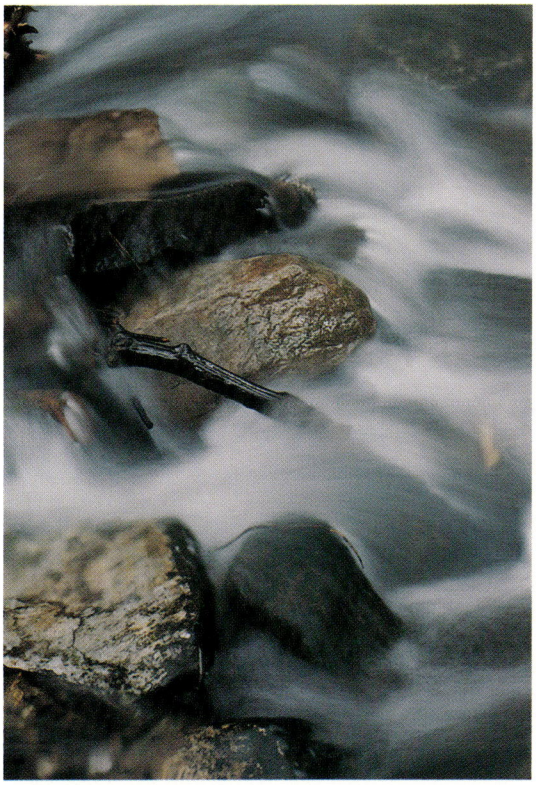

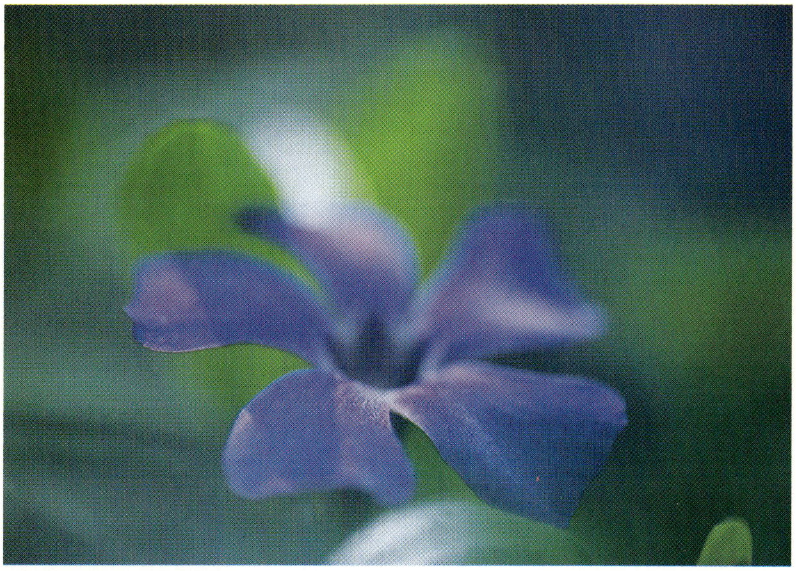

49. Accurate color rendition and excellent image definition are two qualities of Polaroid's High Definition Slide Film. It is an E-6 processible emulsion presently available only in Europe. Polaroid HD 100 @ ISO 100.

48. Using a neutral density filter reduced light to a point where one-half second exposure is required. Then the water is rendered as a soft, smooth blur. Kodachrome 64 Professional @ ISO 80.

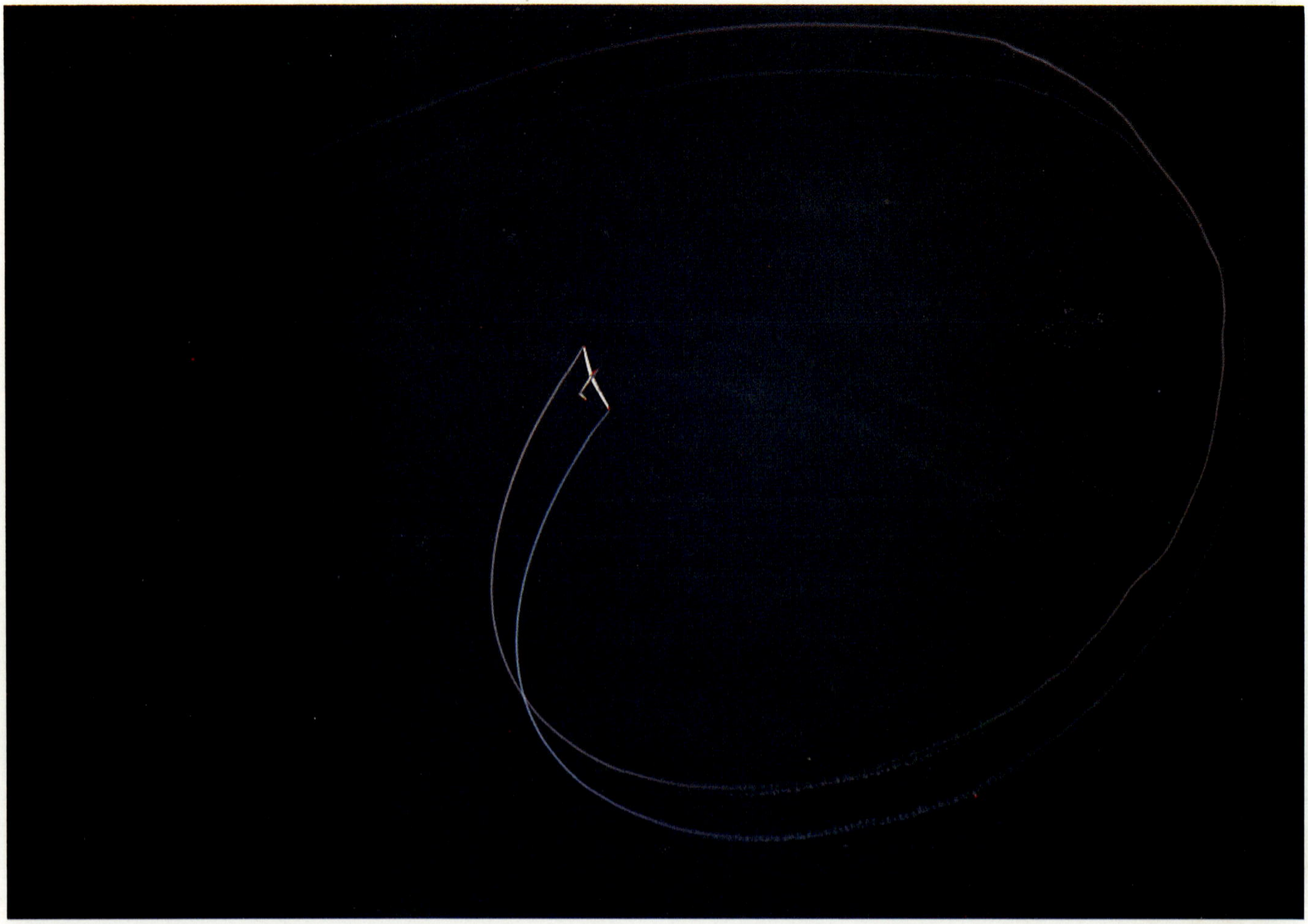

50. Open areas lacking detail often show grain. The very fine grain of Fujichrome 100 RD rated at ISO 100 is readily apparent in the clean sky areas of this image. Fujichrome 100 RD @ ISO 100.

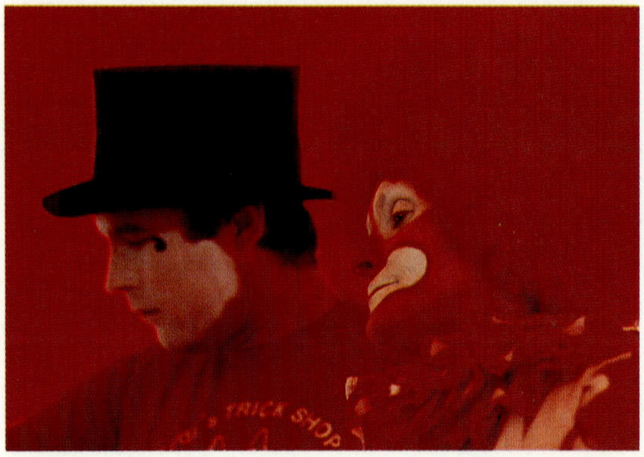

51. Agfachromes reproduce red with particular fidelity. Agfachrome RS 100 Professional @ ISO 100.

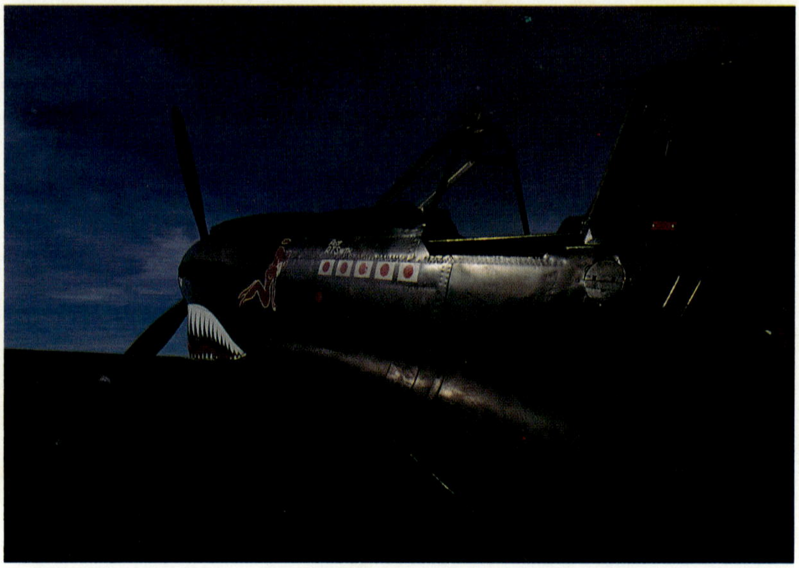

52. Polaroid Presentation Chrome is a specialized film intended for photography of CRT screens. Used as a general purpose film, its extreme contrast and fine grain are clearly visible. Polaroid Presentation Chrome @ ISO 100.

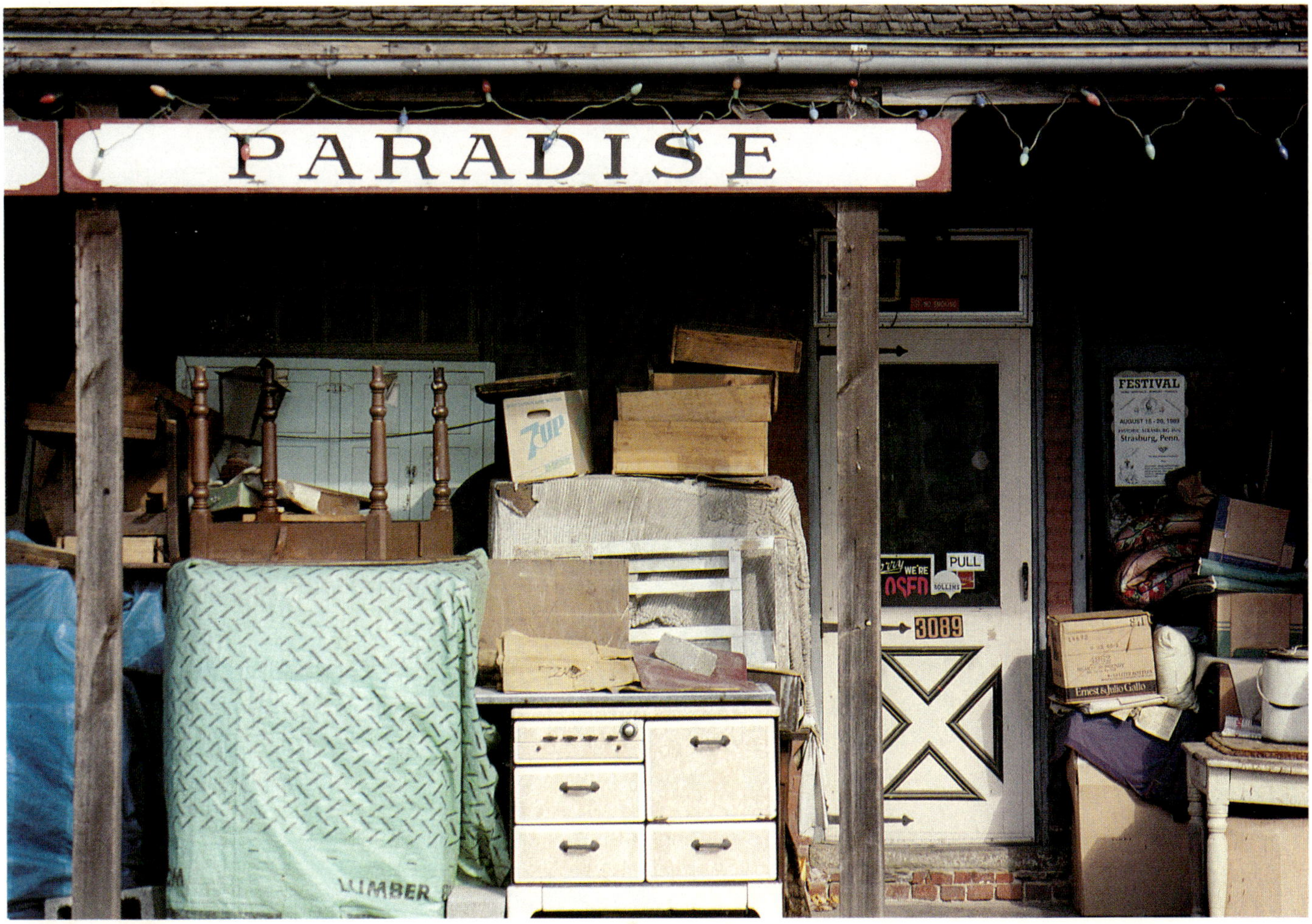

53. Copying a color slide onto low-contrast Fujichrome Slide Duplicating Film helps preserve the fine pictorial qualities of the original. Fujichrome Slide Duplicating Film CDU @ ISO 12.

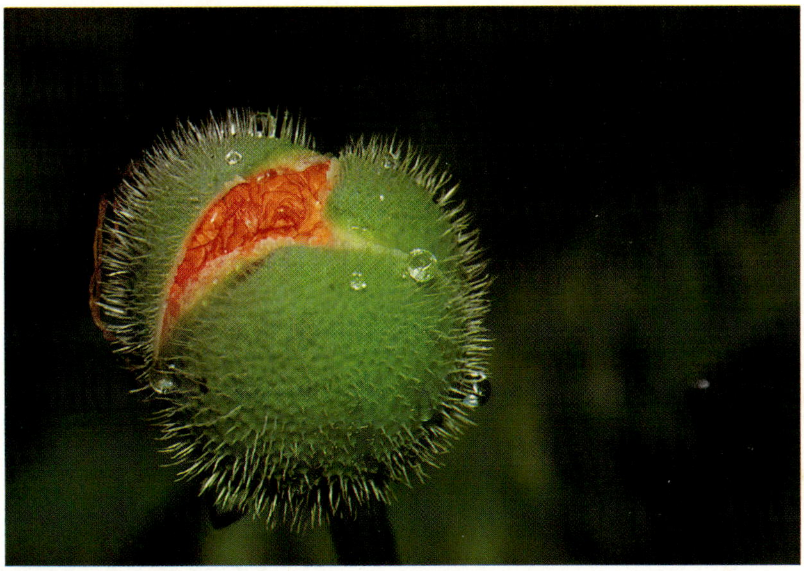

54. A UV filter absorbed excess blue in the open shadows, allowing more accurate color rendition. Konicacolor Super SR 100 @ ISO 100.

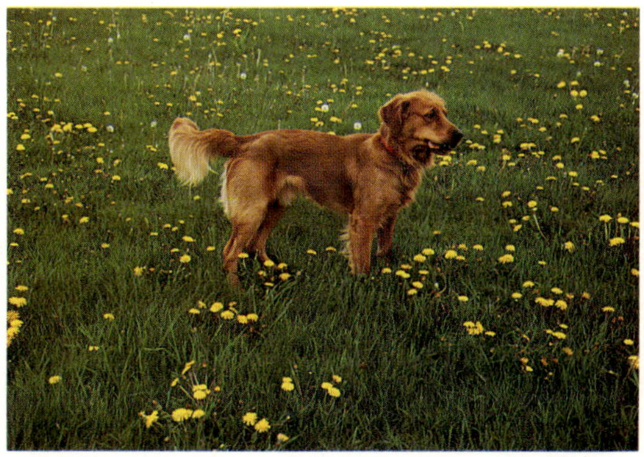

55. Usually color can be deepened by underexposing slide film or overexposing print film. Ektapress 100 @ ISO 80.

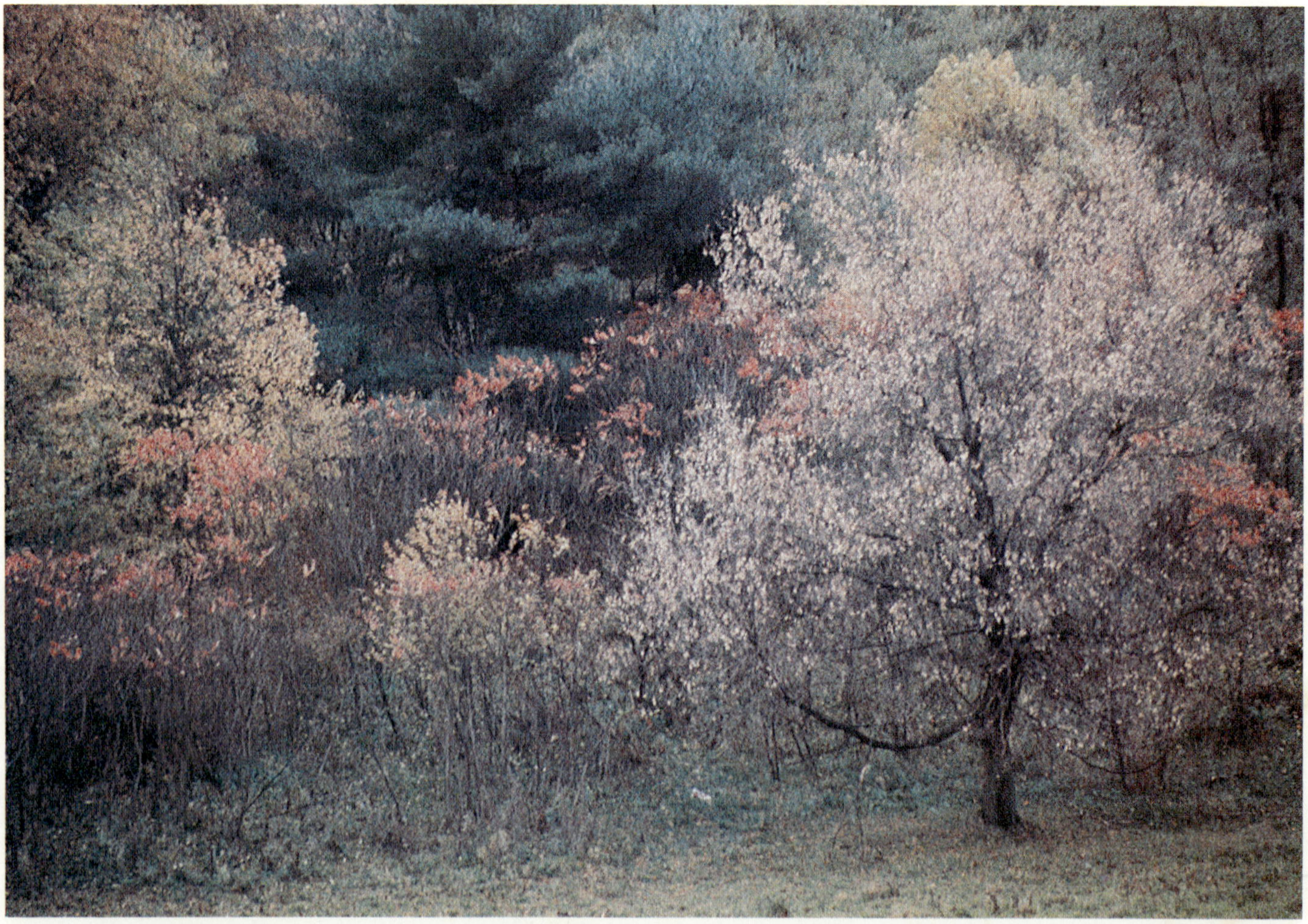

56. Ultra high-speed color film push-processed one or two f-stops generates an image having a pointilistic appearance. Agfachrome RS 1000 Professional @ ISO 4000.

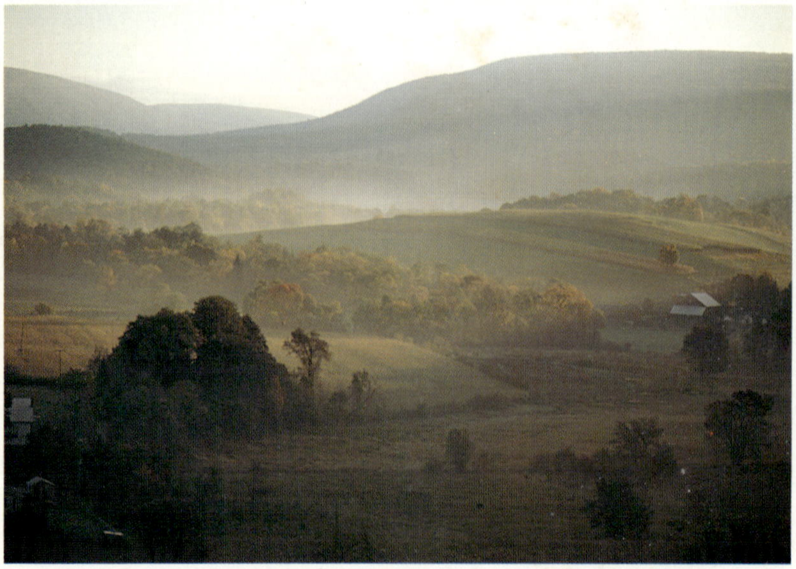

57. Combining soft focus and warming filters underscored the nostalgic look of this early morning landscape. Ektachrome 100 HC @ ISO 100.

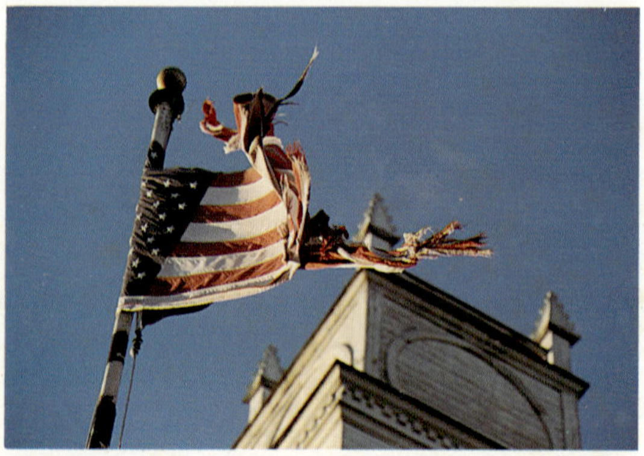

58. Ektachrome 64 Professional produced an image with soft color and wide contrast range. Ektachrome 64 Professional @ ISO 64.

Processing: Standard E-6
Size(s): 35mm *Roll Lengths:* 24, 36 exposures
Long Rolls Available? Yes
Other Sizes Available: 120
Diffuse RMS Granularity Value: 10
Resolving Power
High Contrast 1000:1 = 125 ll/mm
Low Contrast 1.6:1 = 60 ll/mm plus
Exposure Latitude: +1/2, –1 f-stop
Contrast Characteristics: Normal to slightly high
Film Structure: Conventional
Color Balance: Daylight
Color Bias: Neutral to slightly warm
Recommended Applications: Superior film for still-life, landscapes, portraiture, macro
Special Uses: Excellent choice when very fine grain and convenience of E-6 processing is needed
Reciprocity (recommended shutter-speed range): 1/1000–1 second
Exposure Correction Table:

Exposure time:	1/10000	1/1000	1/100	1/10	1
Correction:	none	none	none	none	+1/2 f-stop
Filtration (CC):	none	none	none	none	CC05R

Storage, Keeping Properties, and Special Handling: Normal
Availability: Photo specialty stores, some discount outlets
In-Use Evaluation: This relatively slow fine grained film yields extremely brilliant colors of high fidelity. In situations where speed is not of utmost importance, HC 50 would be an excellent choice.

Designation: **Ektachrome 100 HC M**
Manufactured by: Eastman Kodak Co., USA
Film Type: Very fine grain 35mm color transparency
Speed: ISO 100 *Push and/or Pull Feasible?* Yes
Processing: Standard E-6
Size(s): 35mm *Roll Lengths:* 24, 36 exposures
Long Rolls Available? Yes
Other Sizes Available: 120
Diffuse RMS Granularity Value: 11
Resolving Power
High Contrast 1000:1 = 100 ll/mm
Low Contrast 1.6:1 = 50 ll/mm plus
Exposure Latitude: +3/4, –1 f-stop
Contrast Characteristics: Normal
Film Structure: Conventional
Color Balance: Daylight
Color Bias: Nearly neutral, slightly warm skin tones
Recommended Applications: General amateur photography
Special Uses: None
Reciprocity (recommended shutter-speed range): 1/10000–1/10 second
Exposure Correction Table:

Exposure time:	1/10000	1/1000	1/100	1/10	1
Correction:	none	none	none	none	+1/2 f-stop
Filtration (CC):	none	none	none	none	CC05R

Storage, Keeping Properties, and Special Handling: Normal
Availability: General

In-Use Evaluation: Ektachrome 100HC is an excellent all-around performer. It is highly saturated, has very good contrast qualities, is pushable, and displays tight, even grain. One of the top medium-speed slide films.

Designation: **Ektachrome 100 Plus Professional EPN**
Manufactured by: Eastman Kodak Co., USA
Film Type: Very fine grain 35mm color transparency
Speed: ISO 100 *Push and/or Pull Feasible?* Yes
Processing: Standard E-6
Size(s): 35mm *Roll Lengths:* 36 exposures
Long Rolls Available? Yes
Other Sizes Available: 120, 220, sheets
Diffuse RMS Granularity Value: 11
Resolving Power
High Contrast 1000:1 = 100 ll/mm
Low Contrast 1.6:1 = 50 ll/mm plus
Exposure Latitude: +1, –1 f-stop
Contrast Characteristics: Normal
Film Structure: Conventional
Color Balance: Daylight
Color Bias: Neutral to slightly warm
Recommended Applications: Outdoor landscapes, candid portraiture, general photography
Special Uses: Warmth makes it good choice for flash photography
Reciprocity (recommended shutter-speed range): 1/1000–1 second
Exposure Correction Table:

Exposure time:	1/10000	1/1000	1/100	1/10	1
Correction:	none	none	none	none	+1/2 f-stop
Filtration (CC):	none	none	none	none	CC05R

Storage, Keeping Properties, and Special Handling: Professional film. Refrigerate until use.
Availability: Photo specialty stores
In-Use Evaluation: The professional version of 100 HC appears a bit sharper and more color-saturated with cleaner whites and neutral colors. Pastels reproduce very well in spite of high saturation. A top choice for critical applications

Designation: **Ektachrome 100 Professional EPN**
Manufactured by: Eastman Kodak Co., USA
Film Type: Fine grain 35mm color transparency
Speed: ISO 100 *Push and/or Pull Feasible?* Yes
Processing: Standard E-6
Size(s): 35mm *Roll Lengths:* 36 exposures
Long Rolls Available? Yes
Other Sizes Available: 120, 220, sheets
Diffuse RMS Granularity Value: 11
Resolving Power
High Contrast 1000:1 = 100 ll/mm
Low Contrast 1.6:1 = 50 ll/mm
Exposure Latitude: +3/4, –1 f-stop
Contrast Characteristics: Lower than normal
Film Structure: Conventional
Color Balance: Daylight

Color Bias: Cool
Recommended Applications: General photography
Special Uses: Excellent film for Azo dyed fabric
Reciprocity (recommended shutter-speed range): 1/10000–1/10 second
Exposure Correction Table:

Exposure time:	1/10000	1/1000	1/100	1/10	1
Correction:	none	none	none	none	N/R
Filtration (CC):	none	none	none	none	N/R

Storage, Keeping Properties, and Special Handling: Professional film. Refrigerate until use.
Availability: Photo specialty stores
In-Use Evaluation: Until appearance of 100 Plus this film was the professional's E-6 mainstay. It features a cooler color rendition than later emulsions, with softer edges and a bit more grain. It is a useful film if you need to tone down contrast.

Designation: **Ektachrome 100X Professional EPZ**
Manufactured by: Eastman Kodak Co., USA
Film Type: Extremely fine grain 35mm color transparency
Speed: ISO 100 *Push and/or Pull Feasible?* Yes
Processing: Standard E-6
Size(s): 35mm *Roll Lengths:* 36 exposures
Long Rolls Available? No
Other Sizes Available: 120, 220, sheets
Diffuse RMS Granularity Value: 11
Resolving Power
High Contrast 1000:1 = 100 ll/mm
Low Contrast 1.6:1 = 50 ll/mm
Exposure Latitude: +3/4, –1 f-stop
Contrast Characteristics: Normal
Film Structure: Conventional, added yellow and red sensitivity
Color Balance: Daylight
Color Bias: Warm, especially in relation to older Ektachromes
Recommended Applications: Same as Ektachrome 64X
Special Uses: The added speed of the ISO 100 allows this film to be used in more hand-held situations
Reciprocity (recommended shutter-speed range): 1/10000–1/10 second
Exposure Correction Table:

Exposure time:	1/1000	1/100	1/10	1	10
Correction:	none	none	none	+1/2 f-stop	N/R
Filtration (CC):	none	none	none	none	N/R

Storage, Keeping Properties, and Special Handling: Professional film. Refrigerate until use.
Availability: Photo specialty stores
In-Use Evaluation: See Ektachrome 64X. 100X exhibits slightly more grain for the tradeoff of about 1/2 f-stop more film speed.

Designation: **Ektachrome 160 Professional (Tungsten) EPT**
Manufactured by: Eastman Kodak Co., USA
Film Type: Fine grain 35mm color transparency
Speed: ISO 160 *Push and/or Pull Feasible?* Yes
Processing: Standard E-6
Size(s): 35mm *Roll Length:* 36 exposures
Long Rolls Available? Yes

Other Sizes Available: 120
Diffuse RMS Granularity Value: 13
Resolving Power
High Contrast 1000:1 = 125 ll/mm
Low Contrast 1.6:1 = 50 ll/mm
Exposure Latitude: +3/4, –1 f-stop
Contrast Characteristics: Normal
Film Structure: Conventional
Color Balance: Tungsten 3200K
Color Bias: Nearly neutral
Recommended Applications: General indoor existing-light photography
Special Uses: Outdoor existing-light night work, sports
Reciprocity (recommended shutter-speed range): 1/10000–1/10 second
Exposure Correction Table:

Exposure time:	1/10000	1/1000	1/100	1/10	1
Correction:	none	none	none	none	1/2 f-stop
Filtration (CC):	CC025G	CC05M	none	none	CC025M

Storage, Keeping Properties, and Special Handling: Professional film. Refrigerate until use.
Availability: Photo specialty stores
In-Use Evaluation: This film displays good color characteristics, moderate grain, and is pushable at least one stop. Rated at ISO 320, it can be used hand-held in relatively good night-lighting situations.

Designation: **Ektachrome 160 (Tungsten) ET**
Manufactured by: Eastman Kodak Co., USA
Film Type: Fine to medium grain 35mm color transparency
Speed: ISO 160 *Push and/or Pull Feasible?* Yes
Processing: Standard E-6
Size(s): 35mm *Roll Lengths:* 24, 36 exposures
Long Rolls Available? No
Other Sizes Available: No
Diffuse RMS Granularity Value: 13
Resolving Power
High Contrast 1000:1 = 100 ll/mm
Low Contrast 1.6:1 = 50 ll/mm
Exposure Latitude: +3/4, –1 f-stop
Contrast Characteristics: Normal
Film Structure: Conventional
Color Balance: Tungsten 3200K
Color Bias: Neutral to slightly cool (blue)
Recommended Applications: General tungsten illumination, including regular house lights
Special Uses: See 160 Professional
Reciprocity (recommended shutter-speed range): 1/10000–1/10 second
Exposure Correction Table:

Exposure time:	1/10000	1/1000	1/100	1/10	1
Correction:	none	none	none	none	+1/2 f-stop
Filtration (CC):	none	none	none	none	CC025M

Storage, Keeping Properties, and Special Handling: Normal
Availability: Photo specialty stores, some discount outlets
In-Use Evaluation: Amateur version of Ektachrome 160 (Tungsten) Professional. Virtually identical performance.

Designation: **Ektachrome 200 Professional EPD**
Manufactured by: Eastman Kodak Co., USA
Film Type: Medium-grain 35mm color transparency
Speed: ISO 200 *Push and/or Pull Feasible?* Yes
Processing: Standard E-6
Size(s): 35mm *Roll Lengths:* 36 exposures
Long Rolls Available? Yes
Other Sizes Available: 120, 220, sheets
Diffuse RMS Granularity Value: 13
Resolving Power:
High Contrast 1000:1 = 125 ll/mm
Low Contrast 1.6:1 = 50 ll/mm
Exposure Latitude: +3/4, –1 f-stop
Contrast Characteristics: Lower than normal
Film Structure: Conventional
Color Balance: Daylight
Color Bias: Cool (blue)
Recommended Applications: General photography
Special Uses: None
Reciprocity (recommended shutter-speed range): 1/10000–1/10 second
Exposure Correction Table:

Exposure time:	1/10000	1/1000	1/100	1/10	1
Correction:	none	none	none	none	+1/2 f-stop
Filtration (CC)	none	none	none	none	none

Storage, Keeping Properties, and Special Handling: Professional film. Refrigerate until use.
Availability: Photo specialty stores
In-Use Evaluation: A good, if not outstanding, general-purpose film that is useful if extra speed is required.

Designation: **Ektachrome 200 ED**
Manufactured by: Eastman Kodak Co., USA
Film Type: Medium-grain 35mm color transparency
Speed: ISO 200 *Push and/or Pull Feasible?* Yes
Processing: Standard E-6
Size(s): 35mm *Roll Lengths:* 24, 36 exposures
Long Rolls Available? No
Other Sizes Available: None
Diffuse RMS Granularity Value: 13
Resolving Power
High Contrast 1000:1 = 125 ll/mm
Low Contrast 1.6:1 = 50 ll/mm
Exposure Latitude: +3/4, –1 f-stop
Contrast Characteristics: Low
Film Structure: Conventional
Color Balance: Daylight
Color Bias: Cool (blue)
Recommended Applications: General photography
Special Uses: None
Reciprocity (recommended shutter-speed range): 1/10000–1/10 second
Exposure Correction Table:

Exposure time:	1/10000	1/1000	1/100	1/10	1
Correction:	none	none	none	none	+1/2 f-stop
Filtration (CC):	none	none	none	none	CC05M

Storage, Keeping Properties, and Special Handling: Normal
Availability: General
In-Use Evaluation: Amateur version of Ektachrome 200 Professional. Similar performance.

Designation: **Ektachrome 400X Professional EPL**
Manufactured by: Eastman Kodak Co., USA
Film Type: 35mm color transparency
Speed: ISO 400 *Push and/or Pull Feasible?* Yes
Processing: Standard E-6
Size(s): 35mm *Roll Lengths:* 36 exposures
Long Rolls Available? No
Other Sizes Available: None
Diffuse RMS Granularity Value: 17
Resolving Power
High Contrast 1000:1 = 90 ll/mm
Low Contrast 1.6:1 = 45 ll/mm
Exposure Latitude: +1, –1 f-stop
Contrast Characteristics: Normal
Film Structure: Conventional, added yellow and red sensitivity
Color Balance: Daylight
Color Bias: Warm, especially in relation to older Ektachromes
Recommended Applications: All available light photography
Special Uses: High shutter speeds and reasonable f-stop combination for hand-held photography
Reciprocity (recommended shutter-speed range): 1/10000–1/10 second
Exposure Correction Table:

Exposure time:	1/1000	1/100	1/10	1/1	10
Correction:	none	none	none	+1/2 f-stop	N/R
Filtration (CC):	none	none	none	none	N/R

Storage, Keeping Properties, and Special Handling: Professional film. Refrigerate until use.
Availability: Photo specialty stores
In-Use Evaluation: This new film is an overall improvement over its predecessor. The film is made to tight ±1/6 f-stop tolerances. It features rich colors, improved grain, and warmer color balance—making it a much more useable emulsion.

Designation: **Ektachrome 400HC EL**
Manufactured by: Eastman Kodak Co., USA
Film Type: 35mm color transparency
Speed: ISO 400 *Push and/or Pull Feasible?* Yes
Processing: Standard E-6
Size(s): 35mm *Roll Lengths:* 24, 36 exposures
Long Rolls Available? No
Other Sizes Available: 120
*Diffuse RMS Granularity Value:*17
Resolving Power
High Contrast 1000:1 = 90 ll/mm,
Low Contrast 1.6:1 = 45 ll/mm
Exposure Latitude: +1, –1 f-stop
Contrast Characteristics: Slightly high
Film Structure: Conventional
Color Balance: Daylight

Color Bias: Neutral to slightly warm
Recommended Applications: See Ektachrome 400X
Special Uses: See Ektachrome 400 X
Reciprocity (recommended shutter-speed range): 1/10000–1/10 second
Exposure Correction Table:

Exposure time:	1/1000	1/100	1/10	1	10
Correction:	none	none	none	+1/2	N/R
Filtration (CC):	none	none	none	none	N/R

Storage, Keeping Properties, and Special Handling: Normal
Availability: Photo specialty stores, mass market outlets
In-Use Evaluation: Amateur version of Ektachrome 400X and similar in performance. Color balance appears a bit warmer than its pro sibling.

Designation: **Ektachrome P800/1600 EES**
Manufactured by: Eastman Kodak Co., USA
Film Type: Medium to coarse grain 35mm color transparency
Speed: ISO 400–3200 *Push and/or Pull Feasible?* Yes
Processing: E-6P
Size(s): 35mm *Roll Lengths:* 36 exposures
Long Rolls Available? No
Other Sizes Available: No
Diffuse RMS Granularity Value: EI 400 = 17, EI 800 = 22, EI 1600 = 28, EI 3200 = 36
Resolving Power

High Contrast	1000:1	@ EI 400/800 = 80 ll/mm
		@ EI 800/1600 = 40 ll/mm
Low Contrast	1.6:1	@ EI 400/800 = 40 ll/mm
		@ EI 1600/3200 = 32 ll/mm

Exposure Latitude: +1/2, –1 f-stop
Contrast Characteristics: High
Film Structure: Conventional
Color Balance: Daylight
Color Bias: Similar to Ektachrome 400
Recommended Applications: Available-light high-speed photography
Special Uses: Allows use of high shutter speeds and small apertures
Reciprocity (recommended shutter-speed range): 1/10000–1/10 second
Exposure Correction Table:

Exposure time:	1/1000	1/100	1/10	1	10
Correction:	none	none	none	N/R	N/R
Filtration (CC):	none	none	none	N/R	N/R

Storage, Keeping Properties, and Special Handling: Professional film. Refrigerate until use.
Availability: Photo specialty stores
In-Use Evaluation: P800/1600 performs better than slower slide films that are pushed to P800/1600's speed ratings. Also, this film is designed to generate richer D-Max (more density in shadow areas). It is the fastest Kodak slide film and transparencies project surprisingly well.

Designation: **Fujichrome Velvia RVP**
Manufactured by: Fuji Film Co., Japan
Film Type: Ultra fine grain 35mm color transparency
Speed: ISO 50 *Push and/or Pull Feasible?* Yes
Processing: Standard E-6

Size(s): 35mm *Roll Lengths:* 24, 36 exposures
Long Rolls Available? Yes
Other Sizes Available: 120, 220, sheets
Diffuse RMS Granularity Value: 9
Resolving Power
High Contrast 1000:1 = 160 ll/mm
Low Contrast 1.6:1 = 80 ll/mm
Exposure Latitude: +1/2, –1 f-stop
Contrast Characteristics: Normal to slightly high
Film Structure: Conventional, with 15 percent decrease in thickness, reducing light scattering and increasing sharpness. DIRs used for first time in slide film, providing greater edge sharpness (acutance).
Color Balance: Daylight
Color Bias: Neutral to warm
Recommended Applications: A film for the most critical applications where intense color, extreme sharpness, and micro grain are required
Special Uses: Spectacular results on overcast days. Color reproduction enhanced greatly.
Reciprocity (recommended shutter-speed range): 1/1000–1 second
Exposure Correction Table:

Exposure time:	1/4000	1/1000	1/100	1/10	1	10
Correction:	none	none	none	none	none	+ 1/2 f-stop
Filtration (CC):	none	none	none	none	none	CC 07.5M

Storage, Keeping Properties, and Special Handling: Professional film. Refrigerate until use.
Availability: Photo specialty stores
In-Use Evaluation: Designed for fashion, commercial, advertising, nature, still-life, and other demanding applications, Velvia represents the latest technological tour de force in the film designers' art. It offers the user the highest color saturation of any color film. Colors literally leap off the slide. For some this is not always a desirable quality, however. Additionally, Velvia is not a film that will record a scene with unerring color accuracy for it has a distinctive palette. For sheer impact and super quality rivaling Kodachrome 25 but with one full f-stop more sensitivity, Velvia is surely a film to be seriously considered.

Designation: **Fujichrome 50 Professional D RFP**
Manufactured by: Fuji Film Co., Japan
Film Type: Extremely fine grain 35mm color transparency
Speed: ISO 50 *Push and/or Pull Feasible?* Yes
Processing: Standard E-6
Size(s): 35mm *Roll Lengths:* 36 exposures
Long Rolls Available? Yes
Other Sizes Available: 120, sheets
Diffuse RMS Granularity Value: 11
Resolving Power
High Contrast 1000:1 = 125 ll/mm
Low Contrast 1.6:1 = 50 ll/mm
Exposure Latitude: +3/4, –1 f-stop
Contrast Characteristics: Normal to slightly high
Film Structure: Conventional, twin-crystal grain structure
Color Balance: Daylight

Color Bias: Nearly neutral
Recommended Applications: All professional-caliber photography, especially outdoor scenics
Special Uses: Excellent film choice where extreme enlargements are required.
Reciprocity (recommended shutter-speed range): 1/4000–1 second
Exposure Correction Table:

Exposure time:	1/4000	1/1000	1/100	1/10	1
Correction:	none	none	none	none	+ 1/2 f-stop
Filtration (CC):	none	none	none	none	CC07.5M

Storage, Keeping Properties, and Special Handling: Professional film. Refrigerate until use.
Availability: Photo specialty stores
In-Use Evaluation: Before Velvia's appearance, RFP was the quality champ in Fuji's arsenal of films. It renders blues, reds, and yellows very accurately. It also handles wide tonal variations well, recording detail in deep shadows as well as in highlights. The film has been optimized for photomechanical reproduction with the addition of extra base density, so it delivers splendid separations for book and magazine illustration purposes. Of all its positives, this film is especially noted for the way it handles greens. Pros will tell you no other film does quite as good a job. Fujichrome 50 appears to be less contrasty than its Kodachrome rivals.

Designation: **Fujichrome 50 RF**
Manufactured by: Fuji Film Co., Japan
Film Type: Extremely fine grain 35mm color transparency
Speed: ISO 50 *Push and/or Pull Feasible?* Yes
Processing: Standard E-6
Size(s): 35mm *Roll Lengths:* 24, 36 exposures
Long Rolls Available? No
Other Sizes Available: Only in the professional version
Diffuse RMS Granularity Value: 11
Resolving Power
High Contrast 1000:1 = 125 ll/mm
Low Contrast 1.6:1 = 50 ll/mm
Exposure Latitude: +3/4, –1 f-stop
Contrast Characteristics: Normal
Film Structure: Conventional; improved grain structure increases sharpness
Color Balance: Daylight
Color Bias: Neutral
Recommended Applications: Excellent general-purpose film for occasions where high film sensitivity is not required
Special Uses: Well suited for photomechanical reproduction
Reciprocity (recommended shutter-speed range): 1/4000–1 second
Exposure Correction Table:

Exposure time:	1/4000	1/1000	1/100	1/10	1
Correction:	none	none	none	none	+1/2 f-stop
Filtration (CC)	none	none	none	none	CC07.5 M

Storage, Keeping Properties, and Special Handling: Normal
Availability: Photo specialty stores, some discount outlets
In-Use Evaluation: With an excellent mix of grain, color reproduction, manageable contrast, and neutral color balance, RF offers virtually

the same fine performance as does its professional mate, RFP, but at a substantially lower price.

Designation: **Fujichrome 64 T Professional**
Manufactured by: Fuji Film Co., Japan
Film Type: Extremely fine grain 35mm color transparency
Speed: ISO 64 *Push and/or Pull Feasible?* Yes
Processing: Standard E-6
Size(s): 35mm *Roll Lengths:* 36 exposures
Long Rolls Available? Yes
Other Sizes Available: 120, sheets
Diffuse RMS Granularity Value: 11
Resolving Power
High Contrast 1000:1 = 125 ll/mm
Low Contrast 1.6:1 = 50 ll/mm
Exposure Latitude: +3/4, –1 f-stop
Contrast Characteristics: Normal
Film Structure: Conventional
Color Balance: Tungsten 3100K
Color Bias: Neutral
Recommended Applications: Copy or regular photography under tungsten illumination
Special Uses: Photography using long exposure times
Reciprocity (recommended shutter-speed range): 1/15–4 second
Exposure Correction Table:

Exposure time:	1/4000	1/100	1/10	1	10
Correction:	N/R	N/R	none	none	none
Filtration (CC):	N/R	N/R	none	none	none

Note: exposures of 20 seconds require about +1/3 f-stop exposure correction; 30 seconds CC02.5B and +2/3 f-stop increase in exposure.
Storage, Keeping Properties, and Special Handling: Professional film. Refrigerate until use.
Availability: Photo specialty stores
In-Use Evaluation: This artificial-light (3100K) film is designed to be used with exposures longer than 1/15 second. It records wide tonal ranges extremely well and is chromatically neutral. About as fine grained as Fujichrome 50, it will deliver neutral color and is not overly contrasty. It reproduces all color faithfully with green being especially well rendered. Fujichrome 64 T is less contrasty than its nearest rival, Kodachrome 40.

Designation: **Fujichrome 100 RDP**
Manufactured by: Fuji Film Co., Japan
Film Type: Very fine grain 35mm color transparency
Speed: ISO 100 *Push and/or Pull Feasible?* Yes
Processing: Standard E-6
Size(s): 35mm *Roll Lengths:* 36 exposures
Long Rolls Available? Yes
Other Sizes Available: 120, 220, sheets
Diffuse RMS Granularity Value: 11
Resolving Power
High Contrast 1000:1 = 125 ll/mm
Low Contrast 1.6:1 = 50 ll/mm

Exposure Latitude: +1/2, –1 f-stop
Contrast Characteristics: Normal to slightly high
Film Structure: Conventional
Color Balance: Daylight
Color Bias: Neutral to slightly cool
Recommended Applications: All-purpose film for a wide range of professional assignments
Special Uses: An especially good film for natural light candid portraiture or street photography
Reciprocity (recommended shutter-speed range): 1/4000–1 second
Exposure Correction Table:

Exposure time:	1/4000	1/100	1/10	1	10
Correction:	none	none	none	none	+1/2 f-stop
Filtration (CC):	none	none	none	none	CC07.5M

Storage, Keeping Properties, and Special Handling: Professional film. Refrigerate until use.
Availability: Photo specialty stores
In-Use Evaluation: Vibrant color, a hallmark of Fuji films, is this emulsion's forte. Adjustments to color balance have resulted in a film that is less red than its predecessors. Instead, the current RDP is cooler but still able to deliver healthy skin tones. Field testers report this film is especially effective with nature and landscape subjects when combined with a polarizing or Tiffen enhancing filter. Like Fujichrome 50 Professional, 100 Pro is formulated to produce outstanding color separations.

Designation: **Fujichrome 100 RD**
Manufactured by: Fuji Film Co., Japan
Film Type: Very fine grain 35mm color transparency
Speed: ISO 100 *Push and/or Pull Feasible?* Yes
Processing: Standard E-6
Size(s): 35mm *Roll Lengths:* 24, 36 exposures
Long Rolls Available? Yes
Other Sizes Available: None
Diffuse RMS Granularity Value: 11
Resolving Power
High Contrast 1000:1 = 125 ll/mm
Low Contrast 1.6:1 = 50 ll/mm
Exposure Latitude: +1/2, –1 f-stop
Contrast Characteristics: Normal to slightly high
Film Structure: Conventional
Color Balance: Daylight
Color Bias: Neutral to slightly cool
Recommended Applications: All general photography
Special Uses: Especially suited for subjects in which green is the dominant color
Reciprocity (recommended shutter-speed range): 1/4000–1 second
Exposure Correction Table:

Exposure time:	1/1000	1/100	1/10	1	10
Correction:	none	none	none	none	+1/2 f-stop
Filtration (CC):					

Storage, Keeping Properties, and Special Handling: Normal
Availability: Photo specialty stores, some discount outlets
In-Use Evaluation: Essentially similar to RDP in all major areas of

performance, RD represents a somewhat lower-price alternative to the professional RDP. Fujichromes have become more widely available recently due to a more aggressive marketing strategy. This film differs from its pro version in that it appears to produce somewhat brighter highlights due in part to lower base density.

Designation: **Fujichrome 400 Professional RHP**
Manufactured by: Fuji Film Co., Japan
Film Type: Fine grain 35mm color transparency
Speed: ISO 400 *Push and/or Pull Feasible?* Yes
Processing: Standard E-6
Size(s): 35mm *Roll Lengths:* 36 exposures
Long Rolls Available? No
Other Sizes Available: None
Diffuse RMS Granularity Value: 16
Resolving Power
High Contrast 1000:1 = 125 ll/mm
Low Contrast 1.6:1 = 40 ll/mm
Exposure Latitude: +1/2, –1 f-stop
Contrast Characteristics: Normal to slightly high
Film Structure: Conventional, except has dual blue-sensitive layers
Color Balance: Daylight
Color Bias: Neutral to cool
Recommended Applications: Available-light or bad weather photography, sports, and action
Special Uses: Useful when high shutter speeds are required
Reciprocity (recommended shutter-speed range): 1/4000–1 second
Exposure Correction Table:

Exposure time:	1/4000	1/1000	1/100	1/10	1
Correction:	none	none	none	none	none
Filtration (CC):	none	none	none	none	none

Note: Exposure of 10 seconds requires increase of +1/2 f-stop and addition of CC 02.5 Y filter.
Storage, Keeping Properties, and Special Handling: Professional film. Refrigerate until use.
Availability: Photo specialty stores
In-Use Evaluation: Fujichrome 400 Professional is an especially fine grained film for its speed rating. It also records images shot under tungsten light without filters with a pleasant yellowish glow. In high contrast, this film will reveal a wealth of image detail, with moderate grain and good contrast.

Designation: **Fujichrome 400 RH**
Manufactured by: Fuji Film Co., Japan
Film Type: Fine grain 35mm color transparency
Speed: ISO 400 *Push and/or Pull Feasible?* Yes
Processing: Standard E-6
Size(s): 35mm *Roll Lengths:* 24, 36 exposures
Long Rolls Available? No
Other Sizes Available: None
Diffuse RMS Granularity Value: 16
Resolving Power
High Contrast 1000:1 = 125 ll/mm
Low Contrast 1.6:1 = 40 ll/mm

Exposure Latitude: +1/2, –1 f-stop
Contrast Characteristics: Normal to slightly high
Film Structure: Conventional
Color Balance: Daylight
Color Bias: Neutral to slightly cool
Recommended Applications: General photography where high-speed film is required
Special Uses: Sports and action, animals
Reciprocity (recommended shutter-speed range): 1/4000–1 second
Exposure Correction Table:

Exposure time:	1/4000	1/1000	1/100	1/10	1
Correction:	none	none	none	none	none
Filtration (CC):	none	none	none	none	none

Storage, Keeping Properties, and Special Handling: Normal
Availability: Photo specialty stores, some discount outlets
In-Use Evaluation: Same comments as pertain to RHP. Lower retail price.

Designation: **Fujichrome P 1600 Professional RSP11**
Manufactured by: Fuji Film Co., Japan
Film Type: Medium to coarse grain 35mm color transparency
Speed: ISO 1600–3200 *Push and/or Pull Feasible?* Yes
Processing: Standard E-6P
Size(s): 35mm *Roll Lengths:* 36 exposures
Long Rolls Available? No
Other Sizes Available: None
Diffuse RMS Granularity Value: 23
Resolving Power
High Contrast 1000:1 = 80 ll/mm
Low Contrast 1.6:1 = 40 ll/mm
Exposure Latitude: +1/2, –3/4 f-stop
Contrast Characteristics: Normal to slightly high
Film Structure: Conventional
Color Balance: Daylight
Color Bias: Neutral
Recommended Applications: Ultra high-speed film intended for "available-darkness" photography
Special Uses: Use film in daylight with very high shutter speeds and small f-stops; f-16 @ 1/2000 second is a typical exposure in bright sunlight.
Reciprocity (recommended shutter-speed range): 1/4000–1 second
Exposure Correction Table:

Exposure time:	1/4000	1/1000	1/100	1/10	1
Correction:	none	none	none	none	none
Filtration (CC):	none	none	none	none	none

Storage, Keeping Properties, and Special Handling: Professional film. Refrigerate until use.
Availability: Photo specialty stores
In-Use Evaluation: Although this film is designed to deliver best results at either ISO 1600 or 3200 it can also be successfully exposed at ISO 800. Even though grain structure is pronounced, it is relatively smooth and uniform. Color appears much like Fujichrome 400. Fujichrome P 1600 features the deepest blacks of any of the ultra high speed slide films (highest D-Max.). In high-contrast lighting, shadows tend to block up, losing significant detail.

Designation: **Fujichrome Duplicating Film Improved Type CDU**
Manufactured by: Fuji Film Co., Japan
Film Type: Extremely fine grain 35mm color transparency
Speed: ISO 10 *Push and/or Pull Feasible?* Yes
Processing: Standard E-6
Size(s): 35mm *Roll Lengths:* None
Long Rolls Available? Yes
Other Sizes Available: Various size sheets
Diffuse RMS Granularity Value: 9
Resolving Power
High Contrast 1000:1 = 125 ll/mm
Low Contrast 1.6:1 = 50 ll/mm
Exposure Latitude: +1, –1 f-stop
Contrast Characteristics: Slightly high
Film Structure: Conventional, but employing all the latest technology
Color Balance: Tungsten
Color Bias: Neutral
Recommended Applications: Used for reproducing all kinds of duplicates from original color transparencies. It is also a useful emulsion for the production of color separations.
Special Uses: Adaptable for the production of color transparencies from artwork or photographic prints
Reciprocity (recommended shutter-speed range): 1/1000–20 seconds
Exposure Correction Table:

Exposure time:	1/1000	1/100	1/10	1	10	20
Correction:	none	none	none	none	none	none
Filtration (CC):	none	none	none	none	none	none

Storage, Keeping Properties, and Special Handling: Professional film. Refrigerate until use.
Availability: Photo specialty stores
In-Use Evaluation: An exceptional film for its intended purpose, CDU equals or exceeds the performance of any other duplicating film on the market. Has exceptional color saturation for a duplicating film.

Designation: **Fujichrome Internegative Film IT-N**
Manufactured by: Fuji Film Co., Japan
Film Type: 35mm color transparency
Speed: ISO 1 (est.; no exact numbers are published)
Push and/or Pull Feasible? Yes
Processing: Standard C-41 or similar chemistry
Size(s): 35mm *Roll Lengths:* None
Long Rolls Available? Yes, 30.5 meters (100 ft.)
Other Sizes Available: None
Diffuse RMS Granularity Value: 3
Resolving Power
High Contrast 1000:1 = 160 ll/mm
Low Contrast 1.6:1 = 80 ll/mm
Exposure Latitude: +1/2, –1 f-stop
Contrast Characteristics: Normal
Film Structure: Conventional
Color Balance: Tungsten
Color Bias: Neutral

Recommended Applications: Designed to produce high definition internegatives from color transparencies or color art.
Special Uses: Can be used as a very slow pictorial film
Reciprocity (recommended shutter-speed range): 5–20 seconds
Exposure Correction Table:

Exposure time:	1/100	1/10	5	10	20
Correction::	none	none	none	none	none
Filtration (CC):	none	none	none	none	none

Storage, Keeping Properties, and Special Handling: Professional film. Refrigerate until use.
Availability: Photo specialty stores, professional stock houses
In-Use Evaluation: Perhaps the finest grain color transparency film made, IT-N is finding a growing audience of professional users. Its high color, purity, and extreme sharpness make it a useful professional tool. Because of special mask, film produces extremely sharp internegatives.

Designation: **Scotchchrome 100**
Manufactured by: 3M Photo Products Division, USA
Film Type: Very fine grain 35mm color transparency
Speed: ISO 100 *Push and/or Pull Feasible?* Yes
Processing: Standard E-6
Size(s): 35mm *Roll Lengths:* 24, 36 exposures
Long Rolls Available? No
Other Sizes Available: None
Diffuse RMS Granularity Value: 13
Resolving Power
High Contrast 1000:1 = 94 ll/mm
Low Contrast 1.6:1 = 40 ll/mm plus
Exposure Latitude: +3/4, –1 f-stop
Contrast Characteristics: Normal
Film Structure: Conventional
Color Balance: Daylight
Color Bias: Neutral to warm
Recommended Applications: Good general-purpose film
Special Uses: Particularly well suited for fall colors
Reciprocity (recommended shutter-speed range): 1/10000–1/10 second
Exposure Correction Table:

Exposure time:	1/10000	1/1000	1/100	1/10	1
Correction:	none	none	none	none	+2/3 f-stop
Filtration (CC):	none	none	none	none	CC05Y

Storage, Keeping Properties, and Special Handling: Normal
Availability: In some discount outlets under Scotch label, in mass market outlets under a variety of house brand names
In-Use Evaluation: A good quality medium-speed film with slightly lower contrast than its competitors

Designation: **Scotchchrome 400**
Manufactured by: 3M Photo Products Division, USA
Film Type: Fine grain 35mm color transparency
Speed: ISO 400 *Push and/or Pull Feasible?* Yes
Processing: Standard E-6
Size(s): 35mm *Roll Lengths:* 24, 36 exposures
Long Rolls Available? No
Other Sizes Available: None

Diffuse RMS Granularity Value: ISO 400, 20; 800, 21
Resolving Power
High Contrast 1000:1 = 89 ll/mm
Low Contrast 1.6:1 = 40 ll/mm plus
Exposure Latitude: +3/4, –1 f-stop
Contrast Characteristics: Normal
Film Structure: Conventional
Color Balance: Daylight
Color Bias: Neutral
Recommended Applications: Good film for situations where speed is an important consideration
Special Uses: Designed to be push-processed up to two stops without significant quality loss
Reciprocity (recommended shutter-speed range): 1/10000–1/10 second
Exposure Correction Table:

Exposure time:	1/10000	1/1000	1/100	1/10	1
Correction:	none	none	none	none	none
Filtration (CC):	none	none	none	none	none

Note: 10 seconds exposure requires compensation of +1 f-stop and CC05Y filter.
Storage, Keeping Properties, and Special Handling: Normal
Availability: Some discount outlets, mainly under private labels
In-Use Evaluation: Some feel this emulsion is the best of the ISO-400 speed films. Good blend of pictorial qualities. Moderate contrast helps preserve detail in shadows and highlights.

Designation: **Scotchchrome 1000**
Manufactured by: 3M Photo Products Division, USA
Film Type: Coarse grain 35mm color transparency
Speed: ISO 1000 *Push and/or Pull Feasible?* Yes
Processing: Standard E-6
Size(s): 35mm *Roll Lengths:* 24, 36 exposures
Long Rolls Available? No
Other Sizes Available: None
Diffuse RMS Granularity Value: approx. 30
Resolving Power
High Contrast 1000:1 = 67 ll/mm
Low Contrast 1.6:1 = 25 ll/mm plus
Exposure Latitude: +1/2, –1 f-stop
Contrast Characteristics: Normal to slightly low
Film Structure: Conventional
Color Balance: Daylight
Color Bias: Neutral to slightly warm
Recommended Applications: Available-light photography where speed is essential
Special Uses: Grain effects
Reciprocity (recommended shutter-speed range): 1/10000–1 second
Exposure Correction Table:

Exposure time:	1/1000	1/100	1/10	1	10
Correction:	none	none	none	none	none
Filtration (CC):					

Storage, Keeping Properties, and Special Handling: Normal
Availability: Photo specialty stores
In-Use Evaluation: The grainiest of all slide films, 1000 can be used to

create impressionistic renditions of common scenes particularly when it is pushed to ISO 2000 or beyond. Flowers and bold scenics are especially good subjects for this treatment.

Designation: **Scotchchrome 640T**
Manufactured by: 3M Photo Products Division, USA
Film Type: Medium-grain 35mm color transparency
Speed: ISO 640 *Push and/or Pull Feasible?* Yes
Processing: Standard E-6
Size(s): 35mm *Roll Lengths:* 36 exposures
Long Rolls Available? No
Other Sizes Available: None
Diffuse RMS Granularity Value: 29
Resolving Power
High Contrast 1000:1 = 85 ll/mm
Low Contrast 1.6:1 = 40 ll/mm
Exposure Latitude: +1/2, –1 f-stop
Contrast Characteristics: Normal to slightly high
Film Structure: Conventional
Color Balance: Daylight
Color Bias: Neutral
Recommended Applications: Available tungsten-light photography, concerts, and sports events where flash is not permitted
Special Uses: Creation of pointilistic effects when push-processed to higher ISOs
Reciprocity (recommended shutter-speed range): 1/1000–1 /10 second
Exposure Correction Table:

Exposure time:	1/1000	1/100	1/10	1	10
Correction:	none	none	none	none	none
Filtration (CC):	none	none	none	none	none

Note: Can be exposed up to 100 seconds. At 10-second exposure, +2/3 f-stop and CC10 B filter, at 100 seconds +1-1/2 f-stop and CC015M filter required
Storage, Keeping Properties, and Special Handling: Normal
Availability: Photo specialty stores
In-Use Evaluation: The fastest tungsten balanced film available, 640T delivers images that are grainy but interesting visually. They have much the same look as do slides from Scotch 1000 but are noticeably sharper.

Designation: **Agfacolor XRG 100**
Manufactured by: Agfa Geveart AG, Germany
Film Type: 35mm color negative
Speed: ISO 100 *Push and/or Pull Feasible?* Yes
Processing: Standard C-41 or equivalent
Size(s): 35mm *Roll Lengths:* 12, 24, 36 exposures
Long Rolls Available? No
Other Sizes Available: 120
Diffuse RMS Granularity Value: 6
Resolving Power
High Contrast 1000:1 = 100 ll/mm
Low Contrast 1.6:1 = 50 ll/mm
Exposure Latitude: +3, –2 f-stops
Contrast Characteristics: Normal to slightly high

Film Structure: Conventional
Color Balance: Daylight
Color Bias: Neutral
Recommended Applications: All general photography
Special Uses: None
Reciprocity (recommended shutter speed range): 1/1000–1 second
Exposure Correction Table:

Exposure time:	1/1000	1/100	1/10	1	10
Correction:	none	none	none	+1/2 f-stop	+1 f-stop
Filtration (CC):	none	none	none	025B	050B

Storage, Keeping Properties, and Special Handling: Professional film. Refrigerate until use.
Availability: Photo specialty stores
In-Use Evaluation: An excellent general purpose film that has seen continual improvement. Rich color, neutral balance, and wide exposure latitude make it very forgiving in use.

Designation: **Agfacolor XRG 200**
Manufactured by: Agfa Geveart AG, Germany
Film Type: 35mm color negative
Speed: ISO 200 *Push and/or Pull Feasible?* Yes
Processing: Standard C-41 or equivalent
Size(s): 35mm *Roll Lengths:* 12, 24, 36 exposures
Long Rolls Available? No
Other Sizes Available: 120
Diffuse RMS Granularity Value: 6
Resolving Power
High Contrast 1000:1 = 130 ll/mm
Low Contrast 1.6:1 = 50 ll/mm
Exposure Latitude: +3, –2 f-stop
Contrast Characteristics: Normal to slightly high
Film Structure: Conventional, but with latest technological innovations
Color Balance: Daylight
Color Bias: Neutral
Recommended Applications: All general photo assignments where extra speed of an ISO-200 film would prove useful
Special Uses: None
Reciprocity (recommended shutter-speed range): 1/10000–1/2 second
Exposure Correction Table:

Exposure time:	1/10000	1/1000	1/100	1/10	1	10
Correction:	none	none	none	none	+1/2 f-stop	+2 f-stops
Filtration;	none	none	none	none	none	none

Storage, Keeping Properties, and Special Handling: Normal
Availability: Photo specialty stores, some mass market outlets
In-Use Evaluation: A very good medium-speed film with almost the same performance parameters as its ISO-100 stablemate. Very low apparent grain for a film of its speed.

Designation: **Agfacolor Ultra Professional**
Manufactured by: Agfa Geveart AG, Germany
Film Type: 35mm color negative
Speed: ISO 50 *Push and/or Pull Feasible?* Yes
Processing: Standard C-41 or equivalent
Size(s): 35mm *Roll Lengths:* 36 exposures

Long Rolls Available? No
Other Sizes Available: 120
Diffuse RMS Granularity Value: 4.3
Resolving Power
High Contrast 1000:1 = 140 ll/mm
Low Contrast 1.6:1 = 50 ll/mm plus
Exposure Latitude: +3, –2 f-stop
Contrast Characteristics: Normal to slightly high
Film Structure: Latest technological innovations
Color Balance: Daylight
Color Bias: Neutral
Recommended Applications: Critical professional work where color contrast is important
Special Uses: Use on overcast days to enhance color reproduction
Reciprocity (recommended shutter-speed range): 1/10000–1/2 second
Exposure Correction Table:

Exposure time:	1/10000	1/1000	1/100	1/10	1	10
Correction:	none	none	none	none	+1/2 f-stop	+2 f-stops
Filtration (CC):	none	none	none	none	none	none

Storage, Keeping Properties, and Special Handling: Professional film, but no need to refrigerate
Availability: Photo specialty stores
In-Use Evaluation: This film is reported to be one of the two most color-saturated print films available today. Its overexposure latitude is exceptional, indicating it is probably underrated. Using the film at ISO 80–100 seems about right. Its combination of high color contrast, sharpness, and great acutance produce negatives that have visual snap, that is, excellent color differentiation and tonal separation.

Designation: **Agfacolor XRS 100 Professional**
Manufactured by: Agfa Geveart AG, Germany
Film Type: 35mm color negative
Speed: ISO 100 *Push and/or Pull Feasible?* Yes
Processing: Standard C-41 or equivalent
Size(s): 35mm *Roll Lengths:* 36 exposures
Long Rolls Available? No
Other Sizes Available: 120
Diffuse RMS Granularity Value: 6
Resolving Power
High Contrast 1000:1 = 130 ll/mm
Low Contrast 1.6:1 = 50 ll/mm
Exposure Latitude: +3, –2 f-stop
Contrast Characteristics: Normal to slightly high
Film Structure: Conventional, with latest technical innovations: multiple grain crystals, enhanced edge effects, reduced color infringement, and better acutance
Color Balance: Daylight
Color Bias: Neutral
Recommended Applications: All professional photo assignments
Special Uses: Very useful with subjects exhibiting an unusually wide contrast range
Reciprocity (recommended shutter-speed range): 1/10000–1/2 second
Exposure Correction Table:

Exposure time:	1/10000	1/1000	1/100	1/10	1	10
Correction:	none	none	none	none	+1/2 f-stop	+2 f-stops
Filtration (CC):	none	none	none	none	none	none

Storage, Keeping Properties, and Special Handling: Professional film, but no need to refrigerate.
Availability: Photo specialty stores
In-Use Evaluation: Reproduces a variety of skin tones very accurately. Chromatically neutral so it produces natural color rendition. This is a very good general-purpose film that is forgiving of rather large exposure errors.

Designation: **Agfacolor XRG 100**
Manufactured by: Agfa Geveart AG, Germany
Film Type: 35mm color negative
Speed: ISO 100 *Push and/or Pull Feasible?* Yes
Processing: Standard C-41 or equivalent
Size(s): 35mm *Roll Lengths:* 15, 27, 36 exposures
Long Rolls Available? No
Other Sizes Available: None
Diffuse RMS Granularity Value: 4.5
Resolving Power
High Contrast 1000:1 = 130 ll/mm
Low Contrast 1.6:1 = 50 ll/mm
Exposure Latitude: +3, –2 f-stop
Contrast Characteristics: Normal
Film Structure: Conventional
Color Balance: Daylight
Color Bias: Neutral
Recommended Applications: All general photography where print film is appropriate.
Special Uses: See XRS 100
Reciprocity (recommended shutter-speed range): 1/10000–1/2 second
Exposure Correction Table:

Exposure time:	1/10000	1/1000	1/100	1/10	1	10
Correction:	none	none	none	none	+1/2 f-stop	+2 f-stops
Filtration (CC):	none	none	none	none	none	none

Storage, Keeping Properties, and Special Handling: Normal
Availability: Photo specialty stores, some larger discount outlets
In-Use Evaluation: A general-purpose medium-speed print film that is right up there in terms of performance. More color neutrality than some others. Because of its slightly lower contrast, it is well suited for many nature and close-up assignments as well as child and classic female portraiture.

Designation: **Agfacolor Optima 125 Professional**
Manufactured by: Agfa Geveart AG, Germany
Film Type: 35mm color negative
Speed: ISO 125 *Push and/or Pull Feasible?* Yes
Processing: Standard C-41 or equivalent
Size(s): 35mm *Roll Lengths:* 36 exposures
Long Rolls Available? No
Other Sizes Available: 120
Diffuse RMS Granularity Value: 4

Resolving Power

High Contrast 1000:1 = 150 ll/mm
Low Contrast 1.6:1 = 50 ll/mm plus

Exposure Latitude: +3, –1-1/2 f-stop
Contrast Characteristics: Normal to slightly low
Film Structure: Conventional, with enhanced color couplers, edge effects for better edge sharpness, and superior highlight to shadow contrast reproduction. Color response very neutral and realistic.
Color Balance: Daylight
Color Bias: Neutral
Recommended Applications: All professional photo applications
Special Uses: Especially useful where accurate color rendition is essential
Reciprocity (recommended shutter-speed range): 1/10000–1/2 second
Exposure Correction Table:

Exposure time:	1/10000	1/1000	1/100	1/10	1	10
Correction:	none	none	none	none	+1/2 f-stop	+2 f-stops

Storage, Keeping Properties, and Special Handling: Normal
Availability: Photo specialty stores
In-Use Evaluation: Optima is the middle film of Agfa's lineup of professional color print films. Each of the three offers users different color saturation and contrast qualities in addition to different ISOs. Optima lies in the middle between the super color Ultra and soft working Portrait and as such might be the most useful. Production tolerances seem very tight so the extra cost of this product might very well be justified.

Designation: **Agfacolor Optima 200 OPT 200**
Manufactured by: Agfa Geveart AG, Germany
Film Type: 35mm color negative
Speed: ISO 200 *Push and/or Pull Feasible?* Yes
Processing: Standard C-41 or equivalent
Size(s): 35mm *Roll Lengths:* 12, 24, 36 exposures
Long Rolls Available? No
Other Sizes Available: 120
Diffuse RMS Granularity Value: 5
Resolving Power:

High Contrast 1000:1 = 80 ll/mm
Low Contrast 1.6:1 = 45 ll/mm

Exposure Latitude: +3, –2 f-stop
Contrast Characteristics: Normal to high
Film Structure: Conventional, with latest technological innovations
Color Balance: Daylight
Color Bias: Neutral
Recommended Applications: All general photo assignments, professional and advanced amateur
Special Uses: None
Reciprocity (recommended shutter-speed range): 1/10000–1/2 second
Exposure Correction Table:

Exposure time:	1/10000	1/1000	1/100	1/10	1	10
Correction:	none	none	none	none	+1 f-stop	+2 f-stops

Storage, Keeping Properties, and Special Handling: Normal
Availability: Photo specialty stores, some mass market outlets
In-Use Evaluation: In all essential respects, a performance twin of Optima

125, but updated to fit into the highly competitive ISO 200 market niche.

Designation: **Agfacolor Portrait 160 Professional**
Manufactured by: Agfa Geveart AG, Germany
Film Type: 35mm color negative
Speed: ISO 160 *Push and/or Pull Feasible?* Yes
Processing: Standard C-41 or equivalent
Size(s): 35mm *Roll Lengths:* 36 exposures
Long Rolls Available? No
Other Sizes Available: 120
Diffuse RMS Granularity Value: 3.5
Resolving Power:
High Contrast 1000:1 = 150 ll/mm
Low Contrast 1.6:1 = 50 ll/mm plus
Exposure Latitude: +2-1/2 , –1 f-stop plus
Contrast Characteristics: Low
Film Structure: Conventional, enhanced with latest technology
Color Balance: Daylight
Color Bias: Neutral
Recommended Applications: Designed for portrait work, delicate skin tones reproduced extremely well
Special Uses: Can reduce harsh contrast in brightly lit direct sun. Tones down harsh colors in contrasty light.
Reciprocity (recommended shutter-speed range): 1/10000–1/2 second
Exposure Correction Table:

Exposure time:	1/10000	1/1000	1/100	1	10
Correction:	none	none	none	+ 1/2 f-stop	+ 2 f-stops

Storage, Keeping Properties, and Special Handling: Normal, even though a professional film
Availability: Photo specialty stores
In-Use Evaluation: Soft, understated color suitable for many commercial uses. Film does an excellent job reproducing pastels lit by direct sun or front-firing electronic strobes. Some tests indicate film not as fast as its rated ISO160. You may like the film's interpretation of skin tone which is very different than most other more color saturated and thus, flashier medium-speed print films. Compared to its low-contrast competitors, Vericolor 111, Konica 160S, and HG 400, Portrait 160 displays the softest, most restrained color palette.

Designation: **Agfacolor XRS 200**
Manufactured by: Agfa Geveart AG, Germany
Film *Film Type:* 35mm color negative
Speed: ISO 200 *Push and/or Pull Feasible?* Yes
Processing: Standard C-41 or equivalent
Size(s): 35mm *Roll Lengths:* 24, 36 exposures
Long Rolls Available? No
Other Sizes Available: 120, sheets
Diffuse RMS Granularity Value: 6
Resolving Power
High Contrast 1000:1 = 130 ll/mm
Low Contrast 1.6:1 = 50 ll/ mm
Exposure Latitude: +3 plus , –2 f-stops
Contrast Characteristics: Normal to slightly high

Film Structure: Conventional
Color Balance: Daylight
Color Bias: Neutral
Recommended Applications: Situations where accurate color reproduction, wide tonal range, and extra speed are of prime importance.
Special Uses: Useful for image-making where very accurate color is important
Reciprocity (recommended shutter-speed range): 1/10000–1/2 second
Exposure Correction Table:

Exposure time:	1/10000	1/1000	1/100	1/10	1	10
Correction:	none	none	none	none	+1/2 f-stop	+2 f-stops

Storage, Keeping Properties, and Special Handling: Normal
Availability: Photo specialty stores
In-Use Evaluation: Much the same film as XRS 100 but with somewhat more grain, reduced sharpness, and one full f-stop more film speed.

Designation: **Agfacolor XRS 400**
Manufactured by: Agfa Geveart AG, Germany
Film Type: 35mm color negative
Speed: ISO 400 *Push and/or Pull Feasible?* Yes
Processing: Standard C-41 or equivalent
Size(s): 35mm *Roll Lengths:* 24, 36 exposures
Long Rolls Available? No
Other Sizes Available: 120
Diffuse RMS Granularity Value: 7
Resolving Power
High Contrast 1000:1 = 100 ll/mm
Low Contrast 1.6:1 = 50 ll/mm
Exposure Latitude: +3, –2 f-stop
Contrast Characteristics: Normal
Film Structure: Conventional
Color Balance: Daylight
Color Bias: Neutral
Recommended Applications: Low-light photography, action, sports
Special Uses: High shutter speeds or small apertures in bright light
Reciprocity (recommended shutter-speed range): 1/10000–1/2 second
Exposure Correction Table:

Exposure time:	1/10000	1/1000	1/100	1	10
Correction:	none	none	none	+1 f-stop	+2 f-stops
Filtration:	none	none	none	none	none

Storage, Keeping Properties, and Special Handling: Does not need refrigeration
Availability: Photo specialty stores
In-Use Evaluation: Only slight increase in grain over slower XRS films. Good balance of pictorial characteristics. Very usable at ISOs of 200 through 800 with standard processing.

Designation: **Agfacolor XRG 400**
Manufactured by: Agfa Geveart AG, Germany
Film Type: 35mm color negative
Speed: ISO 400 *Push and/or Pull Feasible?* Yes
Processing: Standard C-41 or equivalent
Size(s): 35mm *Roll Lengths:* 24, 36 exposures

Long Rolls Available? No
Other Sizes Available: None
Diffuse RMS Granularity Value: 7
Resolving Power
High Contrast 1000:1 = 100 ll/mm
Low Contrast 1.6:1 = 50 ll/mm
Exposure Latitude: +3, –2 f-stops
Contrast Characteristics: Normal
Film Structure: Conventional
Color Balance: Daylight
Color Bias: neutral
Recommended Applications: General low-light photography
Special Uses: None
Reciprocity (recommended shutter-speed range): 1/10000–1 second
Exposure Correction Table:

Exposure time:	1/10000	1/1000	1/100	1	10
Correction:	none	none	none	+1 f-stop	+2 f-stops

Storage, Keeping Properties, and Special Handling: Normal
Availability: Photo specialty stores, some discount outlets
In-Use Evaluation: See XRS 400

Designation: **Agfacolor XRS 1000**
Manufactured by: Agfa Geveart AG, Germany
Film Type: 35mm color negative
Speed: ISO 1000 *Push and/or Pull Feasible?* Yes
Processing: Standard C-41 or equivalent
Size(s): 35mm *Roll Lengths:* 24, 36 exposures
Long Rolls Available? No
Other Sizes Available: 120
Diffuse RMS Granularity Value: 8
Resolving Power
High Contrast 1000:1–100 ll/mm
Low Contrast 1.6:1 = 40 ll/mm
Exposure Latitude: +3, –2 f-stops
Contrast Characteristics: Normal to slightly high
Film Structure: Conventional, but three added layers, for a total of 16, to build in extra sensitivity
Color Balance: Daylight
Color Bias: Neutral
Recommended Applications: Nighttime, action sports, high shutter speed
Special Uses: Useful with small maximum aperture long telephoto lens
Reciprocity (recommended shutter-speed range): 1/10000–1/2 second
Exposure Correction Table:

Exposure time:	1/10000	1/1000	1/100	1	10
Correction:	none	none	none	+1/2 f-stop	+2 f-stops

Storage, Keeping Properties, and Special Handling: No refrigeration required
Availability: Photo specialty stores
In-Use Evaluation: Uses include available light, short exposure times, telephoto and macro, indoor hand-held or night photography. Surprisingly fine grained for such a fast film. Typical bright sun exposure is f-11 @ 1/2000 sec. Even under overcast conditions, exposures of f-8 @ 1/500 can be expected.

Designation: **Fujicolor Realia**
Manufactured by: Fuji Film Co., Japan
Film Type: 35mm color negative
Speed: ISO 100 *Push and/or Pull Feasible?* Yes
Processing: Standard C-41 or equivalent
Size(s): 35mm *Roll Lengths:* 24, 36 exposures
Long Rolls Available? No
Other Sizes Available: 120, 200, sheets
*Diffuse RMS Granularity Value:*4
Resolving Power
High Contrast 1000:1 = 125 ll/mm
Low Contrast 1.6:1 = 60 ll/mm plus
Exposure Latitude: +3, –2 f-stops
Contrast Characteristics: Normal to slightly low
Film Structure: Conventional, twin grain structure, enhanced DIRs, thin emulsion
Color Balance: Daylight
Color Bias: Neutral
Recommended Applications: All general professional and amateur photography
Special Uses: Fourth added cyan dye layer allows the film to reproduce colors with a high degree of accuracy. Subjects shot under fluorescent lights, long the bane of color shooters, are reproduced much more naturally than with just about any other film.
Reciprocity (recommended shutter-speed range): 1/4000–1/10 second
Exposure Correction Table:

Exposure time:	1/4000	1/1000	1/100	1/10	1	10
Correction:	none	none	none	none	+1/2 f-stop	+1 f-stop
Filtration (CC):	none	none	none	none	none	none

Storage, Keeping Properties, and Special Handling: Professional film. Refrigerate until use.
Availability: Photo specialty stores
In-Use Evaluation: This film features an excellent blend of qualities. It is extremely sharp for a medium-speed emulsion, shows virtually no grain all the way up to 16 × 20 inch enlargements, yields pleasing colors especially skin tones under many different light sources, and is very easy to print. It must be ranked among the few very best color print films in use today.

Designation: **Fujicolor Super HG 100 CN**
Manufactured by: Fuji Film Co., Japan
Film Type: 35mm color negative
Speed: ISO 100 *Push and/or Pull Feasible?* Yes
Processing: Standard C-41 or equivalent
Size(s): 35mm *Roll Lengths:* 12, 24, 36 exposures
Long Rolls Available? No
Other Sizes Available: 120
Diffuse RMS Granularity Value: 5
Resolving Power
High Contrast 1000:1 = 125 ll/mm
Low Contrast 1.6:1 = 63 ll/mm
Exposure Latitude: +3, –2 f-stops
Contrast Characteristics: Normal to slightly high

Film Structure: Conventional, twin grain structure, enhanced DIRs, thin emulsion
Color Balance: Daylight
Color Bias: Neutral to slightly warm
Recommended Applications: All general photography
Special Uses: Useful to brighten up colors on overcast days
Reciprocity (recommended shutter speed range): 1/4000–1 second
Exposure Correction Table:

Exposure time:	1/4000	1/1000	1/1	1/10	1	10
Correction:	none	none	none	none	+1/2 f-stop	+1 f-stop

Storage, Keeping Properties, and Special Handling: Normal
Availability: Photo specialty stores, supermarkets, mass merchandising outlets
In-Use Evaluation: Fuji's amateur medium-speed print film is finding increasing acceptance among knowledgeable photographers. It produces bright, extremely saturated colors that are accurate, has moderate overall contrast, and is very printable with consistent results.

Designation: **Fujicolor Super HG 200**
Manufactured by: Fuji Film Co., Japan
Film Type: 35mm color negative
Speed: ISO 200 *Push and/or Pull Feasible?* Yes
Processing: Standard C-41 or equivalent
Size(s): 35mm *Roll Lengths:* 12, 24, 36 exposures
Long Rolls Available? No
Other Sizes Available: 120
Diffuse RMS Granularity Value: 5
Resolving Power
High Contrast 1000:1 = 100 ll/mm
Low Contrast 1.6:1 = 50 ll/mm plus
Exposure Latitude: +3, –2 f-stops
Contrast Characteristics: Normal to slightly high
Film Structure: Conventional, twin grain structure, enhanced DIRs, thin emulsion
Color Balance: Daylight
Color Bias: Almost neutral
Recommended Applications: All general photography
Special Uses: To be used when high color saturation and extra speed of an ISO 200 film is required
Reciprocity (recommended shutter-speed range): 1/4000–1 second
Exposure Correction Table:

Exposure time:	1/4000	1/1000	1/100	1/10	1	10
Correction:	none	none	none	none	+1/2 f-stop	+1 f-stop
Filtration (CC):	none	none	none	none	none	none

Storage, Keeping Properties, and Special Handling: Normal
Availability: Photo specialty stores, many mass market outlets, supermarkets
In-Use Evaluation: Very high color saturation, yielding brilliant colors. Low granularity and high degree of sharpness lead to excellent negatives.

Designation: **Fujicolor Super HG 400 CH**
Manufactured by: Fuji Film Co., Japan
Film Type: 35mm color negative
Speed: ISO 400 *Push and/or Pull Feasible?* Yes
Processing: Standard C-41 or equivalent
Size(s): 35mm *Roll Lengths:* 12, 24, 36 exposures
Long Rolls Available? No
Other Sizes Available: 120
Diffuse RMS Granularity Value: 5
Resolving Power
High Contrast 1000:1 = 100 ll/mm
Low Contrast 1.6:1 = 50 ll/mm plus
Exposure Latitude: +3, –2 f-stops
Contrast Characteristics: Normal to slightly high
Film Structure: Conventional, twin grain structure, enhanced DIRs, thin emulsion
Color Balance: Daylight
Color Bias: Almost neutral
Recommended Applications: General photography, night, action, and sports
Special Uses: High-quality prints from negatives using high shutter speeds and small apertures
Reciprocity (recommended shutter-speed range): 1/4000–1/10 second
Exposure Correction Table:

Exposure time:	1/4000	1/1000	1/100	1/10	1	10
Correction:	none	none	none	none	+1/2 f-stop	+1 f-stop

Storage, Keeping Properties, and Special Handling: Normal
Availability: Photo specialty stores, mass market outlets, some supermarkets
In-Use Evaluation: An exceptionally good 400-speed film with many of the qualities of much slower films of supposedly finer grain. An excellent choice for low-contrast lighting. The film's extra bright color compared to other fast negative films will liven up dull or overcast scenes.

Designation: **Fujicolor 400 Professional HG NHG**
Manufactured by: Fuji Film Co., Japan
Film Type: 35mm color negative
Speed: ISO 400 *Push and/or Pull Feasible?* Yes
Processing: Standard C-41 or equivalent
Size(s): 35mm *Roll Lengths:* 36 exposures
Long Rolls Available? No
Other Sizes Available: 120
Diffuse RMS Granularity Value: 5
Resolving Power
High Contrast 1000:1 = 100 ll/mm plus
Low Contrast 1.6:1 = 50 ll/mm plus
Exposure Latitude: +3, –2 f-stops
Contrast Characteristics: Normal to low
Film Structure: Conventional, twin grain structure, enhanced DIRs, thin emulsion
Color Balance: Daylight
Color Bias: Neutral

Recommended Applications: A particularly good film for portraiture
Special Uses: Tones down excessively high contrast, as in direct sunlight, weddings, bar mitzvahs, or Confirmations
Reciprocity (recommended shutter speed range): 1/4000–1/10 second
Exposure Correction Table:

Exposure time:	1/4000	1/1000	1/100	1/10	1	10
Correction:	none	none	none	none	+1/2 f-stop	+1-1/2 f-stop

Storage, Keeping Properties, and Special Handling: Normal
Availability: Photo specialty stores
In-Use Evaluation: Chromatically neutral with excellent skin tones. Soft contrast enhances skin tone reproduction. Colors are saturated but not as much as with some other Fuji print films. Grain structure looks more like ISO 100 than ISO 400.

Designation: **Fujicolor Super HG 1600 II CU**
Manufactured by: Fuji Film Co., Japan
Film Type: 35mm color negative
Speed: ISO 160 *Push and/or Pull Feasible?* Yes
Processing: Standard C-41 or equivalent
Size(s): 35mm *Roll Lengths:* 24, 36 exposures
Long Rolls Available? No
Other Sizes Available: None
Diffuse RMS Granularity Value: 10
Resolving Power:
High Contrast 1000:1 = 90 ll/mm plus
Low Contrast 1.6:1 = 40 ll/mm
Exposure Latitude: +3, –2 f-stops
Contrast Characteristics: Normal to slightly high
Film Structure: Conventional, twin grain structure, enhanced DIRs, thin emulsion
Color Balance: Daylight
Color Bias: Neutral to slightly warm
Recommended Applications: All available-light situations, high shutter speeds
Special Uses: Makes possible the use of ultra high shutter speeds such as 1/8000 second
Reciprocity (recommended shutter-speed range): 1/4000–1/10 second
Exposure Correction Table:

Exposure time:	1/4000	1/1000	1/100	1/10	1	10
Correction:	none	none	none	none	+1/2 f-stop	+1 f-stop
Filtration (CC):	none	none	none	none	none	none

Storage, Keeping Properties, and Special Handling: Professional film. Refrigerate until use.
Availability: Photo specialty stores
In-Use Evaluation: An excellent film especially at its speed. Very good color and contrast. Tight, smooth grain that is apparent but not particularly disturbing. In fact, grain can be used to create special effects such as textures and abstracts.

Designation: **Fujicolor 160 Professional S NSP**
Manufactured by: Fuji Film Co., Japan
Film Type: 35mm color negative
Speed: ISO 160 *Push and/or Pull Feasible?* Yes
Processing: Standard C-41 or equivalent
Size(s): 35mm *Roll Lengths:* 36 exposures

Long Rolls Available? No
Other Sizes Available: 120, 220, sheets
Diffuse RMS Granularity Value: 5
Resolving Power
High Contrast 1000:1 = 113 ll/mm plus
Low Contrast 1.6:1 = 55 ll/mm
Exposure Latitude: +3, –2 f-stops
Contrast Characteristics: Medium
Film Structure: Conventional, twin grain structure, enhanced DIRs, thin emulsion
Color Balance: Daylight
Color Bias: Neutral
Recommended Applications: Professional commercial and advertising
Special Uses: Especially suited for portraiture, excellent skin tone rendition
Reciprocity (recommended shutter-speed range): 1/4000–1/15 second
Exposure Correction Table:

Exposure time:	1/4000	1/1000	1/100	1/10	1	10
Correction:	none	none	none	none	+1/2 f-stop	+1 f-stop
Filtration (CC):	none	none	none	none	none	none

Storage, Keeping Properties, and Special Handling: Professional film. Refrigerate until use.
Availability: Photo specialty stores
In-Use Evaluation: This film is designed for exposures of 1/15 second or faster. Its low to medium contrast is useful for applications such as studio portraiture, product photography, and fashion. Especially broad tonal range.

Designation: **Ilfocolor HR100**
Manufactured by: Ilford Ltd., Great Britain
Film Type: 35mm color negative
Speed: ISO 100 *Push and/or Pull Feasible?* Yes
Processing: Standard C-41 processing
Size(s): 35mm *Roll Lengths:* 24, 36 exposures
Long Rolls Available? None
Other Sizes Available: No
Diffuse RMS Granularity Value: 4
Resolving Power
High Contrast 1000:1 = 100 ll/mm
Low Contrast 1.6:1 = 50 ll/mm
Exposure Latitude: +3, –2 f-stops
Contrast Characteristics: Neutral
Film Structure: Conventional
Color Balance: Daylight
Color Bias: N/A
Recommended Applications: General-purpose photography
Special Uses: None
Reciprocity (recommended shutter-speed range): 1/4000–10 seconds
Exposure Correction Table:

Exposure time:	1/1000	1/100	1/10	1	10
Correction:	See Fujicolor Super HG 100				

Storage, Keeping Properties, and Special Handling: Normal
Availability: Occasionally available in large film supply outlets under other brand names
In-Use Evaluation: This film is distributed in England and on the

Continent but some does slip into this country under other names. It is probably made by Fuji and repackaged by Ilford. The film is up with the best of the medium-speed negative films.

Designation: **Kodacolor Gold 100 Plus GA**
Manufactured by: Eastman Kodak Co., USA
Film Type: 35mm color negative
Speed: ISO 100 *Push and/or Pull Feasible?* Yes
Processing: Standard C-41 or equivalent
Size(s): 35mm *Roll Lengths:* 12, 24, 36 exposures
Long Rolls Available? No
Other Sizes Available: 120
Diffuse RMS Granularity Value: 6
Resolving Power
High Contrast 1000:1 = 100 ll/mm
Low Contrast 1.6:1 = 50 ll/mm
Exposure Latitude: +3, –2 f-stops
Contrast Characteristics: Normal to slightly high
Film Structure: Conventional, T-Grain structure, enhanced DIRs, thin emulsion
Color Balance: Daylight
Color Bias: Nearly neutral with warm, pleasing skin tones
Recommended Applications: All general photography
Special Uses: None
Reciprocity (recommended shutter-speed range): 1/10000–1/10 second
Exposure Correction Table:

Exposure time:	1/10000	1/1000	1/100	1/10	1	10
Correction:	none	none	none	none	+1 f-stop	N/R
Filtration (CC):	none	none	none	CC20 Y	N/R	N/R

Storage, Keeping Properties, and Special Handling: Normal
Availability: Virtually everywhere
In-Use Evaluation: An excellent general purpose film intended for mass consumption. Improved Plus version features increased overall color saturation and even better grain structure.

Designation: **Kodacolor Gold 200 Plus GB**
Manufactured by: Eastman Kodak Co., USA
Film Type: 35mm color negative
Speed: ISO 200 *Push and/or Pull Feasible?* Yes
Processing: Standard C-41 or equivalent
Size(s): 35mm *Roll Lengths:* 12, 24, 36 exposures
Long Rolls Available? No
Other Sizes Available: 120
Diffuse RMS Granularity Value: 6
Resolving Power
High Contrast 1000:1 = 100 ll/mm
Low Contrast 1.6:1 = 50 ll/mm
Exposure Latitude: +3, –1-1/2 f-stops
Contrast Characteristics: Normal to slightly high
Film Structure: Conventional, T-Grain structure, enhanced DIRs, thin emulsion
Color Balance: Daylight
Color Bias: Neutral
Recommended Applications: All general photography
Special Uses: None

Reciprocity (recommended shutter-speed range): 1/10000–1/10 second
Exposure Correction Table:

Exposure time:	1/10000	1/1000	1/100	1/10	1
Correction:	none	none	none	none	+1 f-stop
Filtration (CC):	none	none	none	none	CC20 Y

Storage, Keeping Properties, and Special Handling: Normal
Availability: Virtually everywhere
In-Use Evaluation: Much the same as its ISO 100 speed version but at one f-stop greater sensitivity. Color saturation is even more noticeable than in ISO 100 version.

Designation: **Kodacolor Gold 400 Plus GC**
Manufactured by: Eastman Kodak Co., USA
Film Type: 35mm color negative
Speed: ISO 400 *Push and/or Pull Feasible?* Yes
Processing: Standard C-41 or equivalent
Size(s): 35mm *Roll Lengths:* 12, 24, 36 exposures
Long Rolls Available? No
Other Sizes Available: 120
Diffuse RMS Granularity Value: 7
Resolving Power
High Contrast 1000:1 = 100 ll/mm
Low Contrast 1.6:1 = 50 ll/mm
Exposure Latitude: +3, –2 f-stops
Contrast Characteristics: Normal to slightly high
Film Structure: Conventional, T-Grain structure, enhanced DIRs, thin emulsion.
Color Balance: Daylight
Color Bias: Neutral, warmish but pleasant skin tones
Recommended Applications: All general photography needs where high-film speed is essential
Special Uses: Will deliver believable color under a variety of artificial light sources
Reciprocity (recommended shutter-speed range): 1/10000–1/10 second
Exposure Correction Table:

Exposure time:	1/10000	1/1000	1/100	1/10	1	10
Correction:	none	none	none	none	none	none
Filtration (CC):	none	none	none	none	none	none

Storage, Keeping Properties, and Special Handling: Normal
Availability: Photo specialty stores, discount, supermarkets, many other outlets
In-Use Evaluation: This film like many in the highly competitive ISO 400 category is being constantly revised and improved. In its present form Kodacolor Gold 400 Plus exhibits excellent grain, highly saturated colors, controllable contrast, and warm healthy skin tones. Its excellent overall performance and wide availability make it a favorite of point and shooters as well as serious amateurs.

Designation: **Kodacolor Gold 1600**
Manufactured by: Eastman Kodak Co., USA
Film Type: 35mm color negative
Speed: ISO 1600 *Push and/or Pull Feasible?* Yes
Processing: Standard C-41 or equivalent
Size(s): 35mm *Roll Lengths:* 12, 24, 36 exposures
Long Rolls Available? No

Other Sizes Available: None
Diffuse RMS Granularity Value: 11
Resolving Power
High Contrast 1000:1 = 80 ll/mm
Low Contrast 1.6:1 = 40 ll/mm
Exposure Latitude: +3, –1-1/2 f-stops
Contrast Characteristics: Normal to slightly high
Film Structure: Conventional, T-Grain structure, enhanced DIRs, thin emulsion
Color Balance: Daylight
Color Bias: Almost neutral
Recommended Applications: Situations where very fast film is required
Special Uses: Hand-held extremely low light environment where electronic flash is not desirable
Reciprocity (recommended shutter-speed range): 1/10000–1/10 second
Exposure Correction Table:

Exposure time:	1/10000	1/1000	1/100	1/10	1
Correction:	none	none	none	none	+1/2 f-stop
Filtration (CC):	none	none	none	none	none

Storage, Keeping Properties, and Special Handling: Normal
Availability: Some general merchandising stores, photo specialty shops
In-Use Evaluation: Much improved over Kodacolor VR 1000, this film has very good color and more-than-acceptable blacks. In fact, it looks like a medium-speed film of only five or ten years ago. The film is designed to deliver reasonable color under a wide range of lighting conditions, natural or artificial.

Designation: **Ektar 25**
Manufactured by: Eastman Kodak Co., USA
Film Type: 35mm color negative
Speed: ISO 25 *Push and/or Pull Feasible?* Yes
Processing: Standard C-41 or equivalent
Size(s): 35mm *Roll Lengths:* 24, 36 exposures
Long Rolls Available? No
Other Sizes Available: 120
Diffuse RMS Granularity Value: 4
Resolving Power
High Contrast 1000:1 = 200 ll/mm
Low Contrast 1.6:1 = 80 ll/mm
Exposure Latitude: +3, –2 f-stops
Contrast Characteristics: Normal to slightly high
Film Structure: Conventional, T-Grain structure, enhanced DIRs, thin emulsion
Color Balance: Daylight
Color Bias: Warm
Recommended Applications: Professional commercial, fashion, advertising, and nature
Special Uses: Occasions where utmost quality is required and film speed is not critical
Reciprocity (recommended shutter-speed range): 1/10000–1 second
Exposure Correction Table:

Exposure time:	1/10000	1/1000	1/100	1/10	1	10
Correction:	none	none	none	none	none	none
Filtration (CC):	none	none	none	none	none	none

Storage, Keeping Properties, and Special Handling: Professional film. Refrigerate until use.
Availability: Photo specialty stores, some discount outlets
In-Use Evaluation: A film of exceptional quality, Ektar 25 is the finest grained, highest color saturation print film offered by Kodak. It is perhaps the sharpest color film available with the possible exception of Kodachrome 25 and Fujichrome Velvia (both transparency materials). If you can live with ISO 25, this is the film to use to produce super quality negatives.

Designation: **Ektar 125 CW**
Manufactured by: Eastman Kodak Co., USA
Film Type: 35mm color negative
Speed: ISO 125 *Push and/or Pull Feasible?* Yes
Processing: Standard C-41 or equivalent
Size(s): 35mm *Roll Lengths:* 24, 36 exposures
Long Rolls Available? No
Other Sizes Available: 120
Diffuse RMS Granularity Value: 5
Resolving Power
High Contrast 1000:1 = 160 ll/mm
Low Contrast 1.6:1 = 80 ll/mm
Exposure Latitude: +3, –2 f-stops
Contrast Characteristics: Normal
Film Structure: T-Grain structure, enhanced DIRs, thin emulsion
Color Balance: Daylight
Color Bias: Neutral
Recommended Applications: All general photography, amateur and professional.
Special Uses: High quality, but over two stops more speed than Ektar 25. Can be used for wider range of subjects.
Reciprocity (recommended shutter-speed range): 1/10000–1 second
Exposure Correction Table:

Exposure time:	1/10000	1/1000	1/100	1/10	1	10
Correction:	none	none	none	none	none	none
Filtration (CC):	none	none	none	none	none	none

Storage, Keeping Properties, and Special Handling: Normal
Availability: Photo specialty stores, discount outlets
In-Use Evaluation: This medium-speed film shares many of the performance parameters of Ektar 25 but its speed make it a more usable emulsion. Grain is tight and very sharp—typical of Kodak's contemporary technology. Exposure latitude makes the use of this film possible with all types of cameras.

Designation: **Ektar 1000 CJ**
Manufactured by: Eastman Kodak Co., USA
Film Type: 35mm color negative
Speed: ISO 1000 *Push and/or Pull Feasible?* Yes
Processing: Standard C-41 or equivalent
Size(s): 35mm *Roll Lengths:* 24, 36 exposures
Long Rolls Available? No
Other Sizes Available: 120
Diffuse RMS Granularity Value: 11

Resolving Power
High Contrast 1000:1 = 80 ll/mm
Low Contrast 1.6:1 = 40 ll/mm
Exposure Latitude: +3, –2 f-stops
Contrast Characteristics: Normal to slightly high
Film Structure: T-Grain structure, enhanced DIRs, thin emulsion
Color Balance: Daylight
Color Bias: Warm
Recommended Applications: All high-speed assignments, such as action, and sports
Special Uses: Low-light work, night shooting
Reciprocity (recommended shutter-speed range): 1/10000–1 second
Exposure Correction Table:

Exposure time:	1/10000	1/1000	1/100	1/10	1	10
Correction:	none	none	none	none	none	+1 f-stop

Storage, Keeping Properties, and Special Handling: Professional film. Refrigerate until use.
Availability: Photo specialty stores, discount outlets
In-Use Evaluation: A very good high-speed film that produces fine color prints up to 8 × 10 and, in some instances, 11 × 14.

Designation: **Ektar 25 Professional PHR**
Manufactured by: Eastman Kodak, Co. USA
Film Type: 35mm color negative
Speed: ISO 25 *Push and/or Pull Feasible?* Yes
Processing: Standard C-41 or equivalent
Size(s): 35mm *Roll Lengths:* 36 exposures
Long Rolls Available? No
Other Sizes Available: 120
Diffuse RMS Granularity Value: 4
Resolving Power
High Contrast 1000:1 = 200 ll/mm
Low Contrast 1.6:1 = 80 ll/mm
Exposure Latitude: +3, –2 f-stops
Contrast Characteristics: Normal to slightly high
Film Structure: T-Grain structure, enhanced DIRs, thin emulsion
Color Balance: Daylight
Color Bias: Neutral
Recommended Applications: All critical professional applications where film speed is not a prime consideration
Special Uses: An excellent choice for still-life product work and formal portraiture
Reciprocity (recommended shutter-speed range): 1/10000–1/10 second
Exposure Correction Table:

Exposure time:	1/1000	1/100	1/10	1	10
Correction:	none	none	none	none	none
Filtration (CC):	none	none	none	none	none

Storage, Keeping Properties, and Special Handling: Professional film. Refrigerate until use.
Availability: Photo specialty stores
In-Use Evaluation: Most of the same comments hold for Ektar 25 Professional as for the amateur version. This variant is probably manufactured to higher tolerances, making it a better choice for scientific or exacting color matching applications.

Designation: **Vericolor III Professional Type S VPS**
Manufactured by: Eastman Kodak Co., USA
Film Type: 35mm color negative
Speed: ISO 160 *Push and/or Pull Feasible?* Yes
Processing: Standard C-41 or equivalent
Size(s): 35mm *Roll Lengths:* 36 exposures
Long Rolls Available? No
Other Sizes Available: 120, 220
Diffuse RMS Granularity Value: 6
Resolving Power
High Contrast 1000:1 = 100 ll/mm
Low Contrast 1.6:1 = 40 ll/mm
Exposure Latitude: +3, –2 f-stops
Contrast Characteristics: Normal to slightly low
Film Structure: T-Grain structure, enhanced DIRs, thin emulsion
Color Balance: Daylight
Color Bias: Neutral
Recommended Applications: Commercial, advertising, and fashion
Special Uses: Especially useful for studio portraits
Reciprocity (recommended shutter-speed range): 1/10000–1/10 second
Exposure Correction Table:

Exposure time:	1/10000	1/1000	1/100	1	10
Correction:	none	none	none	+1/2 f-stop	+1 f-stop

Storage, Keeping Properties, and Special Handling: Professional film. Refrigerate until use.
Availability: Photo specialty stores
In-Use Evaluation: The standby of professionals, this latest Vericolor is designed to produce soft contrast and is very forgiving of exposure errors. It is a special favorite of wedding photographers and portraitists.

Designation: **Vericolor 400 Professional VPH**
Manufactured by: Eastman Kodak Co., USA
Film Type: 35mm color negative
Speed: ISO 400 *Push and/or Pull Feasible?* Yes
Processing: Standard C-41 or equivalent
Size(s): 35mm *Roll Lengths:* 36 exposures
Long Rolls Available? Yes
Other Sizes Available: 120, 220, sheets
Diffuse RMS Granularity Value: 6
Resolving Power
High Contrast 1000:1–100 ll/mm
Low Contrast 1.6:1 = 50 ll/mm
Exposure Latitude: +3, –2 f-stops
Contrast Characteristics: Normal
Film Structure: Conventional, T-Grain structure, enhanced DIRs, thin emulsion
Color Balance: Daylight
Color Bias: Neutral
Recommended Applications: All situations where high-speed color film is required
Special Uses: Very tight grain structure enables user to apply this film to situations where a slower film was once needed

Reciprocity (recommended shutter speed range): 1/10000–1/30 second
Exposure Correction Table:

Exposure time:	1/1000	1/100	1/10	1	10
Correction:	none	none	none	+1/2 f-stop	+1 f-stop
Filtration (CC):	none	none	none	none	none

Storage, Keeping Properties, and Special Handling: Professional film. Refrigerate until use.
Availability: Photo specialty stores
In-Use Evaluation: This film incorporates all the latest innovations in film design. It produces rich color that is not overly contrasty or prone to blocking up in highlights or shadow areas. One of the finest grained films in the crowded high-speed color print film group.

Designation: **Ektapress Gold 100 Professional PPA**
Manufactured by: Eastman Kodak Co., USA
Film Type: 35mm color negative
Speed: ISO 100 *Push and/or Pull Feasible?* Yes
Processing: Standard C-41 or equivalent
Size(s): 35mm *Roll Lengths:* 36 exposures
Long Rolls Available? Yes
Other Sizes Available: 120, sheets
Diffuse RMS Granularity Value: 6
Resolving Power
High Contrast 1000:1 = 100 ll/mm
Low Contrast 1.6:1 = 50 ll/mm
Exposure Latitude: +3, –2 f-stops
Contrast Characteristics: Normal
Film Structure: Conventional, T-Grain structure, enhanced DIRs
Color Balance: Daylight
Color Bias: Neutral
Recommended Applications: Designed for photojournalists who needed a professional film that does not need refrigeration
Special Uses: Can be push-processed with better results than most color print films
Reciprocity (recommended shutter-speed range): 1/10000–1/10 second
Exposure Correction Table:

Exposure time:	1/10000	1/1000	1/100	1/10	1	10
Correction:	none	none	none	none	+1/2 f-stop	+1 f-stop
Filtration (CC):	none	none	none	none	none	none

Storage, Keeping Properties, and Special Handling: Professional film, but needs no refrigeration
Availability: Photo specialty stores, professional stock houses
In-Use Evaluation: Very sharp film with very stable chemistry that can accommodate typical professsional handling. Its modest contrast and high color saturation make it a good choice for many professional applications under relatively good lighting or with the use of flash.

Designation: **Ektapress 400 Gold Professional PPB**
Manufactured by: Eastman Kodak Co., USA
Film Type: 35mm color negative
Speed: ISO 400 *Push and/or Pull Feasible?* Yes
Processing: Standard C-41 or equivalent
Size(s): 35mm *Roll Lengths:* 36 exposures

Long Rolls Available? Yes
Other Sizes Available: 120
Diffuse RMS Granularity Value: 7
Resolving Power
High Contrast 1000:1 = 100 ll/mm
Low Contrast 1.6:1 = 50 ll/mm
Exposure Latitude: +3, –3 f-stops
Contrast Characteristics: Normal to high
Film Structure: T-Grain structure, enhanced DIRs
Color Balance: Daylight
Color Bias: Neutral
Recommended Applications: Location film for use in low-illumination situations
Special Uses: Can be push-processed
Reciprocity (recommended shutter-speed range): 1/10000–1/10 second
Exposure Correction Table:

Exposure time:	1/10000	1/1000	1/100	1/10	1	10
Correction:	none	none	none	none	+1/2 f-stop	+1 f-stop
Filtration (CC):	none	none	none	none	none	none

Storage, Keeping Properties, and Special Handling: Professional film, but does not need refrigeration
Availability: Photo specialty stores, professional stock houses
In-Use Evaluation: An excellent high-speed film with good keeping qualities. This film is specially designed to be push-processed up to two f-stops and still deliver results acceptable to working pros.

Designation: **Ektapress 1600 Gold Professional PPC**
Manufactured by: Eastman Kodak Co., USA
Film Type: 35mm color negative
Speed: ISO 1600 *Push and/or Pull Feasible?* Yes
Processing: Standard C-41 or equivalent
Size(s): 35mm *Roll Lengths:* 36 exposures
Long Rolls Available? No
Other Sizes Available: 120
Diffuse RMS Granularity Value: 21
Resolving Power
High Contrast 1000:1 = 80 ll/mm
Low Contrast 1.6:1 = 40 ll/mm
Exposure Latitude: +3, –2 f-stops
Contrast Characteristics: Normal to slightly high
Film Structure: Conventional, T-Grain structure, enhanced DIRs
Color Balance: Daylight
Color Bias: Neutral, warm skin tones
Recommended Applications: Available-light work, sports, action
Special Uses: Can be push-processed up to 2 f-stops with acceptable images
Reciprocity (recommended shutter-speed range): 1/10000–1/10 second
Exposure Correction Table:

Exposure time:	1/10000	1/1000	1/100	1/10	1	10
Correction:	none	none	none	none	+1/2 f-stop	+1 f-stop
Filtration (CC):	none	none	none	none	none	none

Storage, Keeping Properties, and Special Handling: Professional film, but does not need refrigeration
Availability: Photo specialty stores, some professional stock houses

In-Use Evaluation: One of the very best ultra high speed color films, Ektapress 1600 exhibits tight grain for a film this fast. It also shows good blacks with little or no color tinging.

Designation: **Konica Impresa 50 IMP50**
Manufactured by: Konishiroku Photo Co., Japan
Film Type: 35mm color negative
Speed: ISO 50 *Push and/or Pull Feasible?* No
Processing: Standard C-41 or equivalent
Size(s): 35mm *Roll Lengths:* 24, 36 exposures
Long Rolls Available? No
Other Sizes Available: 120
Diffuse RMS Granularity Value: 4
Resolving Power
High Contrast 1000:1 = 130 ll/mm
Low Contrast 1.6:1 = 80 ll/mm
Exposure Latitude: +2, –1-1/2 f-stops
Contrast Characteristics: Medium
Film Structure: Conventional, multigrain structure, enhanced DIRs, thin emulsion
Color Balance: Daylight
Color Bias: Neutral to slightly warm
Recommended Applications: General professional applications
Special Uses: Film can be helpful when very accurate color rendition is required
Reciprocity (recommended shutter speed range): 1/10000–1 second
Exposure Correction Table:

Exposure time:	1/10000	1/1000	1/100	1/10	1	10
Correction:	none	none	none	none	none	by test
Filtration (CC):	none	none	none	none	none	by test

Storage, Keeping Properties, and Special Handling: Professional film. Refrigerate until use.
Availability: Photo specialty stores, under house names
In-Use Evaluation: Outstanding film rivaling the best print emulsions in terms of sharpness and grain structure. In addition, contrast is very well controlled and balanced. Many colors usually difficult to reproduce are handled easily by this excellent film.

Designation: **Konica Color Super SR 100**
Manufactured by: Konishiroku Photo Co., Japan
Film Type: 35mm color negative
Speed: ISO 100 *Push and/or Pull Feasible?* Yes
Processing: Standard C-41 or equivalent
Size(s): 35mm *Roll Lengths:* 12, 24, 36 exposures
Long Rolls Available? No
Other Sizes Available: 120
Diffuse RMS Granularity Value: 4
Resolving Power
High Contrast 1000:1 = 100 ll/mm
Low Contrast 1.6:1 = 50 ll/mm
Exposure Latitude: +3, –2 f-stops
Contrast Characteristics: Normal
Film Structure: Conventional, twin grain structure, enhanced DIRs, thin emulsion

Color Balance: Daylight
Color Bias: Neutral
Recommended Applications: All general photography normally assigned to a medium-speed film
Special Uses: None
Reciprocity (recommended shutter speed range): 1/10000–1 second
Exposure Correction Table:

Exposure time:	1/10000	1/1000	1/100	1/10	1	10
Correction:	none	none	none	none	none	+1 f-stop

Storage, Keeping Properties, and Special Handling: Normal
Availability: Photo specialty stores, under many house names
In-Use Evaluation: Like its faster brand mates, this film has been improved in every performance dimension. It features better color saturation, finer grain, higher acutance, and lower contrast. It is fully competitive with better known makes.

Designation: **Konica Color Super SR 200**
Manufactured by: Konishiroku Photo Co., Japan
Film Type: 35mm color negative
Speed: ISO 200 *Push and/or Pull Feasible?* Yes
Processing: Standard C-41 or equivalent
Size(s): 35mm *Roll Lengths:* 12, 24, 36 exposures
Long Rolls Available? No
Other Sizes Available: 120
Diffuse RMS Granularity Value: 4
Resolving Power
High Contrast 1000:1 = 100 ll/mm
Low Contrast 1.6:1 = 50 ll/mm
Exposure Latitude: +3, –2 f-stops
Contrast Characteristics: Medium to low
Film Structure: Conventional, twin grain structure, enhanced DIRs, thin emulsion
Color Balance: Daylight
Color Bias: Neutral
Recommended Applications: All general photographic assignments where an ISO-200 film's speed would be advantageous
Special Uses: Moderate contrast makes this film a good choice for studio, wedding, or portrait work
Reciprocity (recommended shutter-speed range): 1/10000–1 second
Exposure Correction Table:

Exposure time:	1/10000	1/1000	1/100	1/10	1	10
Correction:	none	none	none	none	none	+1 f-stop
Filtration (CC):	none	none	none	none	none	none

Storage, Keeping Properties, and Special Handling: Normal
Availability: Photo specialty stores, also found under many house brand names
In-Use Evaluation: An excellent, thoroughly modern print film with ISO 100–like performance. May be available at substantial discounts.

Designation: **Konica Color SR-G 160 Professional**
Manufactured by: Konishiroku Photo Co., Japan
Film *Film Type:* 35mm color negative
Speed: ISO 160 *Push and/or Pull Feasible?* Yes
Processing: Standard C-41 or equivalent

Size(s): 35mm *Roll Lengths:* 24, 36 exposures
Long Rolls Available? Yes
Other Sizes Available: 120, 220
Diffuse RMS Granularity Value: 4
Resolving Power
High Contrast 1000:1 = 100 ll/mm
Low Contrast 1.6:1 = 50 ll/mm
Exposure Latitude: +3-1/2, –2-1/2 f-stops
Contrast Characteristics: Normal
Film Structure: Conventional, twin grain structure, enhanced DIRs, thin emulsion
Color Balance: Daylight
Color Bias: Neutral
Recommended Applications: Moderate contrast and open, bright—but not overly harsh—color makes it useful for many product, commercial and formal portrait assignments
Special Uses: Suitable for extremely short exposure times, as fast as 1/10000 second
Reciprocity (recommended shutter-speed range): 1/10000–1/10 second
Exposure Correction Table:

Exposure time:	1/10000	1/1000	1/100	1/10	1	10
Correction:	none	none	none	none	by test	by test

Storage, Keeping Properties, and Special Handling: Professional film. Refrigerate until use.
Availability: Photo specialty stores
In-Use Evaluation: A professional-quality film manufactured with tight tolerances. Its performance dimensions make it suitable for studio flash or location work. I have found it to have an exceptionally wide exposure latitude of almost seven stops.

Designation: **Konica Color Super SR 400**
Manufactured by: Konishiroku Photo Co., Japan
Film Type: 35mm color negative
Speed: ISO 400 *Push and/or Pull Feasible?* Yes
Processing: Standard C-41 or equivalent
Size(s): 35mm *Roll Lengths:* 12, 24, 36 exposures
Long Rolls Available? No
Other Sizes Available: 120
Diffuse RMS Granularity Value: 5
Resolving Power
High Contrast 1000:1 = 100 ll/mm
Low Contrast 1.6:1 = 50 ll/mm
Exposure Latitude: +3, –2 f-stops
Contrast Characteristics: Normal
Film Structure: Conventional, twin grain structure, enhanced DIRs, thin emulsion, recently improved
Color Balance: Daylight
Color Bias: Neutral
Recommended Applications: All low-light situations. Useful where high shutter speeds are required.
Special Uses: Formulated to reproduce colors more naturally than other print films under fluorescent lights
Reciprocity (recommended shutter-speed range): 1/4000–1 second

Exposure Correction Table:

Exposure time:	1/4000	1/1000	1/100	1/10	1	10
Correction:	none	none	none	none	none	+1 f-stop
Filtration (CC):	none	none	none	none	none	CC010C

Storage, Keeping Properties, and Special Handling: Normal
Availability: Photo specialty stores, under house brand names
In-Use Evaluation: Like other Super SR films, 400 is a significant improvement over its previous incarnation, SRV 400. It is finer grained, with better color response. An outstanding performer at ISO 400. And it may be found at substantial discounts.

Designation: **Konica Color SR-G 3200**
Manufactured by: Konishiroku Photo Co., Japan
Film Type: 35mm color negative
Speed: ISO 3200 *Push and/or Pull Feasible?* Yes
Processing: Standard C-41 or equivalent
Size(s): 35mm *Roll Lengths:* 24, 36 exposures
Long Rolls Available? No
Other Sizes Available: 120
Diffuse RMS Granularity Value: 11
Resolving Power
High Contrast 1000:1 = 70 ll/mm plus
Low Contrast 1.6:1 = 35 ll/mm
Exposure Latitude: +3, –1-1/2 f-stops
Contrast Characteristics: Normal to slightly high
Film Structure: Conventional, multi-grain structure, enhanced DIRs, thin emulsion
Color Balance: Daylight
Color Bias: Neutral
Recommended Applications: All available-light, night photography, extra fast shutter speeds up to 1/10000 second, long telephoto
Special Uses: Astro photography without use of motor-driven clock drives
Reciprocity (recommended shutter-speed range): 1/10000–1 second
Exposure Correction Table:

Exposure time:	1/10000	1/1000	1/100	1/10	1	10
Correction:	none	none	none	none	none	by test
Filtration (CC):	none	none	none	none	none	by test

Storage, Keeping Properties, and Special Handling: Professional film. Refrigerate until use.
Availability: Photo specialty stores, under house names
In-Use Evaluation: SR-G 3200 represents the world's fastest color film, print or slide. As such, it offers some unique possibilities. It is surprisingly sharp and displays reasonable color and grain for its speed. When employed to make moderate enlargements, it is sometimes hard to distinguish it from some slower films.

Designation: **Polaroid One Film**
Manufactured by: Polaroid Corp., USA
Film Type: 35mm color negative
Speed: ISO 200 *Push and/or Pull Feasible?* Yes
Processing: Standard C-41 or equivalent
Size(s): 35mm *Roll Lengths:* 12, 24, 36 exposures
Long Rolls Available? No

Other Sizes Available: None
Diffuse RMS Granularity Value: 4
Resolving Power
High Contrast 1000:1 = 100 ll/mm
Low Contrast 1.6:1 = 50 ll/mm
Exposure Latitude: +3, –2 f-stops
Contrast Characteristics: Normal
Film Structure: Conventional, multigrain structure, enhanced DIRs, thin emulsion
Color Balance: Daylight
Color Bias: Neutral
Recommended Applications: General photography
Special Uses: None
Reciprocity (recommended shutter-speed range): 1/10000–1 second
Exposure Correction Table:

Exposure time:	1/10000	1/1000	1/100	1/10	1	10
Correction:	none	none	none	none	none	+1 f-stop

Storage, Keeping Properties, and Special Handling: Normal
Availability: Mass market outlets, supermarkets
In-Use Evaluation: A very good general-purpose film with a wide exposure latitude (thus its name). Made for Polaroid by Konica.

Designation: **Scotchcolor HR 100**
Manufactured by: 3M Photo Products Division, USA
Film Type: 35mm color negative
Speed: ISO 100 *Push and/or Pull Feasible?* Yes
Processing: Standard C-41 or equivalent
Size(s): 35mm *Roll Lengths:* 12, 24, 36 exposures
Long Rolls Available? No
Other Sizes Available: None
Diffuse RMS Granularity Value: 5
Resolving Power
High Contrast 1000:1 = 100 ll/mm
Low Contrast 1.6:1 = 50 ll/mm
Exposure Latitude: +3, –2 f-stops
Contrast Characteristics: Normal
Film Structure: Conventional, twin grain structure, enhanced DIRs, thin emulsion
Color Balance: Daylight
Color Bias: Neutral to warm
Recommended Applications: General-purpose photography
Special Uses: Very long exposures up to two minutes
Reciprocity (recommended shutter-speed range): 1/1000–1/10 second
Exposure Correction Table:

Exposure time:	1/10000	1/1000	1/100	1/10	1	10
Correction:	none	none	none	none	+1	+1-1/2 f-stops
Filtration (CC):	none	none	none	none	CC05R	CC15R +CC05Y

Storage, Keeping Properties, and Special Handling: Normal
Availability: Sometimes available in discount stores under Scotch brand, many other times under house brand such as K-Mart's Focal.
In-Use Evaluation: A film that has been improved in virtually all aspects of performance and now can hold its own with almost any other ISO-100 color print film.

Designation **Scotchcolor Print 160**
Manufactured by: 3M Photo Products Division, USA
Film Type: 35mm color negative
Speed: ISO 160 *Push and/or Pull Feasible?* Yes
Processing: Standard C-41 or equivalent
Size(s): 35mm *Roll Lengths:* 12, 24, 36 exposures
Long Rolls Available? No
Other Sizes Available: 120
Diffuse RMS Granularity Value: 7
Resolving Power
High Contrast 1000:1 = 100 ll/mm
Low Contrast 1.6:1 = 50 ll/mm
Exposure Latitude: +3, –2 f-stops
Contrast Characteristics: Medium
Film Structure: Conventional, but with latest technological innovations
Color Balance: Daylight
Color Bias: Neutral
Recommended Applications: All general photo assignments with special application to studio, weddings, and other professional assignments
Special Uses: Neutral color balance and moderate contrast make the film especially suitable for studio portraiture
Reciprocity (recommended shutter-speed range): 1/1000–1/2 second
Exposure Correction Table:

Exposure time:	1/10000	1/1000	1/100	1/10	1
Correction:	none	none	none	none	N/R

Storage, Keeping Properties, and Special Handling: Normal
Availability: Photo specialty stores, some mass market outlets
In-Use Evaluation: Scotch film intended for the professional market. Capable performer featuring very good blend of characteristics. Price should be very competitive with other ISO 160 pro films.

Designation: **Scotchcolor HR 200**
Manufactured by: 3M Photo Products Division, USA
Film Type: 35mm color negative
Speed: ISO 200 *Push and/or Pull Feasible?* Yes
Processing: Standard C-41 or equivalent
Size(s): 35mm *Roll Lengths:* 12, 24, 36 exposures
Long Rolls Available? No
Other Sizes Available: None
Diffuse RMS Granularity Value: 6
Resolving Power
High Contrast 1000:1 = 98 ll/mm
Low Contrast 1.6:1 = 40 ll/mm plus
Exposure Latitude: +3, –2 f-stops
Contrast Characteristics: Normal
Film Structure: Conventional, twin grain structure, enhanced DIRs, thin emulsion
Color Balance: Daylight
Color Bias: Neutral to warm
Recommended Applications: General-purpose photography
Special Uses: Same as HR 100
Reciprocity (recommended shutter-speed range): 1/10000–1/10 second

Exposure Correction Table:

Exposure time:	1/10000	1/1000	1/100	1/10	1	10
Correction:	none	none	none	none	+1-1/2 f-stops	+1-1/2 f-stops
Filtration (CC):	none	none	none	none	CC 5R+CC 5Y	CC 10R+CC 20Y

Storage, Keeping Properties, and Special Handling: Normal
Availability: Same as other Scotch print films
In-Use Evaluation: Virtually the same as HR100 except one full stop faster

Designation: **Scotchcolor HR 400**
Manufactured by: 3M Photo Products Division, USA
Film Type: 35mm color negative
Speed: ISO 400 *Push and/or Pull Feasible?* Yes
Processing: Standard C-41 or equivalent
Size(s): 35mm *Roll Lengths:* 12, 24, 36 exposures
Long Rolls Available? No
Other Sizes Available: None
Diffuse RMS Granularity Value: 7
Resolving Power
High Contrast 1000:1 = 85 ll/mm
Low Contrast 1.6:1 = 40 ll/mm
Exposure Latitude: +3, –1-1/2 f-stops
Contrast Characteristics: Normal
Film Structure: Conventional, twin grain structure, enhanced DIRs, thin emulsion
Color Balance: Daylight
Color Bias: Neutral to warm
Recommended Applications: General-purpose photography
Special Uses: Good fall color reproduction
Reciprocity (recommended shutter-speed range): 1/10000–1/10 second
Exposure Correction Table:

Exposure time:	1/10000	1/1000	1/100	1/10	1
Correction:	none	none	none	none	+1-1/2 f-stop
Filtration (CC):	none	none	none	none	CC 05Y

Storage, Keeping Properties, and Special Handling: Normal
Availability: Discount outlets, under many house brands
In-Use Evaluation: A competitive 400-speed film that prints like Kodacolor. Available at substantially lower prices than competition.

Designation: **Agfapan APX 25 Professional**
Manufactured by: Agfa Geveart AG, Germany
Film Type: 35mm black-and-white negative
Speed: ISO 25 *Push and/or Pull Feasible?* Yes
Processing: Standard developers, Refinal, Rodinol, D-76, etc.
Size(s): 35mm *Roll Lengths:* 24, 36 exposures
Long Rolls Available? Yes
Other Sizes Available: 120
Diffuse RMS Granularity Value: 7
Resolving Power
High Contrast 1000:1 = 200 ll/mm
Low Contrast 1.6:1 = 100 ll/mm
Exposure Latitude: +3, –2 f-stops
Contrast Characteristics: Varies with exposure and developing
Film Structure: Conventional

Color Balance: N/A
Color Bias: N/A
Recommended Applications: General-purpose photography, particularly still-life, figures
Special Uses: High-magnification enlargements, fine detail
Reciprocity (recommended shutter-speed range): 1/1000–1 second
Exposure Correction Table:

Exposure time:	1/1000	1/100	1/10	1/2	1	10	100
Correction:	none	none	none	none	+1	+1-1/2	+2 f-stops
Development correction (%):	none	none	none	none	none	none	none

Storage, Keeping Properties, and Special Handling: Professional film. Refrigerate until use.
Availability: Photo specialty stores
In-Use Evaluation: A very sharp, extremely fine grain film that produces fine salon-quality prints at very high magnifications. This film is in the top rank among the slow group.

Designation: **Agfapan APX 100 Professional**
Manufactured by: Agfa Geveart AG, Germany
Film Type: 35mm black-and-white negative
Speed: ISO 100 *Push and/or Pull Feasible?* Yes
Processing: Standard developers, Rodinol, D-76, etc.
Size(s): 35mm *Roll Lengths:* 20, 36 exposures
Long Rolls Available? Yes
Other Sizes Available: 120, sheets
Diffuse RMS Granularity Value: 9
Resolving Power
High Contrast 1000:1 = 150 ll/mm
Low Contrast 1.6:1 = 70 ll/mm
Exposure Latitude: +3, –2 f-stop
Contrast Characteristics: Varies with exposure and developing
Film Structure: Conventional
Color Balance: N/A
Color Bias: N/A
Recommended Applications: General-purpose photography
Special Uses: None
Reciprocity (recommended shutter-speed range): 1/1000–1/2 second
Exposure Correction Table:

Exposure time:	1/1000	1/100	1/10	1	10	100
Correction:	none	none	none	+1	+2	+3 f-stops
Development correction (%):	none	none	none	–10	–25	–35

Storage, Keeping Properties, and Special Handling: Professional film. Refrigerate until use.
Availability: Photo specialty stores
In-Use Evaluation: An excellent medium-speed film with wonderfully fine and sharp-edged grain. This emulsion is sometimes available at discounted prices.

Designation: **Agfapan 400 Professional**
Manufactured by: Agfa Geveart AG, Germany
Film Type: 35mm black-and-white negative
Speed: ISO 400 *Push and/or Pull Feasible?* Yes

Processing: Standard developers, Rodinol, D-76, etc.
Size(s): 35mm *Roll Lengths:* 36 exposures
Long Rolls Available? Yes
Other Sizes Available: 120, sheets
Diffuse RMS Granularity Value: 14
Resolving Power
High Contrast 1000:1 = 110 ll/mm
Low Contrast 1.6:1 = 50 ll/mm plus
Exposure Latitude: +3, –2 f-stops
Contrast Characteristics: Varies with exposure and developing
Film Structure: Conventional
Color Balance: N/A
Color Bias: N/A
Recommended Applications: General-purpose photography
Special Uses: Available-light photojournalism
Reciprocity (recommended shutter-speed range): 1/1000–1/2 second
Exposure Correction Table:

Exposure time:	1/1000	1/100	1/10	1	10	100
Correction:	none	none	none	+1	+2-1/2	+3-1/2 stops
Development correction (%):	none	none	none	–10	–25	–35

Storage, Keeping Properties, and Special Handling: Professional film. Refrigerate until use.
Availability: Photo specialty stores
In-Use Evaluation: A very high quality speed film that has been a well-kept secret in Europe for many years. It is the equal of many of the better known ISO 400 black-and-white films. Best results seem to come from rating this emulsion at ISO 250–320.

Designation: **Neopan 400 Professional**
Manufactured by: Fuji Film Co., Japan
Film Type: 35mm black-and-white negative
Speed: ISO 400 *Push and/or Pull Feasible?* Yes
Processing: Standard developers, Rodinol, D-76, etc.
Size(s): 35mm *Roll Lengths:* 36 exposures
Long Rolls Available? Yes
Other Sizes Available: 120
Diffuse RMS Granularity Value: 9 (est.)
Resolving Power
High Contrast 1000:1 = 100 ll/mm
Low Contrast 1.6:1 = 50 ll/mm
Exposure Latitude: +2, –2 f-stops
Contrast Characteristics: Varies with exposure and developing
Film Structure: Conventional
Color Balance: N/A
Color Bias: N/A
Recommended Applications: General-purpose photography
Special Uses: For use where extra speed is required
Reciprocity (recommended shutter-speed range): 1/4000–1/10 second
Exposure Correction Table:

Exposure time:	1/4000	1/1000	1/100	1/10	1
Correction:	none	none	none	none	+1 f-stop

Storage, Keeping Properties, and Special Handling: Professional film. Refrigerate until use.

Availability: Some photo specialty stores
In-Use Evaluation: An unusually fine grained film for its speed, this emulsion is forgiving of exposure and processing errors. Unfortunately not generally available.

Designation: **Neopan 1600 Professional**
Manufactured by: Fuji Film Co., Japan
Film Type: 35mm black-and-white negative
Speed: ISO 1600 *Push and/or Pull Feasible?* Yes
Processing: Standard developers, ID-11, Rodinol, D-76, etc.
Size(s): 35mm *Roll Lengths:* 36 exposures
Long Rolls Available? Yes
Other Sizes Available: None
Diffuse RMS Granularity Value: 11 (est.)
Resolving Power
High Contrast 1000:1 = 70 ll/mm plus
Low Contrast 1.6:1 = 35 ll/mm plus
Exposure Latitude: +2, –1 f-stop
Contrast Characteristics: Varies with exposure and developing
Film Structure: Conventional
Color Balance: N/A
Color Bias: N/A
Recommended Applications: General-purpose photography
Special Uses: "Available darkness" photography
Reciprocity (recommended shutter-speed range): 1/4000–1/10 second
Exposure Correction Table:

Exposure time:	1/4000	1/1000	1/100	1/10	1
Correction:	none	none	none	none	+1 f-stop

Storage, Keeping Properties, and Special Handling: Professional film. Refrigerate until use.
Availability: Some photo specialty stores
In-Use Evaluation: This film features an extra high definition layer to increase image sharpness. The result is an outstanding performer for its speed over a very wide ISO range.

Designation: **Pan F**
Manufactured by: Ilford Ltd., Great Britain
Film Type: 35mm black-and-white negative
Speed: ISO 50 *Push and/or Pull Feasible?* Yes
Processing: Standard developers, Rodinol, D-76, ID-11Plus, etc.
Size(s): 35mm *Roll Lengths:* 20, 36 exposures
Long Rolls Available? Yes
Other Sizes Available: 120, sheets
Diffuse RMS Granularity Value: 7 (est.)
Resolving Power
High Contrast 1000:1 = better than 100 ll/mm
Low Contrast 1.6:1 = better than 50 ll/mm
Exposure Latitude: +2, –1 f-stop
Contrast Characteristics: Varies with exposure and developing
Film Structure: Conventional
Color Balance: N/A
Color Bias: N/A
Recommended Applications: General-purpose photography
Special Uses: Architectural details, close up, and nature

Reciprocity (recommended shutter-speed range): 1/1000–1/2 second
Exposure Correction Table:

Exposure time:	1/1000	1/100	1/10	1	10
Correction:	none	none	none	+1/2 f-stop	+1 f-stop
Filtration:	none	none	none	025B	050B

Storage, Keeping Properties, and Special Handling: Professional film. Refrigerate until use.
Availability: Photo specialty stores
In-Use Evaluation: This film is a standard high performance emulsion that is capable of producing extremely high quality enlargements (16 × 20 and beyond).

Designation: **FP-4 Plus**
Manufactured by: Ilford Ltd., Great Britain
Film Type: 35mm black-and-white negative
Speed: ISO 125 *Push and/or Pull Feasible?* Yes
Processing: Standard developers, Rodinol, D-76, ID-11Plus, etc.
Size(s): 35mm *Roll Lengths:* 24, 36 exposures
Long Rolls Available? Yes
Other Sizes Available: 120
Diffuse RMS Granularity Value: 10
Resolving Power
High Contrast 1000:1 = 100 ll/mm
Low Contrast 11.6:1 = 50 ll/mm
Exposure Latitude: +2, –2 f-stops
Contrast Characteristics: Varies with exposure and developing
Film Structure: Conventional
Color Balance: N/A
Color Bias: N/A
Recommended Applications: General-purpose photography
Special Uses: None
Reciprocity (recommended shutter speed range): 1/1000–1 second
Exposure Correction Table:

Exposure time:	1/1000	1/100	1/10	1	10
Correction:	none	none	none	+1/2 f-stop	+1 f-stop
Filtration (CC):	none	none	none	025B	050B

Storage, Keeping Properties, and Special Handling: Professional film. Refrigerate until use.
Availability: Photo specialty stores
In-Use Evaluation: This is an improved version of a standard medium-speed film. The film has an excellent tonal scale with very good separation of middle tones.

Designation: **HP-5 Plus**
Manufactured by: Ilford Ltd., Great Britain
Film Type: 35mm black-and-white negative
Speed: ISO 400 *Push and/or Pull Feasible?* Yes
Processing: Standard developers, Rodinol, D-76, ID-11Plus, etc.
Size(s): 35mm *Roll Lengths:* 24, 36 exposures
Long Rolls Available? Yes
Other Sizes Available: 120
Diffuse RMS Granularity Value: 13
Resolving Power
High Contrast 1000:1 = 100 ll/mm
Low Contrast 1.6:1 = 40 ll/mm

Exposure Latitude: +2, –2 f-stops
Contrast Characteristics: Varies with exposure and developing
Film Structure: Conventional
Color Balance: N/A
Color Bias: N/A
Recommended Applications: General-purpose photography
Special Uses: None
Reciprocity (recommended shutter-speed range): 1/1000–1/2 second
Exposure Correction Table:

Exposure time:	1/1000	1/100	1/10	1	10
Correction:	none	none	none	+1/2 f-stop	+1 f-stop
Filtration (CC):	none	none	none	025B	050B

Storage, Keeping Properties, and Special Handling: Normal
Availability: Photo specialty stores
In-Use Evaluation: A substantially improved version of the already excellent HP5. Film rivals Tri-X and many professionals prefer its softer edge effects that can result in smoother skin tones.

Designation: **Delta 400 Professional**
Manufactured by: Ilford Ltd., Great Britain
Film Type: 35mm black-and-white negative
Speed: ISO 400 *Push and/or Pull Feasible?* Yes
Processing: Standard developers, Rodinol, D-76, ID-11Plus, etc.
Size(s): 35mm *Roll Lengths:* 24, 36 exposures
Long Rolls Available? Yes
Other Sizes Available: 120
Diffuse RMS Granularity Value: 12
Resolving Power
High Contrast 1000:1 = 100 ll/mm
Low Contrast 1.6:1 = 50 ll/mm
Exposure Latitude: +2, –2 f-stops
Contrast Characteristics: Varies with exposure and developing
Film Structure: Conventional
Color Balance: N/A
Color Bias: N/A
Recommended Applications: General-purpose photography
Special Uses: Very usable at elevated EIs
Reciprocity (recommended shutter-speed range): 1/1000–1 second
Exposure Correction Table:

Exposure time:	1/1000	1/100	1/10	1	10
Correction:	none	none	none	+1/2 f-stop	+1 f-stop
Filtration (CC):	none	none	none	025B	050B

Storage, Keeping Properties, and Special Handling: Professional film. Refrigerate until use.
Availability: Photo specialty stores
In-Use Evaluation: This relatively new film seems to produce negatives with greater contrast than other high speed emulsions. Its performance should yield professional quality negatives.

Designation: **XP-2**
Manufactured by: Ilford Ltd., Great Britain
Film Type: 35mm black-and-white negative
Speed: ISO 100–1600 *Push and/or Pull Feasible?* Yes
Processing: C-41 or similar color print film chemistry
Size(s): 35mm *Roll Lengths:* 24, 36 exposures

Long Rolls Available? Yes
Other Sizes Available: 120
Diffuse RMS Granularity Value: variable, depending on ISO rating
Resolving Power
High Contrast 1000:1 = variable depending on ISO rating
Low Contrast 1.6:1 = variable depending on ISO rating
Exposure Latitude: +3, –2 f-stops
Contrast Characteristics: Varies with exposure and developing
Film Structure: Chromogenic black and white
Color Balance: N/A
Color Bias: N/A
Recommended Applications: General-purpose photography
Special Uses: Variable ISO and contrast on same roll without altering processing
Reciprocity (recommended shutter-speed range): 1/1000–1 second
Exposure Correction Table:

Exposure time:	1/1000	1/100	1/10	1	10
Correction:	none	none	none	+1/2 f-stop	+1 f-stop
Filtration (CC):	none	none	none	025B	050B

Storage, Keeping Properties, and Special Handling: Professional film. Refrigerate until use.
Availability: Photo specialty stores
In-Use Evaluation: A chromogenic black-and-white film, XP-2 forms images in much the same way as does a color print film. Black-and-white dyes are used in place of print film's color couplers. This is an exceptional film capable of delivering excellent negatives from ISO 100 to 800 and beyond, with normal development, all on the same roll.

Designation: **Technical Pan TP 2415**
Manufactured by: Eastman Kodak Co., USA
Film Type: 35mm black-and-white negative
Speed: ISO 25 *Push and/or Pull Feasible?* Yes
Processing: Can be developed in standard developers, Rodinol, D-76, ID-11 Plus, etc., but performs best when developed in its own Technidol LC or Technidol Liquid chemistry
Size(s): 35mm *Roll Lengths:* 36 exposures
Long Rolls Available? Yes
Other Sizes Available: 120
Diffuse RMS Granularity Value: 3 (est.)
Resolving Power
High Contrast 1000:1 = 330 ll/mm
Low Contrast 1.6:1 = 150 ll/mm
Exposure Latitude: +2, –2 f-stops
Contrast Characteristics: Varies with exposure and developing
Film Structure: Conventional
Color Balance: N/A
Color Bias: N/A
Recommended Applications: Originally designed as a very high contrast special applications film
Special Uses: General photography when processed in specific developers
Reciprocity (recommended shutter-speed range): 1/1000–1 second

Exposure Correction Table:

Exposure time:	1/1000	1/100	1/10	1	10
Correction:	none	none	none	none	+1/2 f-stop
Development correction (%):	none	none	none	none	–10%

Storage, Keeping Properties, and Special Handling: Professional film. Refrigerate until use.
Availability: Photo specialty stores
In-Use Evaluation: When processed as intended, the film produces ultra-sharp high-contrast negatives. With its custom designed developer and a number of other products, Tech Pan can be used as a full tonal range micro-grain film that delivers more lines of resolution than any other film, color or black-and-white. Grain is all but invisible even in 20 × 24 inch prints. And, because the film has an extended red sensitivity, it produces dramatic sky cloud contrasts in outdoor scenics. In its 35mm version, negatives have the look of medium format.

Designation: **T-Max 100 Professional TMX**
Manufactured by: Eastman Kodak Co., USA
Film Type: 35mm black-and-white negative
Speed: ISO 100 *Push and/or Pull Feasible?* Yes
Processing: Standard developers, Rodinol, D-76, ID-11Plus, and specifically recommended T-Max, T-Max RS developers
Size(s): 35mm *Roll Lengths:* 24, 36 exposures
Long Rolls Available? Yes
Other Sizes Available: 120, sheets
Diffuse RMS Granularity Value: 8
Resolving Power
High Contrast 1000:1 = 200 ll/mm
Low Contrast 1.6:1 = 100 ll/mm
Exposure Latitude: +2, –2 f-stops
Contrast Characteristics: Varies with exposure and developing
Film Structure: Kodak Tabular (T-) Grain technology
Color Balance: N/A
Color Bias: N/A
Recommended Applications: General-purpose photography
Special Uses: With special developing kit, film can be processed into direct B&W positives
Reciprocity (recommended shutter-speed range): 1/10000–1/10 second
Exposure Correction Table:

Exposure time:	1/10000	1/1000	1/100	1/10	1	10
Correction:	none	none	none	none	+1/3	+1/2 f-stop
Filtration (CC):	none	none	none	none	none	none

Storage, Keeping Properties, and Special Handling: Professional film. Refrigerate until use.
Availability: Photo specialty stores
In-Use Evaluation: An exceptional film, T-Max 100 has the performance of Panotomic-X at more than twice the speed. This is not a film for casual users for it is rather unforgiving of imprecise processing techniques. It has rapidly become the black-and-white film by which all others are judged. For best results, use developers specifically formulated for this film.

Designation: **T-Max 400 Professional TMY**
Manufactured by: Eastman Kodak Co., USA
Film Type: 35mm black-and-white negative
Speed: ISO 400 *Push and/or Pull Feasible?* Yes
Processing: Standard developers, Rodinol, D-76, ID-11Plus, T-Max, T-Max RS
Size(s): 35mm *Roll Lengths:* 24, 36 exposures
Long Rolls Available? Yes
Other Sizes Available: 120
Diffuse RMS Granularity Value: 10
Resolving Power
High Contrast 1000:1 = 125 ll/mm
Low Contrast 1.6:1 = 50 ll/mm
Exposure Latitude: +3, –1 f-stop
Contrast Characteristics: Varies with exposure and developing
Film Structure: Same state-of-the-art technology as other T-Max films
Color Balance: N/A
Color Bias: N/A
Recommended Applications: General-purpose photography
Special Uses: Use like a medium-speed film
Reciprocity (recommended shutter-speed range): 1/10000–1/10 second
Exposure Correction Table:

Exposure time:	1/10000	1/1000	1/100	1/10	1	10
Correction:	Same as T-Max 100					
Filtration (CC):	Same as T-Max 100					

Storage, Keeping Properties, and Special Handling: Professional film. Refrigerate until use.
Availability: Photo specialty stores, discount outlets
In-Use Evaluation: Even though this film is a bit fussy about processing, outstanding results are possible with careful darkroom procedures. EIs of up to 3200 are possible with more-than-acceptable results. To retain the inherent fine grain and sharpness of T-Max 400, one should rate it no higher than ISO 800. Tests indicate that with proper care, it should outperform Plus-X or FP-4.

Designation: **Plus-X Pan PX**
Manufactured by: Eastman Kodak Co., USA
Film Type: 35mm black-and-white negative
Speed: ISO 125 *Push and/or Pull Feasible?* Yes
Processing: Standard developers, Rodinol, D-76, ID-11Plus, etc.
Size(s): 35mm *Roll Lengths:* 24, 36 exposures
Long Rolls Available? Yes
Other Sizes Available: 120
Diffuse RMS Granularity Value: 9 (est.)
Resolving Power
High Contrast 1000:1 = 125 ll/mm
Low Contrast 1.6:1 = 50 ll/mm plus
Exposure Latitude: +3, –2 f-stops
Contrast Characteristics: Varies with exposure and developing
Film Structure: Conventional
Color Balance: N/A
Color Bias: N/A
Recommended Applications: General-purpose photography
Special Uses: None

Reciprocity (recommended shutter-speed range): 1/1000–1/10 second
Exposure Correction Table:

Exposure time:	1/1000	1/100	1/10	1	10
Correction:	none	none	none	+1 f-stop	+2 f-stops
Development adjustment (%):	none	none	none	–10	–20

Storage, Keeping Properties, and Special Handling: Normal
Availability: Everywhere
In-Use Evaluation: Kodak's longtime standby in the medium speed category, Plus-X is very tolerant of exposure errors as well as imprecision in processing. Its universal availability makes it attractive to casual black-and-white shooters. All in all, it is still an excellent performer.

Designation: **Tri-X Pan TX**
Manufactured by: Eastman Kodak Co., USA
Film Type: 35mm black-and-white negative
Speed: ISO 400 *Push and/or Pull Feasible?* Yes
Processing: Standard developers, Rodinol, D-76, ID-11Plus, etc.
Size(s): 35mm *Roll Lengths:* 24, 36 exposures
Long Rolls Available? Yes
Other Sizes Available: 120
Diffuse RMS Granularity Value: 12 (est.)
Resolving Power
High Contrast 1000:1 = 100 ll/mm
Low Contrast 1.6:1 = 50 ll/mm
Exposure Latitude: +3, –2 f-stops
Contrast Characteristics: Varies with exposure and developing
Film Structure: Conventional
Color Balance: N/A
Color Bias: N/A
Recommended Applications: General-purpose photography
Special Uses: Typically used as an available-light, photojournalists' film at elevated EIs up to 1600
Reciprocity (recommended shutter-speed range): 1/1000–1/10 second
Exposure Correction Table:

Exposure time:	1/1000	1/100	1/10	1	10
Correction:	none	none	none	+1 f-stop	+2 f-stops
Development adjustment (%):	none	none	none	–10	–20

Storage, Keeping Properties, and Special Handling: Normal
Availability: Everywhere
In-Use Evaluation: Tri-X has gained an almost legendary reputation for its ease of use and processing, adaptability and excellent overall performance. Many knowledgeable users rate the film at EI of about 320 and, with reduced development time, achieve a broader tonal range in a film already boasting a six or seven f-stop spread.

Designation: **T-Max P 3200 Professional TMZ**
Manufactured by: Eastman Kodak Co., USA
Film Type: 35mm black-and-white negative
Speed: ISO 3200 *Push and/or Pull Feasible?* Yes
Processing: Standard developers, Rodinol, D-76, ID-11Plus, etc.
Size(s): 35mm *Roll Lengths:* 36 exposures
Long Rolls Available? No

Other Sizes Available: None
Diffuse RMS Granularity Value: 18
Resolving Power
High Contrast 1000:1 = 1125 ll/mm
Low Contrast 1.6:1 = 50 ll/mm plus
Exposure Latitude: +4, –2 f-stops
Contrast Characteristics: Varies with exposure and developing
Film Structure: T-Grain technology
Color Balance: N/A
Color Bias: N/A
Recommended Applications: Photojournalism, action, extreme low light
Special Uses: Police, military
Reciprocity (recommended shutter-speed range): 1/10000–1 second
Exposure Correction Table:

Exposure time:	1/1000	1/100	1/10	1	10
Correction:	none	none	none	none	+2/3 f-stop
Development adjustment (%):	none	none	none	none	none

Storage, Keeping Properties, and Special Handling: Professional film. Refrigerate until use.
Availability: Photo specialty stores, some mass marketers
In-Use Evaluation: Currently the fastest black-and-white film in the world, P-3200 benefits from the latest advances making it a very good performer even at extended speed ratings. Rated normally (ISO 3200) it looks like a much slower emulsion. Reasonable size enlargements of 8 × 10 or less appear as if they were made from Tri-X or similar negatives.

Designation: **Fort Pan 100**
Manufactured by: Forte, Hungary
Film Type: 35mm black-and-white negative
Speed: ISO 100 *Push and/or Pull Feasible?* Yes
Processing: Wide range of conventional developers
Size(s): 35mm *Roll Lengths:* 36 exposures
Long Rolls Available? No
Other Sizes Available: 120
Diffuse RMS Granularity Value: 9
Resolving Power
High Contrast 1000:1 = 90 ll/mm
Low Contrast 1.6:1 = 50 ll/mm
Exposure Latitude: +2, –2 f-stops
Contrast Characteristics: Normal
Film Structure:
Color Balance: N/A
Color Bias: N/A
Recommended Applications: General-purpose photography
Special Uses: None
Reciprocity (recommended shutter-speed range): 1/1000–1 second
Exposure Correction Table:

Exposure time:	1/1000	1/100	1/10	1
Correction:	none	none	none	none

Storage, Keeping Properties, and Special Handling:
Availability: Photo specialty stores or through GMI Photographic, Inc.
In-Use Evaluation: The films produced in the former Kodak plant in

Hungary feature an exceptionally long tonal scale said to be some 30 percent greater than comparable films. Films should prove popular if quality is maintained from batch to batch. Prices appear very competitive.

Designation: **Fort Pan 200**
Manufactured by: Forte, Hungary
Film Type: 35mm black-and-white negative
Speed: ISO 200 *Push and/or Pull Feasible?* Yes
Processing: Conventional
Size(s): 35mm *Roll Lengths:* 36 exposures
Long Rolls Available? No
Other Sizes Available: 120, 4 × 5, 8 × 10
Diffuse RMS Granularity Value: 6
Resolving Power
High Contrast 1000:1 = 85 ll/mm
Low Contrast 1.6:1 = 50 ll/mm
Exposure Latitude: +2, –2 f-stops
Contrast Characteristics: Medium
Film Structure:
Color Balance: N/A
Color Bias: N/A
Recommended Applications: General-purpose photography
Special Uses: Useful film when extended tonal range is preferred
Reciprocity (recommended shutter-speed range): 1/1000–1 second
Exposure Correction Table:

Exposure time:	1/1000	1/100	1/10	1
Correction:	none	none	none	none

Storage, Keeping Properties, and Special Handling: None
Availability: Photo specialty stores or through GMI Photographic, Inc.
In-Use Evaluation: Same as ISO 100 version, but at double the speed rating

Designation: **Fort Pan 400**
Manufactured by: Forte, Hungary
Film Type: 35mm black-and-white negative
Speed: ISO 400 *Push and/or Pull Feasible?* Yes
Processing: Conventional
Size(s): 35mm *Roll Lengths:* 36 exposures
Long Rolls Available? No
Other Sizes Available: 120, 4 × 5, 8 × 10
Diffuse RMS Granularity Value: 9+
Resolving Power
High Contrast 1000:1 = 80 ll/mm
Low Contrast 1.6:1 = 40 ll/mm
Exposure Latitude: +2, –2 f-stops
Contrast Characteristics: Normal
Film Structure: Conventional, with double emulsion layer
Color Balance: N/A
Color Bias: N/A
Recommended Applications: General-purpose photography
Special Uses: Useful for hand-held, available-light photography, sports, and photojournalism
Reciprocity (recommended shutter-speed range): 1/1000–1 second

Exposure Correction Table:

Exposure time:	1/1000	1/100	1/10	1
Correction:	none	none	none	none

Storage, Keeping Properties, and Special Handling: Normal
Availability: Photo specialty stores or through GMI Photographic, Inc.
In-Use Evaluation: Image characteristics similar to competitive products except for longer tonal range.

Designation: **PolaChrome CS**
Manufactured by: Polaroid Corporation, Photo Products Division, USA
Film Type: Instant color transparency
Speed: ISO 40 *Push and/or Pull Feasible?* No
Processing: Special Polaroid Instant Chemistry and Auto Processor
Size(s): 35mm *Roll Lengths:* 12, 36 exposures
Long Rolls Available? No
Other Sizes Available: None
Diffuse RMS Granularity Value: N/A
Resolving Power
High Contrast 1000:1 = 90 ll/mm
Low Contrast 1.6:1 = 40 ll/mm plus
Exposure Latitude: +3, –2 f-stops
Contrast Characteristics: Overall contrast, medium to high
Film Structure: Similar in some ways to historic color film structures going back into late ninteenth century, PolaChrome employs a three-color (red, blue, green) additive screen coated on the emulsion layer as a repeating color triplet.
Color Balance: Daylight
Color Bias: Nearly neutral
Recommended Applications: General-purpose photography
Special Uses: Use to preview lighting setups on location and in the studio
Reciprocity (recommended shutter-speed range): 1/1000–1/15 second
Exposure Correction Table:

Exposure time:	1/1000	1/100	1/10	1	10
Correction:	none	none	+1/3	1	+2 f-stops
Filtration (CC):	none	none	none	none	none

Storage, Keeping Properties, and Special Handling: Normal
Availability: Photo specialty stores
In-Use Evaluation: Even though PolaChrome CS neither looks as sharp nor delivers the lines per millimeter of other slow color slide films, its unique palette and soft contrast compared to other slide films make it adaptable to general contemplative photography such as landscapes, still-life, and figure studies. It is a film to be experienced, not written about. Two words of caution: the silver surface of the film is very fragile and can be damaged easily, so handle with extreme care; second, the film has a different surface reflectance from other 35mm films, so built in off-the-film camera metering systems will give inaccurate readings when you use any 35mm Polaroid Instant film. There is a chart contained in Appendix G of this book giving meter corrections, but many current cameras are absent. It is best to switch to manual exposure if possible when using this film.

Designation: **High Contrast PolaChrome HCP**
Manufactured by: Polaroid Corporation, Photo Products Division
Film Type: Instant color transparency

Speed: ISO 40 *Push and/or Pull Feasible? No*
Processing: Same as PolaChrome CS
Size(s): 35mm *Roll Lengths:* 12 exposures
Long Rolls Available? No
Other Sizes Available: None
Diffuse RMS Granularity Value: N/A
Resolving Power
High Contrast 1000:1 = 90 ll/mm
Low Contrast 1.6:1 = 40 ll/mm plus
Exposure Latitude: +1/2, –1/2 f-stop
Contrast Characteristics: Very high
Film Structure: Special Polaroid matrix
Color Balance: Daylight
Color Bias: Neutral to slightly warm
Recommended Applications:
Reciprocity (recommended shutter-speed range): 1/1000–1/15 second
Exposure Correction Table:

Exposure time:	1/1000	1/100	1/10	1	10
Correction:	none	none	+1/3	+1	+2 f-stops
Filtration (CC):	none	none	none	none	none

Storage, Keeping Properties, and Special Handling: Normal
Availability: Photo specialty stores
In-Use Evaluation: PolaChrome CS and High Contrast Polachrome are the same films. Only the processing chemistry is different with the reagent being altered for the high-contrast version. This film does a good job reproducing realistic colors under tungsten light. Color is highly saturated but grain (or its Polaroid equivalent) is rather apparent. In certain situations the effect is striking and visually arresting. For instance, fall outdoor scenes lend themselves to this treatment admirably. While not intended for general pictorial work, numbers of experimenters use it regularly.

Designation: **PolaPan**
Manufactured by: Polaroid Corporation, Photo Products Division
Film Type: 35mm instant black-and-white transparency
Speed: ISO 125 *Push and/or Pull Feasible?* No
Processing: Instant processing with autoprocessor
Size(s): 35mm *Roll Lengths:* 12, 36 exposures
Long Rolls Available? No
Other Sizes Available: None
Diffuse RMS Granularity Value: N/A
Resolving Power
High Contrast 1000:1 = 90 ll/mm
Low Contrast 1.6:1 = 40 ll/mm plus
Exposure Latitude: +1, –1/2 f-stop
Film Structure: Same as other 35 mm Polaroid instant films except does not contain a color triplet screen
Color Balance: N/A
Color Bias: N/A
Recommended Applications: General-purpose photography, but contrasty subjects can be problematic
Special Uses: Lighting setup previews
Reciprocity (recommended shutter-speed range): 1/1000–1/15 second

Exposure Correction Table:

Exposure time:	1/1000	1/100	1/10	1	10
Correction:	none	none	+1/3	+1	+2 f-stops
Filtration (CC):	none	none	none	none	none

Storage, Keeping Properties, and Special Handling: Normal
Availability: Photo specialty stores
In-Use Evaluation: This film produces beautiful mid tones and highlights. It is highly useful for advertising transparencies in black-and-white, a newly reinvented form. Up to about 8 × 10, grain is almost nonexistent and high definition prints to 20 × 24 inches are routinely possible.

Designation: **Polaroid Polablue**
Manufactured by: Polaroid Corporation, USA
Film Type: High-contrast blue-and-white negative
Speed: ISO 8 *Push and/or Pull Feasible?* No
Processing: Polaroid processing, via Polaroid autoprocessor
Size(s): 35mm *Roll Lengths:* 12 exposures
Long Rolls Available? No
Other Sizes Available: None
Diffuse RMS Granularity Value: N/A
Resolving Power
High Contrast 1000:1 = 90 ll/mm
Low Contrast 1.6:1 = 40 ll/mm
Exposure Latitude: +1, –1/2 f-stop
Contrast Characteristics: High, especially with some overexposure
Film Structure: 35 mm Polaroid without color screen
Color Balance: N/A
Color Bias: N/A
Recommended Applications: Generates high-contrast blue on white slides for audiovisual presentations
Special Uses: Many special applications to the experimenter
Reciprocity (recommended shutter-speed range): 1/100000–1/15 second
Exposure Correction Table:

Exposure time:	1/1000	1/100	1/10	1	10
Correction:	none	none	+1/3	+1	+2 f-stops
Filtration (CC):	none	none	none	none	none

Storage, Keeping Properties, and Special Handling: Normal
Availability: Photo specialty stores
In-Use Evaluation: PolaBlue is an ortho film that is sensitive to everything except red. To compare it to a conventional black-and-white film, black will reproduce as white and white as blue.

Designation: **PolaGraph HC**
Manufactured by: Polaroid Corporation, Photo Products Division, USA
Film Type: 35mm instant black -and-white transparency
Speed: ISO 400 *Push and/or Pull Feasible?* No
Processing: 35 mm Polaroid via Autoprocessor
Size(s): 35mm *Roll Lengths:* 12 exposures
Long Rolls Available? No
Other Sizes Available: None
Diffuse RMS Granularity Value: N/A
Resolving Power
High Contrast 1000:1 = 90 ll/mm
Low Contrast 1.6:1 = 40 ll/mm plus
Exposure Latitude: +1, –1/2 f-stop

Contrast Characteristics: High
Film Structure: Same as other 35 mm black-and-white Polaroid films
Color Balance: N/A
Color Bias: N/A
Recommended Applications: Produce all high-contrast black-and-white slides
Special Uses: Many graphic arts applications
Reciprocity (recommended shutter-speed range): 1/100000–1/15 second
Exposure Correction Table:

Exposure time:	1/1000	1/100	1/10	1	10
Correction:	none	none	+1/3	+1	+2 f-stops
Filtration (CC):	none	none	none	none	none

Storage, Keeping Properties, and Special Handling: Normal
Availability: Photo specialty stores
In-Use Evaluation: Fast film that gives added contrast to subjects that are flatly lit or in shadows. Especially useful in copying tasks.

Designation: **35 mm Presentation Chrome**
Manufactured by: Polaroid Corporation, Office Products Group, USA
Film Type: 35mm color transparency
Speed: ISO 100 *Push and/or Pull Feasible?* Yes
Processing: E-6 or equivalent
Size(s): 35mm *Roll Lengths:* 36 exposures
Long Rolls Available? No
Other Sizes Available: None
Diffuse RMS Granularity Value: 11 (est.)
Resolving Power
High Contrast 1000:1 = 100 ll/mm plus
Low Contrast 1.6:1 = 50 ll/mm plus
Exposure Latitude: +1/2, –1 f-stop
Contrast Characteristics: Very high
Film Structure: Conventional
Color Balance: Daylight
Color Bias: Nearly neutral
Recommended Applications: Graphics arts, illustration, audiovisual preparation, photography from CRT screens
Special Uses: Can be used to brighten a low-contrast scene
Reciprocity (recommended shutter-speed range): 1/100000–1/10 second
Exposure Correction Table:

Exposure time:	1/1000	1/100	1/10	1	10
Correction:	none	none	none	none	+2/3 f-stop
Filtration (CC):	none	none	none	none	CC05Y

Storage, Keeping Properties, and Special Handling: Professional film. Refrigerate until use.
Availability: Some large photo specialty stores, graphic arts supply houses, or audiovisual suppliers
In-Use Evaluation: Presentation Chrome is a very interesting film not usually found in camera shops. In fact it isn't even supplied by the same division as are other Polaroid films. It is an unusual product because of its extreme contrast compared to other slide films. Not only does it exhibit high overall contrast, but color contrast and saturation are very high as well. It can produce startling effects when used under the right circumstances. Besides having great contrast, the emulsion produces exceptionally clean and vibrant whites.

This is a film that will produce poster-like effects in direct sun and enhance both color and contrast in bad weather or under heavy overcast. Presentation Chrome is very similar and perhaps identical to 3M CRT 100.

Designation: **High Speed Infrared HIE**
Manufactured by: Eastman Kodak Co., USA
Film Type: 35mm black-and-white negative
Speed: ISO 160 (approx.) *Push and/or Pull Feasible?* N/A
Processing: Standard developers: HC110 (Dil B), D-76, ID-11Plus, etc.
Size(s): 35mm *Roll Lengths:* 36 exposures
Long Rolls Available? No
Other Sizes Available: None
Diffuse RMS Granularity Value: None given, but is relatively grainy. Grain on a par with many high speed black-and-white films in the ISO 1000 range.
Resolving Power
High Contrast 1000:1 = 80 ll/mm
Low Contrast 1.6:1 = 30 ll/mm plus
Exposure Latitude: Varies with amount of infrared radiation present.
Contrast Characteristics: High
Film Structure: Conventional black-and-white film, except it is almost totally insensitive to visible light and sensitive to infrared and near infrared
Color Balance: N/A
Color Bias: N/A
Recommended Applications: Aerial, scientific, industrial, environmental, forensic.
Special Uses: Atmospheric haze penetration, abstract art photography, vastly altered tonal relationships
Reciprocity (recommended shutter-speed range): 1/1000–1/10 second
Exposure Correction Table:

Exposure time:	1/1000	1/100	1/10	1	10
Correction:	+1/3 f-stop	none	none	none	none
Filtration (CC):	none	none	none	none	none

Storage, Keeping Properties, and Special Handling: Professional film. Refrigerate until use. Load, unload, and process in total darkness. *Caution:* some plastic developing tanks are not infrared light-tight: use stainless steel only.
Availability: Photo specialty stores
In-Use Evaluation: When used as a pictorial film, infrared produces somewhat unpredictable results because it is difficult to determine the amount of infrared radiation reflecting from different subjects. Strange and bizarre effects are routine with this film. Green foliage reflects a great deal of infrared and with the recommended red 25A filter over the lens it appears as a ghostly white against a much darker, sometimes black sky. Halos sometimes appear around backlit subjects. Even though Kodak recommends ISO 80 as a starting point, many experienced IR users set ISOs up to 200 on their cameras. You may wish to bracket beginning at ISO 50 up. IR work can be frustrating for two reasons. First, work will be done in the dark, both figuratively and literally, and second, it may not always be possible to repeat effects that you find appealing. Use infrared focusing marks on lens barrels of most 35mm lenses. If no IR focus indicators are present add about 4 percent of the focal

length to visible focus distance. If you are shooting at f-11 or smaller, IR focus shift will usually fall within depth of focus so no correction will be needed.

Designation: **Ektachrome Infrared 2236**
Manufactured by: Eastman Kodak Co., USA
Film Type: 35mm color infrared transparency
Speed: ISO 125 (approx.) w/ 25A filter *Push and/or Pull Feasible?* No
Processing: E-4. See Appendix C for list of commercial processing labs
Size(s): 35mm *Roll Lengths:* 36 exposures
Long Rolls Available? No
Other Sizes Available: none
Diffuse RMS Granularity Value: 17; similar to ISO 1000 color slide film.
Resolving Power
High Contrast 1000:1 = 63 ll/mm
Low Contrast 1.6:1 = 30 ll/mm
Exposure Latitude: +1, –1 f-stop
Contrast Characteristics: High compared to conventional slide films
Film Structure: Conventional, except almost insensitive to visible light. Sensitive to infrared and near infrared.
Color Balance: Daylight
Color Bias: N/A
Recommended Applications: Most of the comments for black-and-white infrared are applicable to the color version. The manufacturer recommends the use of a red, green, or orange filter.
Special Uses: Graphic surreal effects possible but somewhat unpredictable
Reciprocity (recommended shutter-speed range): 1/1000–1/10 second
Exposure Correction Table:

Exposure time:	1/1000	1/100	1/10	1	10
Correction:	N/A				

Storage, Keeping Properties, and Special Handling: Professional film. Refrigerate until use. Load and unload in complete darkness.
Availability: Photo specialty stores
In-Use Evaluation: Color equivalent of high-speed infrared, produces weird color effects that some, if not all, photographers find very interesting. Color effects vary with filter in use and volume of reflected IR radiation. Below are some common color shifts with selected filters:

Filter	*Expected Sky Color*	*Foliage*	*Skin Tone*
Green (58)	Deep blue	Magenta	Purple
Red (25A)	Green	Reddish-orange	Yellow-green
Yellow (12)	Blue	Red	Green

Other color filters will give different results.

Unlike high-speed infrared, Ektachrome Infrared needs no focus correction because two of its three emulsion layers are sensitive to visible light.

Designation: **Recording Film 2475 HIE**
Manufactured by: Eastman Kodak Co., USA
Film Type: 35mm black-and-white negative
Speed: ISO 1000 *Push and/or Pull Feasible?* Yes
Processing: Kodak HC-110, DK-50
Size(s): 35mm *Roll Lengths:* 36 exposures

Long Rolls Available? No
Other Sizes Available: None
Diffuse RMS Granularity Value: 20 Plus
Resolving Power
High Contrast 1000:1 = 63 ll/mm
Low Contrast 1.6:1 = 30 ll/mm
Exposure Latitude: +2 ,–2 f-stops
Contrast Characteristics: Varies with exposure and developing
Film Structure: Conventional
Color Balance: N/A
Color Bias: N/A
Recommended Applications: Used for surveillance work, other very low-light situations
Special Uses: Enhanced grain effects
Reciprocity (recommended shutter-speed range): 1/1000–1 second
Exposure Correction Table:

Exposure time:	1/1000	1/100	1/10	1	10
Correction:	none	none	none	+1/2 f-stop	+1 f-stop
Filtration (CC):	none	none	none	none	none

Storage, Keeping Properties, and Special Handling: Professional film. Refrigerate until use.
Availability: Photo specialty stores
In-Use Evaluation: Until the introduction of Fuji Neopan 1600 and Kodak T-Max P3200, Recording Film was the fastest black-and-white film in use. In the opinion of the test panel, it now has limited uses, one of which is low-definition, ultra high grain abstract imagery.

Designation: **Ektachrome Slide Duplicating Film 5071 ESD**
Manufactured by: Eastman Kodak Co., USA
Film Type: 35mm color slide
Speed: ISO 12 *Push and/or Pull Feasible?* Yes
Processing: E-6 or equivalent
Size(s): 35mm *Roll Lengths:* None
Long Rolls Available? Yes
Other Sizes Available: None
Diffuse RMS Granularity Value: less than 5
Resolving Power
High Contrast 1000:1 = 80 ll/mm plus
Low Contrast 1.6:1 = 50 ll/mm
Exposure Latitude: +1, –1-1/2 f-stops
Contrast Characteristics: Low
Film Structure: Conventional
Color Balance: Tungsten 3200K
Color Bias: Neutral to slightly cool
Recommended Applications: Designed for copying original slides with minimum contrast buildup.
Special Uses: Can be used in general photography of relatively static subjects shot under harsh, contrasty light
Reciprocity (recommended shutter-speed range): 1/10–1 second. Manufacturer recommends 1-second exposure with duplicating equipment such as Besler Illumitron.
Exposure Correction Table:

Exposure time:	1	10
Correction::	None required with duplicator	

Storage, Keeping Properties, and Special Handling: Professional film. Refrigerate until use.
Availability: Photo specialty stores
In-Use Evaluation: An excellent low -contrast, high-sharpness film that does a credible job duplicating a variety of types of color slides with reasonable fidelity. No film can duplicate exactly as some contrast buildup and loss of detail is inevitable when going from first to second generation. A daylight version of this film is stocked by many large professional photo supply firms. It is designated Ektachrome SE Duplicating Film SO-366 and is balanced for 5500K and electronic flash. It should be exposed at shutter speeds between 1/10 and 1/1000 second. Otherwise it has the same characteristics as its tungsten mate.

Designation: **Verichrome Slide Film 5072/ SO-279**
Manufactured by: Eastman Kodak Co., USA
Film Type: 35 mm negative
Speed: ISO 3-12 *Push and/or Pull Feasible?* Yes
Processing: C-41 or similar chemistry
Size(s): 35mm *Roll Lengths:* 36 exposures (SO-279)
Long Rolls Available? 100 ft. (5072)
Other Sizes Available: Sheets as Vericolor Print Film 4111
Diffuse RMS Granularity Value: 9
Resolving Power

High Contrast	1000:1 = 250 ll/mm
Low Contrast	1.6:1 = 100 ll/mm

Exposure Latitude: +1, –1 f-stop
Contrast Characteristics: Low
Film Structure: Conventional
Color Balance: Daylight
Color Bias: N/A
Recommended Applications: Intended for producing color transparencies for color negatives and internegatives by direct printing
Special Uses: Generating reverse-text slides for audiovisual presentation. White text on various colored backgrounds.
Reciprocity (recommended shutter-speed range): 1/4–8 seconds
Exposure Correction Table:

Exposure time:	1/1000	1/100	1/10	1	10
Correction:	N/R	N/R	none	none	none
Filtration (CC):	N/R	N/R	none	none	none

Storage, Keeping Properties, and Special Handling: Professional film. Refrigerate until use. Handle and process in complete darkness.
Availability: Photo specialty stores
In-Use Evaluation: Trial times for optimum exposures should be from 1 to 4 seconds. Starting filter packs for all Ektar, Ektapress Gold, and Kodacolor Gold films is CC20M+ CC30Y; for Vericolor III, Vericolor 400: CC20M, CC35Y. Even though it is a negative film developed in normal negative chemistry, this film yields quality slides from negatives or internegatives. The slides produced by this method print very well for they retain almost all of their original sharpness and do not pick up much additional contrast.

Designation: **Photomicrography Film 2483 PCF**
Manufactured by: Eastman Kodak Co., USA
Film Type: 35mm color transparency
Speed: ISO 12–16 *Push and/or Pull Feasible?* Yes
Processing: E-4 Chemistry
Size(s): 35mm *Roll Lengths:* 36 exposures
Long Rolls Available? No
Other Sizes Available: Some sheet sizes
Diffuse RMS Granularity Value: Lower than 9
Resolving Power
High Contrast 1000:1 = 125 ll/mm plus
Low Contrast 1.6:1 = 100 ll/mm plus
Exposure Latitude: +1, –1 f-stop
Contrast Characteristics: Very high
Film Structure: Conventional
Color Balance: Daylight
Color Bias: Marked magenta cast
Recommended Applications: High-definition scientific and microphotography
Special Uses: Experimental and abstract photography
Reciprocity (recommended shutter-speed range): 1/1000–1 second
Exposure Correction Table:

Exposure time:	1/1000	1/100	1/10	1	10
Correction:	N/A				
Filtration (CC):	N/A				

Storage, Keeping Properties, and Special Handling: Professional film. Refrigerate until use.
Availability: Photo specialty stores
In-Use Evaluation: As a pictorial film, photomicrography film has unlimited appeal. It displays spectacular reds and blues and almost normal color distribution when used in conjunction with a CC 20G filter. The filter counteracts almost all the excess magenta exhibited by the unfiltered film. While producing exciting color, exquisite detail, and remarkably rich blacks, the film can go even further with the addition of a Tiffen Enhancing filter. Reds, yellows, and oranges pop out like no other film/filter combination. Bear in mind that this film is not intended to give accurate colors. It is an interpretive tool, and as such should be tried under many different conditions and with numerous subjects.

Note: Film can only be processed in E-4 chemistry.

Designation: **Vericolor III ID Film**
Manufactured by: Eastman Kodak Co., USA
Film Type: 35mm color negative
Speed: ISO 100 *Push and/or Pull Feasible?* No
Processing: C-41 or similar chemistry
Size(s): 35mm *Roll Lengths:* None
Long Rolls Available? Yes
Other Sizes Available: None
Diffuse RMS Granularity Value: 6
Resolving Power
High Contrast 1000:1 = 100 ll/mm
Low Contrast 1.6:1 = 40 ll/mm
Exposure Latitude; +3, –2 f-stops

Contrast Characteristics: Medium
Film Structure: Conventional, but with two separate light sensitive emulsion layers
Color Balance: Daylight
Color Bias: Neutral
Recommended Applications: Especially designed for use in special cameras to produce two different exposures on the same frame containing both portrait and line copy
Special Uses: Use for producing photo ID cards, drivers licenses, and so on
Reciprocity (recommended shutter-speed range): 1/10000–1/10 second
Exposure Correction Table:

Exposure time:	1/10000	1/1000	1/100	1/10	1	10
Correction:	none	none	none	none	N/R	N/R
Filtration (CC):	none	none	none	none	N/R	N/R

Storage, Keeping Properties, and Special Handling: Professional film. Refrigerate until use.
Availability: Photo specialty stores, professional stock houses
In-Use Evaluation: Even though this product has specific commercial applications, experimenters will most surely find ways to apply it to pictorial uses, particularly in view of its unique structure.

Designation: **Verichrome Internegative Film 6011**
Manufactured by: Eastman Kodak Co., USA
Film Type: 35mm color negative
Speed: ISO 1-3 *Push and/or Pull Feasible?* No
Processing: C-41 or similar chemistry
Size(s): 35mm *Roll Lengths:* 36 exposures
Long Rolls Available? Yes
Other Sizes Available: Sheets
Diffuse RMS Granularity Value: 5
Resolving Power

High Contrast	1000:1 = 100 ll/mm
Low Contrast	1.6:1 = 50 ll/mm

Exposure Latitude: +2, –2 f-stops
Contrast Characteristics: Low
Film Structure: Conventional
Color Balance: Tungsten 3200K
Color Bias: Neutral
Recommended Applications: Designed for making internegative from color transparencies
Special Uses: Experimenters use as an extremely high definition, low-contrast film for static subjects. May also be used to generate black-and-white internegatives
Reciprocity (recommended shutter-speed range): 1/10000–1/10 second
Exposure Correction Table:

Exposure time:	Recommended exposure time, 1 second
Correction:	Starting filter pack: CC30M+CC30Y

Storage, Keeping Properties, and Special Handling: Professional film. Refrigerate until use.
Availability: Photo specialty stores
In-Use Evaluation: This film is very versatile, enabling photographers to correct many color imbalances while making high-quality internegatives for later printing. It comes in a variety of sizes.

Designation: **Kodak Slide Duplicating Film Type K 8071**
Manufactured by: Eastman Kodak Co., USA
Film Type: color transparency
Speed: ISO 25 *Push and/or Pull Feasible?* Yes
Processing: E-6 or equivalent
Size(s): 35mm *Roll Lengths:* 24, 36 exposures
Long Rolls Available? Yes
Other Sizes Available: Sheets
Diffuse RMS Granularity Value: 5
Resolving Power
High Contrast 1000:1 = 80 ll/mm plus
Low Contrast 1.6:1 = 50 ll/mm
Exposure Latitude: +3, –2 f-stops
Contrast Characteristics: Varies with exposure and developing
Film Structure: Conventional
Color Balance: Tungsten
Color Bias: Neutral
Recommended Applications: Specially designed for accurate duplication of Kodachrome slides. Holds highlight detail better than general-purpose duplicating films.
Special Uses: Can be used to duplicate Scotchchrome slides
Reciprocity (recommended shutter-speed range): 1/1000–1 second
Exposure Correction Table:

Exposure time:	1/1000	1/100	1/10	1	10
Correction:	none required with duplicator				
Filtration (CC):	none required with duplicator				

Storage, Keeping Properties, and Special Handling: Professional film. Refrigerate until use.
Availability: Photo specialty stores
In-Use Evaluation: Shares most of the performance qualities of Kodak 5071 duping film, except it is more closely matched to Kodachromes. Produces superior results with Kodachrome originals. Starting filter pack is now CC 80Y + CC 40C.

Designation: **Eastman Negative Color Film Type 5247**
Manufactured by: Eastman Kodak Co., USA
Film Type: 35mm color negative film
Speed: ISO 125 *Push and/or Pull Feasible?* Yes
Processing: Processing in ECN-2 chemistry
Size(s): 35mm *Roll Lengths:* none
Long Rolls Available? Yes, 100, 200, 400, 1000 ft.
Other Sizes Available: None
Diffuse RMS Granularity Value: < 5
Resolving Power
High Contrast 1000:1 = 100 ll/mm
Low Contrast 1.6:1 = 50 ll/mm
Exposure Latitude: +3, –2 f-stops
Contrast Characteristics: medium
Film Structure: Conventional
Color Balance: Tungsten
Color Bias: Neutral
Recommended Applications: Motion picture stock
Special Uses: General still photography
Reciprocity (recommended shutter-speed range): 1/10000–1 seconds

Exposure Correction Table:

Exposure time:	1/1000	1/100	1/10	1
Correction:	none	none	none	+1/2 f-stop
Filtration (CC):	none	none	none	CC10Y

Storage, Keeping Properties, and Special Handling: Professional film. Refrigerate until use.

Availability: Photo specialty stores, by special order

In-Use Evaluation: 5347 is a standard motion-picture stock. When both sides of film are perforated with sprocket holes and the film is respooled in 35mm cassettes, it can be used as a normal color negative film. It is unique in that it can be processed to deliver both slides and prints from the same roll. A number of labs offer either slides or prints alone or combinations of slides, negatives, and/or prints. The best of these rival conventional slide and print quality. The film itself is fine grain, medium contrast, and yields excellent color. Even though it is balanced for tungsten light, the processing lab takes care of color when film is exposed in daylight. You should not mix light sources on the same roll, however.

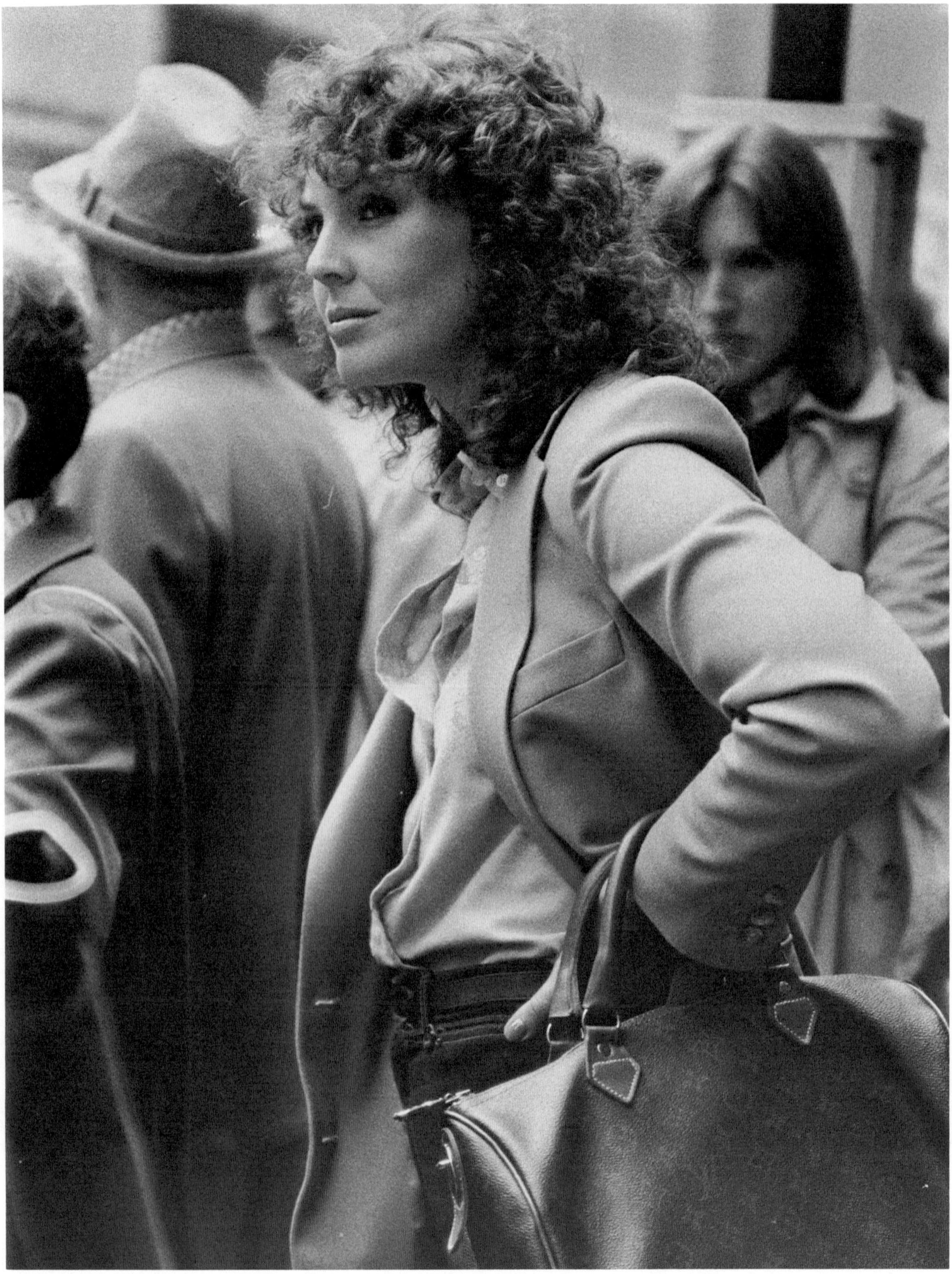

Chapter 7

Film Processing

Sooner or later, nearly every serious photographer becomes interested in the darkroom. There are several good reasons for spending some time learning processing procedures and techniques. But the one that drives most of us to do it, finally, is perhaps the most obvious: It's very satisfying to "make" a photograph, in the literal sense, from beginning to end.

Sending exposed film off to be developed can leave you feeling that you have cut short the creative process. Of course there will be occasions when, because of time pressure or film type, you will have to rely on a commercial processor. But when you can do it yourself, you will have far greater control over the results, in many instances with greater economy, which is an important consideration if you shoot lots of pictures.

Most of us are initiated to the darkroom experience by processing black-and-white films. Traditionally, black-and-white films had greater latitude than color films and therefore were more forgiving both in exposure and processing. Color processing was more exacting and required more care than black-and-white. Fortunately, this is beginning to change as color films and processing are updated and improved.

Developing Black-and-White Film

Developing black-and-white film is reasonably easy and can be done without benefit of a darkroom. Follow the following steps and pointers carefully.

In Room Light

1. Mix the developer, stop bath, and fixer chemicals according to the manufacturer's instructions. Be sure to record the date on a label attached to each bottle. Chemicals weaken with use and time. Photographic chemicals keep better in dark bottles where they are shielded from light. Use brown glass or opaque plastic bottles.
2. Check temperatures of all three chemical solutions. Each should be at 20° C (68° F), average room temperature. Make necessary adjustments before you begin processing, either by running warm or cool water over the outside of the bottles if you have mixed small quantities of the chemicals or, if you have made gallon quantities, by pouring the required volume into graduates and standing them in a water bath at 20° C. In either case, the solutions, once they are brought to the correct temperature, should be kept in a 20° C water bath for the entire processing session. A plastic dishwashing tub is fine for this purpose.

Figure 7.1 Marc Levey. Kodak T-Max 3200 at ISO 6400.

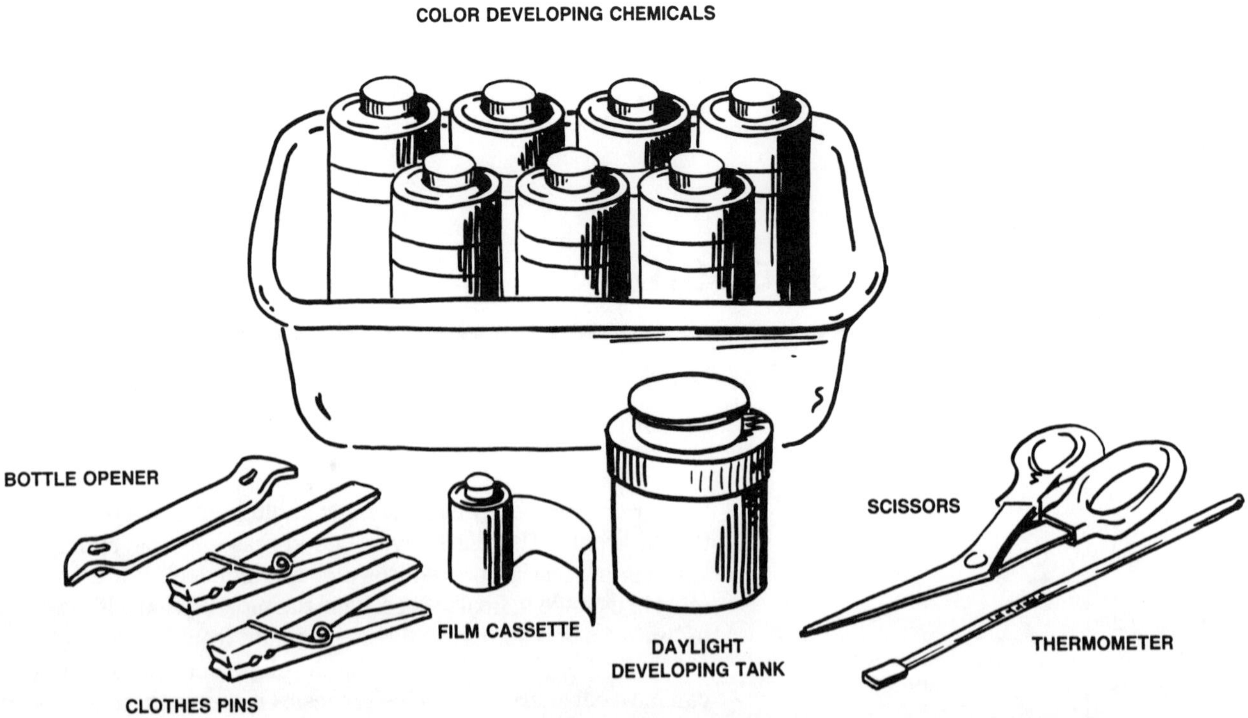

Figure 7.2 This illustration sets out the basic tools needed to process color transparencies or negatives. If a daylight-loading developing tank like the one shown is not used you will need to purchase a lightproof changing bag.

3. If you are working in a room that can be completely darkened, pour the developer into the developing tank. If you use a changing bag, leave the tank empty and go on to step 4.
4. Set, but do not start, the timer for the appropriate developing time. A simple kitchen timer can be used if you put on a bit of masking tape which you can feel in the dark opposite the correct time. (This isn't necessary if you use a changing bag—the timer won't be in the dark.)

In Total Darkness

5. Holding the film cassette by its spool end, pry open the other, flat, end with a bottle-cap opener. Carefully remove the spool of film.
6. With a scissors, square off the lead end of the film.
7. Thread and wind the film onto the reel of the developing tank according to the instructions supplied with the type of tank you own. It's a good idea to practice steps 4, 5, and 6 with a spoiled roll of film until you feel confident enough to repeat them smoothly with your unprocessed film. You will soon learn to keep opener, scissors, reel, developing tank, lid, and timer within easy and familiar reach to avoid fumbling for them in the dark.
8. Gently drop the loaded reel into the chemical filled developing tank with one hand, and, with the other hand, start the timer as the reel reaches the bottom of the tank. (In a changing bag, the developing tank is dry and the timer should not yet be started.) Place the lighttight cover firmly on the tank and remove the tank from the water bath. Tap the bottom of the tank lightly on the table or counter to dislodge any air bubbles that might be clinging to the film.

9. Agitate the tank immediately for 15 seconds, using random motions. Turn on the room lights. (If film was loaded in a changing bag, pour the developing solution through the light-tight opening in the lid of the developing tank. Start the timer as you pour.)
10. Agitate the tank throughout the developing time of the film according to the instructions provided by the manufacturer of the chemicals. In between agitations return the tank to the water bath. Agitation is very important—it moves the exhausted developing solution away from the emulsion and keeps fresh solution flowing around and over the film. Use the recommended agitation times and intervals. Too little agitation can result in underdeveloped film; too much agitation can overdevelop the film. Random motion in agitation will prevent wave marks from forming on the film.
11. At the end of the developing time, pour out the developer.
12. Pour in the correctly measured volume of stop bath. Agitate gently for about 15 seconds and then pour out.
13. Fill in the tank with fixer solution and set the timer for the maximum recommended fixing time according to the manufacturer's instructions.
14. Agitate the developing tank frequently during the fixing period, about 15 seconds at first and about 10 seconds every 60 seconds thereafter. When the timer rings, pour out the fixer.
15. Open the developing tank—the film is no longer light-sensitive—and run a hose from a sink faucet into the core of the developing reel. Be sure the water temperature is adjusted so that it is the same as the chemicals used during development, usually 68°F. Allow the water to run very slowly into the bottom of the developing tank. The film emulsion is soft when wet and can be damaged by too strong a flow of water.

    The washing time will vary depending on whether you use a washing agent. Such agents cut the wash time considerably and save water as well. Follow the manufacturer's instructions. If you use water only, the film must be washed for about 30 minutes. Don't cut the wash time short!
16. Soak the film in a wetting agent as directed by the manufacturer. This will prevent water spots from forming on the film as it dries. You can omit this wash if you choose, but be sure you wipe the water carefully from the film as described in step 17.
17. Gently unwind the film from the reel. Hang it to dry in a dust-free place such as a cabinet or the middle of a room—not near doors or windows. Attach a clip or clothespin to the bottom of the film so that it hangs straight. Wipe it carefully on each side with a Viscose sponge held at a 45° angle.

    Special clips are made for the purpose of hanging film to dry, but plastic or wooden spring-type clothespins work just as well. Don't use ordinary sponges for the wipe however, as they may scratch the film.

If You Are Using Reusable Chemicals

18. Finally, make a record, preferably on a label attached to each bottle, of the number of rolls you have developed and fixed. Trying to exceed the number of rolls that can be successfully processed by chemicals recommended by its manufacturer is poor economy. You risk ruining your film with exhausted chemicals.

Exact and strictly controlled film processing sequences yield negatives ready to be transformed into the final product of your creation; the point of all you have invested in time and energy and money, the result of your equipment, technical know-how, and artistic effort. The transformation of negative to positive is, or ought to be, a creative process in itself.

At this point, you may prefer to take the negatives to a professional finisher. A custom lab will make prints according to your instructions. But, if you want to make the investment in your own darkroom, or if you have access to a darkroom through a club or rental facility, you can give free reign to your creative notions of how that image, captured back when you pressed the shutter-release button, should ultimately appear as a finished photograph.

Because there are so many black-and-white film/developer combinations, it is impossible to cover them all in this chapter. Instead, I have selected several of the more popular combinations along with a few modifications or options.

Most black-and-white films can be exposed at exposure indices other than their rated speed. Doing so requires modified development times. Whenever possible, I have included modifications. The most significant new black-and-white emulsions are Kodak's T-Max rated at ISO 100 to 3200. Processing these films was at first a bit tricky. With the introduction of a T-Max developer specifically formulated for the films, most of the initial processing problems have been greatly reduced or eliminated. With T-Max developer, processing is straightforward if developer temperature is maintained to ± 2° F. Table 7.1 gives some recommended development combinations using T-Max developer mixed as per directions.

Aside from these and other standard developers, there are chemistries that are formulated to give specific effects. They can be grouped into the following basic categories:

General-Purpose Fine-Grain Developers

So-called universal developers work well with a vast range of general-purpose black-and-white films as well as several specialized emulsions, such as black-and-white infrared. They produce fine grain and good contrast, and are usable over a broad temperature range. Such developers include the ubiquitous D-76, FG-7, ID-11 Plus, Rodinol, and many, many more.

High-Energy Developers

These developers increase the speed of film at the expense of grain buildup. They may allow you to increase film speed two or three times normal. Acufine is an example of this type of developer.

Acutance-Type Developers

Acutance types act to reduce the effects of light diffusion and to enhance adjacent-edge sharpness, thus contributing to sharply etched detail from one part of the negative to another. Perceptol is an example of an acutance-type developer. Many fine-grain general-purpose developers claim they improve acutance.

Contrast-Enhancing Developers

Designed to produce maximum contrast, these developers reduce or eliminate middle gray tones by emphasizing development of maximum black and clear portions of the negative. Some of these developers are two-solution types while others are monobaths.

Table 7.1 T-Max Developer

	EI/ISO	68°F/20°C	72°F/22°C	75°F/24°C
T-Max 100	100/200	8.0	7.0	6.5 min.
	400	12.0	10.5	9 min.
	800	NR	NR	10.5 min.
T-Max 400	400/800	7.0	6.5	6 min.
	1600	10.0	9.0	8 min.
	3200	NR	NR	NR
T-Max P 3200	400	7.5	6.5	6 min.
	800	8.0	7.0	6.5 min.
	1600	8.5	7.5	7 min.
	3200	11.5	10.5	9.5 min.
	6400	14.0	12.0	11 min.
		With D-76 1: 1		
T-Max P3200	400	10.5	8.5	7.5 min.
	800	11.0	9.0	8 min.
	1600	11.5	9.5	8.5 min.
	3200	15.0	12.5	10 min.
		Developing Technical Pan Film with Kodak LC Technidol		
Technical Pan	125	11.0	9.0	8 min.
		With T-Max Developer		
Tri-X	400/800	6.0	5.75	5.5 min.
	1600	10.0	9.0	8.5 min.
	3200	NR	NR	9.5 min.
Plus-X	125/250	5.5	5.25	5 min.
	500	NR	NR	9.5 min.

In addition to Kodak black-and-white film, Fuji Neopan, Ilford Pan F, FP4-Plus, HP5 Plus, Delta 400, and Agfa's general-purpose black-and-white films are processable in the developers just mentioned. Also, there are literally dozens of other chemistries on the market from straightforward to exotic. And, like most other things in photography, there is no one best combination. With some experimentation and research, each darkroom worker will come up with a personal favorite.

Color Film Processing

As was mentioned earlier, there was a time when home-processing color film loomed as a nightmare only to be taken on by the self-destructive. Chemistry was difficult to mix, temperature control and processing could only be described as taxing. Today, the situation is much improved and, with some care, color film processing at home is possible, even for the novice.

The main caution in all color darkroom work revolves around temperature control. Both color film and print processing demand tight temperature control in some instances to ±¼° F. Several methods exist to control temperatures to this level of precision. Darkroom devices similar to the Jobo Tempering Box or Color Processor use heating elements in a water bath to maintain solution temperatures. They consist of an emersion-type heating element and precision thermostat. Water temperatures are maintained at proper levels by the unit's heater cycling on and off as the water temperature cools. The next most important item in the darkroom is a laboratory-grade thermometer. Obviously it must be accurate to within ¼° F to be of any use in color work. You should be aware that nasty things happen when temperatures are not maintained to stated tolerances. Wild and unpredictable color shifts are likely, along with loss of film speed and major contrast changes.

If economy is important to you, construct a temperature control box of your own. What you will need is a large plastic tub about 13 × 18 inches and around 6 to 8 inches deep. Purchase a good quality thermometer if you don't have one already. In addition, locate six to eight one-pint capped plastic bottles. The critical unit in this construction is an aquarium heater. Buy the best one you can afford.

To set the unit up for use, fill the tub and bottles with water until the bottles are submerged to at least two-thirds of their length. Turn on the heater and gradually bring the temperature up to the desired solution working temperature. Monitor temperatures of the *chemistry in the bottles* and not the water in the tub. When the desired temperature is arrived at, mark the dial of the heater with the temperature. Setting the dial at this point in the future will get very close with only minor adjustments needed to arrive at the precise temperature again. The procedure described here is slightly more trouble than a commercially built unit but it does work and it saves money.

Processing Color Negatives and Chromogenic Black-and-White Film

Process C-41 and its derivatives are now accepted as the universal process for virtually all color negative and black-and-white chromogenic film processing worldwide. Aside from the need to maintain very stringent temperature and time control over each step in the operation, the procedure is straightforward. The process itself consists of either three or four chemistry stages, with a number of intermediate washes. The entire development sequence including film drying takes only 31½ minutes. This compares very favorably to the old C-22 process with its six chemical steps and a total time of 53 minutes *plus* drying.

Color Transparencies

E-6 Processing

When you've become proficient at black-and-white processing, you may want to try your hand at color. Though it seems intimidating to consider and does require some extra care and precautions, if you follow the instructions exactly, you'll be successful. More than that, you will have the advantage of greater control over the results and at less than half the cost of commercial processing.

The best way to begin is with color-positive (transparency) film which can be processed by E-6 chemistry. These include Ektachrome, Fuji-

Table 7.2 Standard C-41 Film Development

Step	Temperature	Time	Remarks
1. Color developer	100°F (38.7°C) ±1/4°F	3-1/4 min.	See instruction for agitation requirement
2. Bleach	75–04° F (24–40°C)	6-1/2 min.	
3. Wash	75–04° F (24–40°C)	3-1/4 min.	
4. Fix	75–04° F (24–40°C)	6-1/2 min.	
5. Wash	75–04° F (24–40°C)	3-1-/4 min.	
6. Stabilizer	75–04° F (24–40°C)	1-1/2 min.	
7. Film drying	75-110° F (24–44°C)	10–20 min.	
Total processing time = 34-1/2 minutes.			

Besler CN2 and Similar Temperature Variation—Color Developer ±1/4° F

1. Color developer	75° F (23.8 ° F)	6 min.
2. Combined bleach/fix (BLIX)	75° F	9 min.
3. Wash	75–90°F	4 min.
Total processing "wet" time = 29 minutes.		

chrome, Agfachrome, and Scotchchrome. All Kodachrome must be processed by the manufacturers.

You'll need a little more equipment than you already have for black-and-white processing. The extras should include an accurate dial or rod, thermometer such as a Weston, Cibachrome, or Premier, and a timer with both a minute hand and a sweep second hand. You'll also need seven sixteen-ounce storage bottles; the brown glass variety available at pharmacies is best. Put a label on each. Protective waterproof gloves are not imperative if you're careful but, since several of the E-6 chemicals are toxic and can cause skin and eye irritation, better safe than sorry. Another warning here: ***Mix and use these chemicals in a well-ventilated room because the vapors can cause eye and respiratory damage!***

If you find you are doing a lot of E-6 processing, you may eventually want to invest in a motor-based film agitator and a temperature-controlled water bath. These represent an investment that you will soon recoup if you shoot much slide film. They are not essential to start, but do ease the process considerably. Try developing a few rolls first before you decide on these.

E-6 Procedure

Several manufacturers sell E-6 processing kits. They vary only slightly. Whichever you choose, be sure to follow the enclosed instructions for that kit precisely. Tack them up near your work area so you can glance at them easily.

The procedure which follows is for the popular Unicolor kit. (The processing times are the same for any of the E-6 films unless you have pulled or pushed the film to change its effective ASA rating [EI].)

1. Mix the chemicals in the order in which they will be used according to the instructions. Label each bottle with the name of the chemical and the date mixed—these chemicals have a definite shelf life after

which they must be discarded. Put a number on each label, too, according to the sequence in which the chemicals are used. The number should also go on the cap to avoid mixing up caps and contaminating solutions. If you plan to use solutions immediately, mix them at the proper temperature, 38° C (100° F) for inversion-type tanks, 40.5° C (105° F) for motorized agitators.

Pour a little of each solution, after it's mixed, back into the bottle it came in, cap the bottle, shake, and pour the contents back into the graduate. This ensures that you've got all of the chemicals into solution. After that's done, pour the solution into a clean brown glass bottle, cap it, carefully wash the graduate and the thermometer (which doubles as a stirring rod), and go on to the next solution.

2. Place each bottle, after it's filled, into your plastic tub water bath. This should contain water at about 39° C (102° F), assuming you are using an inversion-type tank. *It is critical for the success of this procedure to maintain the temperature of the chemicals at 38°C (100°F) plus or minus one-half degree C (one degree F).* Without a temperature-controlled water bath, you will have to check water-bath temperature frequently and add hot water as the temperature drops. Between steps, take the time to check the temperature of the chemical itself for the next step. You may need to run hot water over the capped bottle to adjust the temperature. Keep some cold water or ice cubes nearby, too, in case you overcompensate and get the water bath or solution temperatures too warm.
3. A changing bag is all you'll need in the way of darkness for E-6 processing. In the bag, load the film onto the reel just as you do for black-and-white film. Place the film in the tank and attach the tank cover. Put the developing tank into the water-bath while you do step 4. Because the tank is empty except for the film, it may float in the water—that's okay.
4. Check again to be sure all solutions and the water bath are at 38° C (100° F). Set your time for 6½ minutes.
5. Start the timer as you pour the first developer, solution number 1, into the tank. Tap the tank twice, gently but firmly, on the table or counter, to dislodge any air bubbles that may be clinging to the film. Agitate the tank for 30 seconds immediately and then for 5 seconds every 30 seconds until the developing time is up.
6. About 10 seconds before the end of the 6½-minute period, begin pouring out the first developer. Use a funnel and pour it back into its brown bottle—it can be used again.
7. Run water at 38° C (100° F), into the baffle top of the developing tank for 2 to 3 minutes. Don't open the tank! Drain tank.
8. Set timer for 2 minutes and start it as you pour in reversal bath, solution number 2. Agitate 5 seconds every 30 seconds. At the end of the two-minute period, pour the reversal bath back into its bottle.
9. Set timer for 6 minutes and start it as you pour in the color developer, solution number 3. Agitate, as in step 5, for the first 30 seconds and then for 5 seconds every 30 seconds thereafter. At the end of the developing time, pour the color developer back into its bottle.

   From the next step on, the solution temperatures are less critical than in the first developer, reversal bath, and color developer steps. Temperatures should be between 32° C (90° F) and 43° C (110° F).
10. Set timer for 1 minute and start it as you pour in stop bath, solution

number 4. Agitate 5 seconds every 30 seconds. Pour the stop bath back into its bottle at the end of the step.

11. Another water rinse, 2 minutes into the baffle top of the developing tank. Don't open the tank! Drain tank.
12. Set timer for 3 minutes and start it as you pour in bleach, solution number 5. Agitate 5 seconds every 30 seconds. At the end of the time period, pour the bleach back into its bottle.
13. Set timer for 2 minutes and start it as you pour in fixer, solution number 6. Agitate the tank 5 seconds every 30 seconds. At the end of the time period, pour the fixer back into its bottle.
14. A third water rinse, for 2 or 3 minutes into the baffle top of the developing tank. The light-tight top of the tank must not be removed yet!
15. Pour in stabilizer, solution number 7, and time for 30 seconds, agitating gently the first 10 to 15 seconds only. Although Unicolor specifies room temperature for this solution, keep it at or near the temperatures of the other solutions so as not to make too drastic a change for the film. At the end of the one-half minute period, pour the stabilizer back into its bottle.
16. The film may now be removed from the tank. Its appearance will give you a jolt—it will look opalescent and no images will be visible. That's normal—your pictures will gradually appear as the film dries. Hang the film, as you do with black-and-white film, in a dust-free place—in the center of a room away from doors and windows, or, better yet, inside a shower stall with the curtain pulled.
17. When the film is dry, you'll want to mount the pictures. There are several ways to accomplish this. Slide mounting gadgets are available for almost any budget. Or you can use the heat-sealed type mount with an electric iron. There are snap mounts, if you can find them. Simplest of all are slip-in mounts. You can save money by mounting only the acceptable pictures. Cut carefully, mount, and enjoy!

The standard E-6 process is much like C-41 and is a great improvement over previous home processing techniques. Its predecessor, E-4, consisted of a nine-chemical, thirteen-step marathon. Taking more than one hour, compared to E-6's 37½ minutes, it was a real turnoff.

Table 7.3 E-6 Color Transparency Film Processing

Step	*Temperature*	*Time*	*Remarks*
1. First Development	100° F/937.8° C) ±0.2° C	7 min.	See instructions for agitation requirements
2. Wash	91–102°F (33–39° C)	2 min.	
3. Reversal Bath	91–102°F (33–39° C)	2 min.	
4. Color Development	100.4° F (38° C) ±0.6° C	6 min.	
5. Conditioner	91–102° F (33–39° C)	2 min.	
6. Bleach	91–102° F (33–39° C)	7 min.	
7. Fix	91–102° F (33–39° C)	4 min.	
8. Wash	91–102° F (33–39° C)	6 min.	
9. Stabilizer	91–102° F (33–39° C)	1/2–1 min.	
10. Dry	91–102° F (33–39° C)	At least 10 minutes	

Table 7.4 E-6 Film Processing (Modifications)

Type of Change	*Effect*	*Time Change in First Development or*	*Temperature Change*
Pull 1/2 f-stop (N-1/2)	Lighten shadows and colors, reduce contrast	–1 minute	–4° F
Push 1/2 f-stop (N + 1/2)	Darken, intensify color, increase effective film speed by 50%	+ 1 minute	+ 4° F
Push 1 f-stop (N + 1)	Same as above except doubles effective film speed	+ 2 minutes	+ 8° F
Push 2 f-stops (N + 2)	Same as above except quadruples effective film speed	+ 5 minutes	+ 12° F

Note: Limited push- and pull-processing for Kodachrome film is now available from a few speciality labs. Consult local camera shop or PSA Chapter for addresses and/or price list.

Kodak Ektachrome Infrared and Photomicrography Film still require processing in the mostly discontinued E-4 chemistry. Consult Appendix C for a list of labs offering E-4 processing.

Be aware that several color chemicals are dangerous if ingested and cause severe skin reactions in some individuals if not neutralized promptly. Color developers, bleach, and fix are the principal culprits. If you are hyper- sensitive to chemicals, you should wear rubber gloves during most of the processing stages involving the previously mentioned chemicals.

Pushing and Pulling

A great advantage of Ektachrome-type films is their ability to be "pulled" or "pushed." Pulling film, overexposing and underdeveloping, reduces contrast and decreases color saturation. Pushing film, underexposing and overdeveloping, effectively increases film speed by one, two, three, or four stops. With Ektachrome available in ASA ratings from 64 to 400, effective exposures from ASA 32 to ASA 6400 are possible (see Table 7.4).

To compensate in the processing for pulling or pushing E-6 films, you need only alter the time for the first developer and, with some processing kits, the color developer. Film pushed more than one stop, however, exhausts the first developer—discard it after one use.

Does It Make Sense to Home Process?

This is not an easy question to answer. For example, I enjoy the adventure of photo discovery in the field as I seek out new subjects to explore. I do not enjoy spending time with chemicals in the dark doing what others do better than I do and in less time. Since I work almost exclusively with color transparency material, straightforward slide processing does not make much sense for me. I have set up several darkrooms in my photographic life from rudimentary bathroom variety to custom-designed, electronically controlled units complete with commercial temperature controls and automated processing machinery. In all instances, I came to realize that the darkroom was not where I could be most creative. However, I also recognize that wonderful and exciting things happen in darkrooms. Many individuals derive great satisfaction from the work they do there. For large numbers of people, it is a wonderful experience to see full color emerging from the haze as slides dry or prints develop.

If you opt for home film processing you should be able to cut processing costs in half. That translates into about five to six dollars per hour for the time you invest in processing.

A case can be made for home processing if, first, you cannot find a consistently good lab close enough and quick enough to make its use practical; second, if you enjoy it; third, if you intend to process film at other than standard times or with exotic chemistry; and fourth, if you want to ensure your film is being processed in the freshest chemicals, using absolutely correct procedures.

In addition to all of this, there is a down side to sending film out to be processed commercially. A small risk always exists that your film will be damaged or lost. Also, random problems at most labs result in substandard results that often can't be corrected. Industry research indicates a well-run processing laboratory should experience a problem rate of less than one-tenth of 1 percent. Additionally, home processing is not without risk itself. Failures do occur to even the most fastidious darkroom experts. The question of whether to process film at home or not ultimately comes down to the question of how fulfilling it is. If processing your own film is enjoyable and satisfying, by all means do it. On the other hand, if it becomes a tedious chore, find a dependable, consistent commercial processor.

Finally, if you intend doing a great deal of black-and-white printing, you should strongly consider developing your own film. Locating a lab that does custom black-and-white is difficult and the results are expensive. Much of the creativity in black-and-white photography begins with precise exposure and processing. Black-and-white film development is not as critical as color, so a relatively simple darkroom arrangement will suffice.

Suppose you decide to have your processing done commercially. How does one find a first-rate lab? First, ask experienced photographers for their recommendations. Next, send away for lists of services found in the back of leading photo magazines. Be careful of labs offering services priced much lower than the competition. You can also inquire at local colleges or tech schools offering photo courses. Try contacting the instructors of black-and-white courses.

Be careful to read advertising carefully. Ads for processing and printing services sometimes imply custom work, which should mean all services are performed by hand rather than by machine. Advertising claiming "individual video analysis" of negatives is not custom work but a semiautomatic procedure whereby negatives are automatically printed after a split-second decision is made about contrast and color by an operator viewing a video screen. The phrase "each roll individually handled" is also misleading. In order to put your film into the automatic processor, your film must be "individually handled" at one point or another, if only to take them out of the mailing envelope.

Let us assume you have narrowed your lab choice down to a few local and one or two mail-order processors. To conduct a meaningful test of the labs, you will need to spend some money. Select representative rolls of unprocessed film and send them to each lab with identical instructions, especially if you are requesting special services like push-processing. If you are only judging print quality, you can use the same frame or frames and send them to all the labs on the list.

Set up a grid as follows:

Film Type	*Non-Standard Processing (if requested)*	*Turn-Around Time*	*Cost*	*Quality*

Comments:

Note: Quality consists of items such as cleanliness, packaging, evenness of exposure, color, mounting overall density, freedom from scratches or other surface defects.

You may find yourself repeating this procedure more than once if the initial tests prove inconclusive. This procedure is useful as you attempt to identify a good commercial black-and-white or color printing facility.

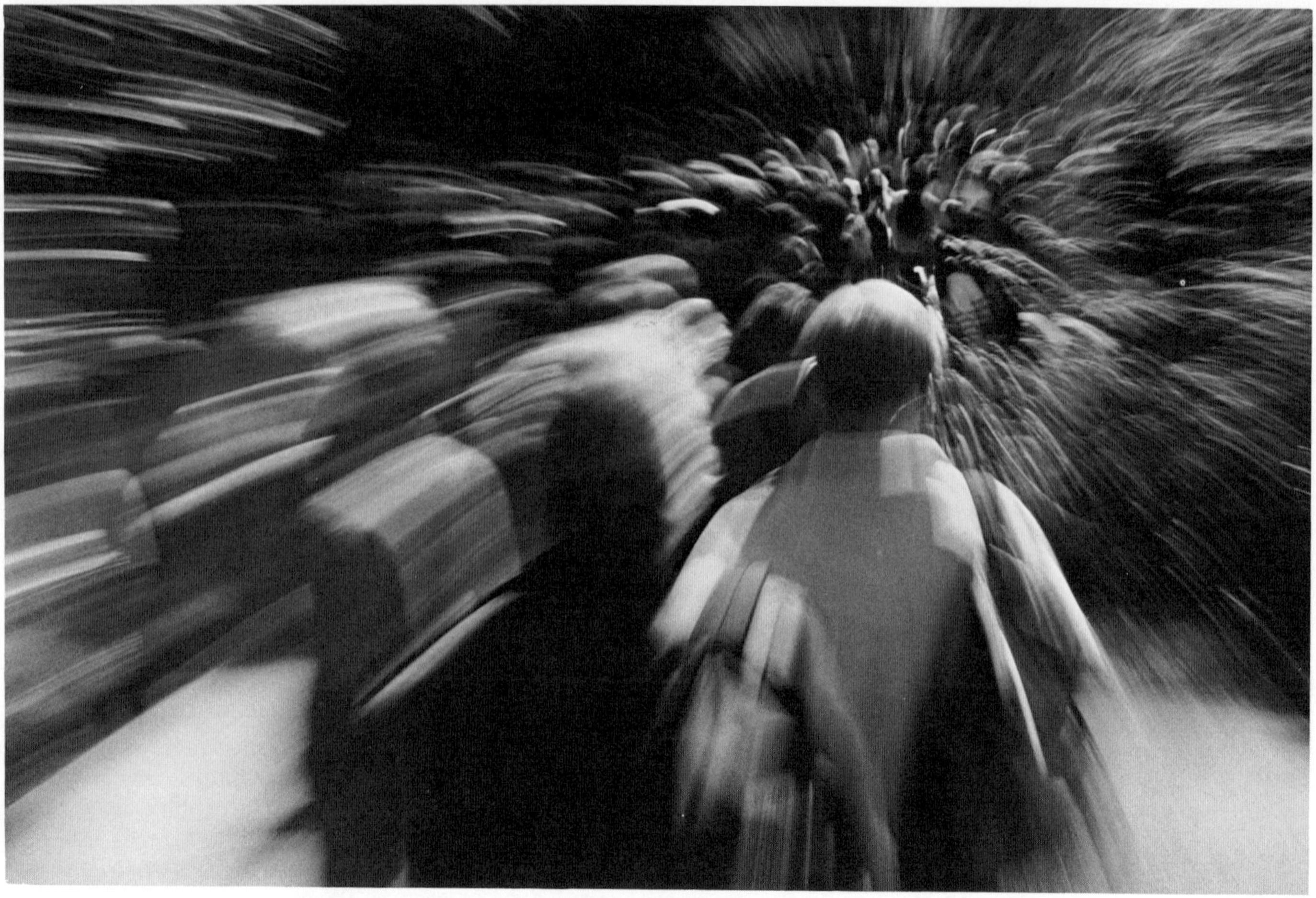

Figure 7.3 Marc Levey. Fiji Neopan ISO 400.

Appendix A

Cleaning Processed Film

Even though modern 35mm film is protected by a tough antiabrasion layer, it is still susceptible to scratches, abrasions, pock marks, and other image-distorting damage. Additionally, film that is composed of a gelatin substrate rich in proteins is a potent arena for the formation and growth of fungus in many forms. High humidity and heat, like that found in your car's glove compartment, are perfect conditions for rapid fungal growth. Fungus attacks raw, exposed, and processed film equally. The best way to combat fungus is to avoid its formation in the first place. Once formed, it is very difficult—if not totally impossible—to eliminate before it does permanent damage. For processed film, the 3M Company has developed a protective coating process called Photoguard that should offer a shield against almost all common fungi. It also coats the film's surface with an extraordinarily tough, clear protective surface on film that is in no way image-distorting. In fact, tests indicate it actually increases light transmission. Unfortunately, this treatment is not available everywhere but it is worth seeking out in a processing lab. For the few extra cents it costs per slide or print, it represents a genuine bargain.

Cleaning slides or negatives is a bit more straightforward than fungus protection. Two problems exist: Skin oils and other foreign matter directly *on* the film surface and dirt or dust particles adhering to the surface of the film via static attraction. Let us consider the latter first. Two methods to combat static electricity seem most effective. Blowing on the film surface with specially filtered, inert pressurized air of the type specifically designated for scientific and optical applications should remove most residual dirt and dust not embedded in the film. You can also brush very lightly on the film surface with a device called a *Staticmaster.* Consisting of an alpha particle generator that in effect neutralizes the charge of dust particles so they may be easily swept away with the Staticmaster's integral camel-hair brush. The watchword here is, softly does it!

For really stubborn dirt, right *in* the film, little can be done. Despite advice to the contrary, dirt embedded in the emulsion layer of the film cannot successfully be removed without degrading the image significantly. How much degradation you are willing to accept is up to you.

When confronted with foreign matter such as fingerprints, remove them as soon as possible. Skin oils will eventually etch themselves into the emulsion layer of the film. When that happens, there is no possibility of removing them.

There are a number of commercial products designed to clean film and dissolve skin oils or petroleum-based substances. All these products, of which Kodak Film Cleaner and Edwal Cleaner are typical, work, after a fashion. You should not expect miracles from them. In all cases, follow the directions faithfully. As a last resort, you can try rewashing the film following the time and temperature procedures used for initial film processing. Be sure to apply a final stabilizer or other conditioning agent to prevent surface spotting.

Appendix B

Labs Offering Negatives, Prints, and Slides From a Single Print Film

Dale Labs
2960 Simms Street
Hollywood, FL 33022-9990

RGB Color Labs
816 N. Highland Avenue
Los Angeles, CA 90038

Seattle Film Works
500 3rd Avenue West
Seattle, WA 98124

For processed negatives, slides, and prints, prices range between $14.00 and $17.50.

Appendix C

Labs Offering E-4 Processing for Color Infrared and Photomicrography Film

Dynacolor Corp.
95 Liberty Street
Springfield, MA 01103

I K Curtis Service, Inc.
2907 Empire Avenue
Burbank, CA 91504

Johns Hopkins University
School of Medicine
Room 100-Pathology
Monument and Wolfe Streets
Baltimore, MD 21205

Criterion Photo Works
119 E. 4th Street
P.O. Box 007
Minden, NE 68959 (Infrared only)

Whites Camera Shop
216 W. 8th Street
P.O. Box 497
Columbia, TN 38401

Kolor Print, Inc.
2121 Thayer Street
Little Rock, AR 72202

Picture This, Inc.
145 W. 28th Street
New York, NY 10001

Light Works Labs
509 N. W. 10th Avenue
Gainesville, FL 32601

Rocky Mt. Film Lab
145 Madison Street
Denver, CO 80206

Appendix D

Kodak Licensed Kodachrome Slide Processing

Grand Canyon Color Lab (M)
4456 East Thomas Road
Phoenix, AZ 85018
602/266-5671

Kolor Print (M)
2121 Thayer Street
Little Rock, AR 72202
501/375-5581

A & I Color Lab (*, P)
933 North Highland Avenue
Hollywood, CA 90038
213/464-8361

Kodalux Processing Service (P, M)
Qualex, Inc.
925 Page Mill Road
Palo Alto, CA 94304
415/493-7200

The New Lab, Inc. (*, P)
10 Cleveland Street
San Francisco, CA 94103
415/431-8555

LaSalle Photo
1700 W. Diversey Avenue
Chicago, IL 60614
312/327-6402

Qualex, Inc. (M)
220 Graceland
Des Plaines, IL 60016
708/827-6141

Kodalux Processing Service (P)
Qualex, Inc.
1 Choke Cherry Road
Rockville, MD 20850
301/670-8626

Qualex, Inc. (M)
616 Dwight Street
Springfield, MA 01103
800/274-1234 or 413/739-2521

Guardian Industries (M)
43045 W. Nine Mile Road
Northville, MI 48167
800/482-0123 (inside MI)
800/521-2034 (outside MI)

Black's Photo
1201 W. Broadway
Minneapolis, MN 55412
612/588-7861

Kodalux Processing Service (*, P)
Qualex, Inc.
16–31 Route 208
Fair Lawn, NJ 07410
201/797-0600

Duggal Downtown
560 Broadway
New York, NY 10012
212/941-7000

New York Film Works (P)
928 Broadway
New York, NY 10010
212/475-5700

Kodalux Processing Service
Qualex, Inc.
1100 East Main Cross Street
Findlay, OH 45840
419/422-7654

Guardian Industries (P, M)
Allentown, PA
800/322-9556 (inside PA)
800/523-9340 (outside PA)

The Color Place (*, P)
1330 Conant Street
Dallas, TX 75207
214/631-7174

Reprinted courtesy of
Kodak Information Center
343 State Street
Rochester, NY 14650-0811
800/242-2424

Kodalux Processing Service (P)
Qualex, Inc.
3131 Manor Way
Dallas, TX 75235
214/352-5900

Qualex, Inc. (M)
6300 Cedar Springs
Dallas, TX 75235
214/351-5101

(*) Offers 120 Processing
(P) Offers Push- and/or Pull-Processing
(M) Offers Movie Processing

Appendix E

Film Care and Protection

Protecting Film Below 32° F

1. Load film indoors, if possible.
2. Kodachrome is less likely to produce static electricity than E-6 films.
3. Outdoors before loading, hold the film leader between your hands to keep pliable.
4. Keep spare film and batteries in an inside coat pocket near your body or in a pocket containing a battery-powered hand warmer.
5. Keep film in their unopened containers.
6. Subzero temperatures can result in film exhibiting some color shifts or speed loss. Bracket exposures!

In Heat Above 85° F

1. Keep film out of closed car trunk or glove compartments. Temperatures can go as high as 140° F.
2. If you take film out from an air-conditioned building or car, allow 30 minutes for the film to come up to outside temperature. This will eliminate the danger of condensation build up on the film surface.
3. Use a cooler to store film, especially those with dry ice or electric types that plug into the car's heater.
4. Warm up times for film stored in a refrigerator is at least one hour. For film stored in the freezer, allow 2–3 hours. Long rolls (100 ft. or longer) require 4 to 6 hours.

In Very High Humidity or Dusty Conditions

1. Pack film in a sealed container with adequate silica gel packets that absorb excess humidity. Use one ounce per 10 rolls of 35mm film.
2. In exceptionally dusty conditions as those encountered in the desert or on the African plain, load and unload film in a changing bag.
3. You can find an inexpensive but effective waterproof and dustproof film storage container in a surplus store. Look for a military ammo box.
4. Process exposed film as soon as practical.

General Film Care

1. Storing excess film in the refrigerator extends useful film life by at least 6 months to one year beyond the expiration date appearing on the film packaging.

2. Freezing unopened film should extend useful film life at least two years and usually much more beyond the expiration date.

   Note: I have used frozen color transparency film five years out of date. I am hard pressed to see any meaningful changes in either film speed or color balance.
3. For maximum permanence of exposed and processed film, dark-store at 75° F at 45 percent relative humidity.
4. High-speed color negative film is most susceptible to deterioration. It should be kept refrigerated before use.
5. Avoid storing slides or negatives in PVC (polyvinyl chloride) or other common plastics. Even rubber bands, adhesive labels, common glues or cements emit harmful gases that can damage emulsions.
6. Choose storage cabinets made of unglued, untreated wood or ones lined with neutral ph paper. Negative File and Light Impressions are two companies that offer several types.
7. If you opt to file negatives or slides in transparent sheets, choose those made from polyethylene triacetate, polypropylene, and some mylars. A number of companies specialize in so-called archive preservation products. They include Light Impressions Corp., Negafile Systems, Inc., Print File, Inc., Visual Horizons, Vue All, Franklin Dist., Leedal, and Conservation Resources Int., Inc.
8. Because background radiation is always present, all sensitized material will deteriorate in time no matter how it is stored. Sensible storage can retard deterioration significantly.
9. Properly processed and stored black-and white negatives should last over 100 years. The longest lasting color transparencies (Kodachromes) will retain their physical characteristics for up to 50 years. Even the most fastidiously processed and stored color negatives will begin to show deterioration after as little as 10 years.
10. Color slides deteriorate as they accumulate projection time. Fujichromes seem to stand up best, followed by Scotchchrome, Ektachrome, Agfachrome, and last Kodachrome. The film most resistant to projection fading is about five times as resistant to fading than the worst (Kodachrome).
11. Kodachrome stores best in the dark, followed by Ektachrome, Fujichrome and then, somewhat below, Agfachrome and Scotchchrome. Properly stored Kodachromes and even Ektachromes and Fujichromes should retain their basic physical qualities for our lifetimes or beyond.
12. Color negative film has less than one-half the stability of slide film. The reason lies in the fact that color containing couplers in negative film are far less stable than the colorless couplers contained in slide film.
13. X-rays are cumulative. The more a film passes through machines, the greater chance of the film being fogged.
14. In the United States, FAA guidelines limit X-ray emissions to 1 milliroentgen. This level is great enough to fog ISO 1000–3200 films. And because radiation is cumulative, several passes through a domestic X-ray machine can fog ISO-400 film. Extensive foreign travel with multiple passes through older, higher powered machines risks fogging film as slow as ISO-50.
15. Try to convince airport security to hand inspect film. Packing it in clear plastic will enhance your chances. In the United States, Canada, and England, you should be able to get film hand-inspected.

However, in Africa, Russia, and much of Western Europe, you can argue all day long with little or no effect. It is only when you present film faster than ISO-1000 that you can expect any success having your film hand-inspected. Countries that are particularly difficult include France (De Gaulle Airport), Belgium, Germany, Italy (especially Da Vinci), Denmark, Switzerland, all of Eastern Europe, and Russia, Kenya, Israel, Nigeria, and South Africa.

16. Sima and Midwest Glove Co. make lead-lined containers and bags. Recently they began to offer bags with extra lead lining to protect film up to ISO-1000. Even so, older high-powered machines in Eastern Europe and the third world will, in a very few passes, emit enough radiation to damage ultra-high-speed emulsion.

17. Polaroid Photoguard is an optically clear surface application to slides, negatives, and prints that will protect delicate material. The product forms an epoxy-tough bond that does three things:

a. It seals the film or print surface, making it impervious to ordinary hazards.

b. The material actually increases light transmission and apparent sharpness.

c. Photoguard absorbs UV radiation to slow color fading.

Appendix F

Equivalent Measures in the Metric System and the English System

Measures of Length					
Metric to English			*English to Metric*		
1 meter	=	39.37 inches	1 inch	=	25.4 millimeters
	=	3.28 feet		=	2.54 centimeters
	=	1.09 yards		=	.03 meter
1 centimeter	=	.39 inch	1 foot	=	.30 meter
1 millimeter	=	.04 inch	1 yard	=	.91 meter
1 kilometer	=	.62 mile	1 mile	=	1.61 kilometers
Measures of Capacity					
1 liter	=	1.06 liquid quarts	1 liquid quart	=	.95 liter
	=	.91 dry quart	1 dry quart	=	1.10 liters
Measures of Weight					
1 gram	=	.04 ounce	1 ounce	=	28.35 grams
1 kilogram	=	2.20 pounds	1 pound	=	.45 kilogram
1 metric ton	=	2,204.62 pounds	1 short ton	=	.91 metric ton

Appendix G

Arresting Motion

The following table gives shutter speeds needed to arrest motion in a slide or print. This table is based on the assumption that a degree of blur in an 8 × 10 inch print made from the whole negative or slide will be no more than .50mm (.020 inch). Generally speaking, blur that is larger than this value will be objectionable to a majority of viewers and will be defined as general unsharpness. Obviously, individual visual acuity varies greatly so take the chart recommendations as minimums. The table also assumes the use of a 50mm lens on a 35mm camera.

Shutter Speed Table (For Lenses of Normal Focal Length)

Speed			*Distance*		*Direction of Motion*		
ml/hr	*kkm/hr*	*Type of Motion*	*ft*	*m*	◄ *or* ►	◤◢ *or* ◥◣	▲ *or* ▼
5	8	Slow walk, hand work, sitting or standing people	12	4	1/500	1/250	1/125
			25	8	1/250	1/125	1/60
			50	16	1/125	1/60	1/30
			100	33	1/60	1/30	1/15
10	16	Fast walk, children and pets playing, horses walking, slow-moving vehicles	12	4	1/1000	1/500	1/250
			25	4	1/500	1/250	1/125
			50	16	1/250	1/125	1/60
			100	33	1/125	1/60	1/30
25	40	Running, sports, very active play, horses running, vehicles moving at a moderate speed	12	4	1/2000	1/1000	1/500
			25	8	1/1000	1/500	1/250
			50	16	1/500	1/250	1/125
			100	33	1/250	1/125	1/60
50	80	Fast-moving vehicles, birds flying, race horses running	12	4	1/2000	1/2000	1/1000
			25	8	1/2000	1/1000	1/500
			50	16	1/500	1/250	1/60
			100	33	1/500	1/250	1/125
			200	66	1/250	1/125	1/60
100	160	Very fast-moving vehicles	12	4	1/8000	1/4000	1/2000
			25	8	1/4000	1/2000	1/1000
			50	16	1/2000	1/1000	1/500
			100	33	1/1000	1/500	1/250
			200	66	1/500	1/250	1/125

If you choose the next slower shutter speed to the one recommended in the chart, you can expect a doubling of the amount of blur. Conversely, choosing the next higher shutter speed halves apparent blur. Wide-angle lenses of 28mm or shorter focal length permit the user of the next slower shutter speed that the one shown in the chart. Using a telephoto lens requires that you increase shutter speeds by the degree of magnification of the telephoto in use as compared to a normal lens. Therefore, if you are using a 200mm lens you will need to use a shutter speed four times faster than the one recommended for use with a normal lens to maintain the same level of blur. Wide-angle lenses of 28mm or wider permit the use of the next slower shutter speed to maintain the same level of blur as a normal lens.

Film choice can have a dramatic effect on picture quality. Selecting a film twice as fast as one that would give acceptable blur at a pre-determined f-stop as indicated in the table will reduce blur by one-half by allowing the use of one faster shutter speed or will maintain the standard amount of blur but increase depth of field by 50 percent. Selecting even faster films will do even more to reduce image blur and therefore apparent image sharpness.

Appendix H

Determining Print Quality from Color Transparencies

When you project a slide or examine it using a lightbox, you are viewing it via transmitted light. These viewing methods will give you a false indication of what a print viewed via reflected light will look like. Think of what neon lights look like when turned off and on. On, they are seen by mostly transmitted light. When turned off, their color is seen only by reflected light.

A more accurate simulation of what to expect from a print generated from a color transparency can be seen by employing the following procedure: Place a sheet of white typing paper on a table or desk with strong, intense light striking it. Hold the transparency as in the illustration and examine it via light reflected from the typing paper. This viewing method should give you a much better idea of the color quality and relative brilliance of a finished print.

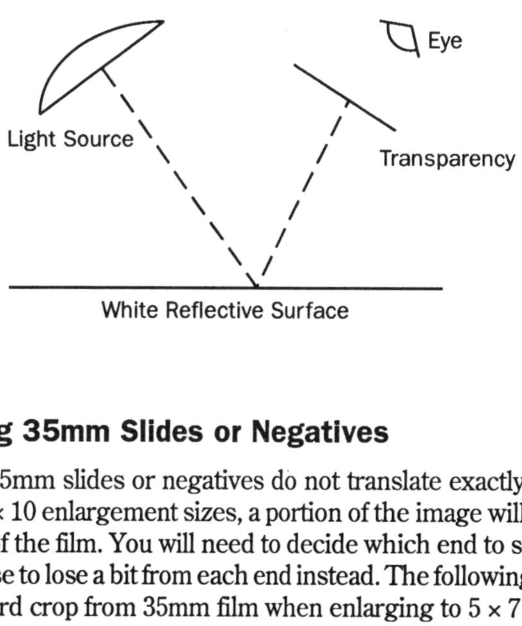

Cropping 35mm Slides or Negatives

Because 35mm slides or negatives do not translate exactly into standard 5 × 7 or 8 × 10 enlargement sizes, a portion of the image will be lost on the long end of the film. You will need to decide which end to sacrifice or you may choose to lose a bit from each end instead. The following illustration is the standard crop from 35mm film when enlarging to 5 × 7 or 8 × 10-inch prints. If you desire to print the entire image you will need to accept a print somewhat less than 7 inches or 10 inches long. Some printers are now offering new sizes, such as 8 × 12 or 11 × 16 inches, to deal with this problem.

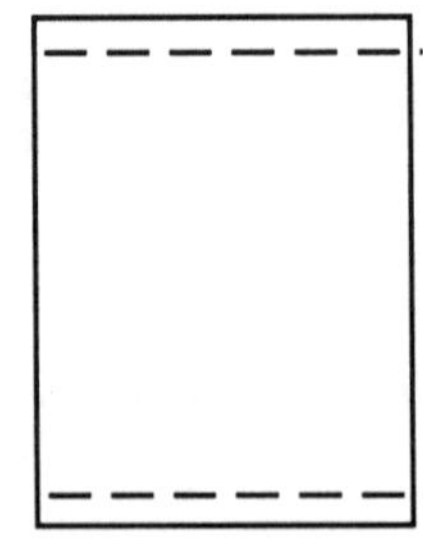

8 × 10 – .910 × 1.137
5 × 7 – .910 × 1.286

Appendix I

Light Source Conversion Nomograph

This nomograph can help you determine which filters, if any, should be used to balance specific color films and light sources.

From right to left, the scales on the chart indicate common light sources and their color temperatures in degrees Kelvin. The scale on the far right denotes the position of three types of color film also on a Kelvin scale.

In the middle between these two scales is another, indicating shift in mired values and pointing to the correct conversion filter to balance film and light source.

Using the nomograph is quite simple. Merely lay a ruler on the chart connecting the prevailing light source you wish to use and the type of film loaded in the camera. The edge of the ruler will cut through the middle scale and identify the correct filter to use. For accurate readings of color temperature, you will need a color temperature meter. If you intend absolute balance between light source and film, Decamired filters, which are finely calibrated, will give you precise balance and, subsequently, accurate color rendition.

Note that the black line on the Nomograph is only for demonstration purposes.

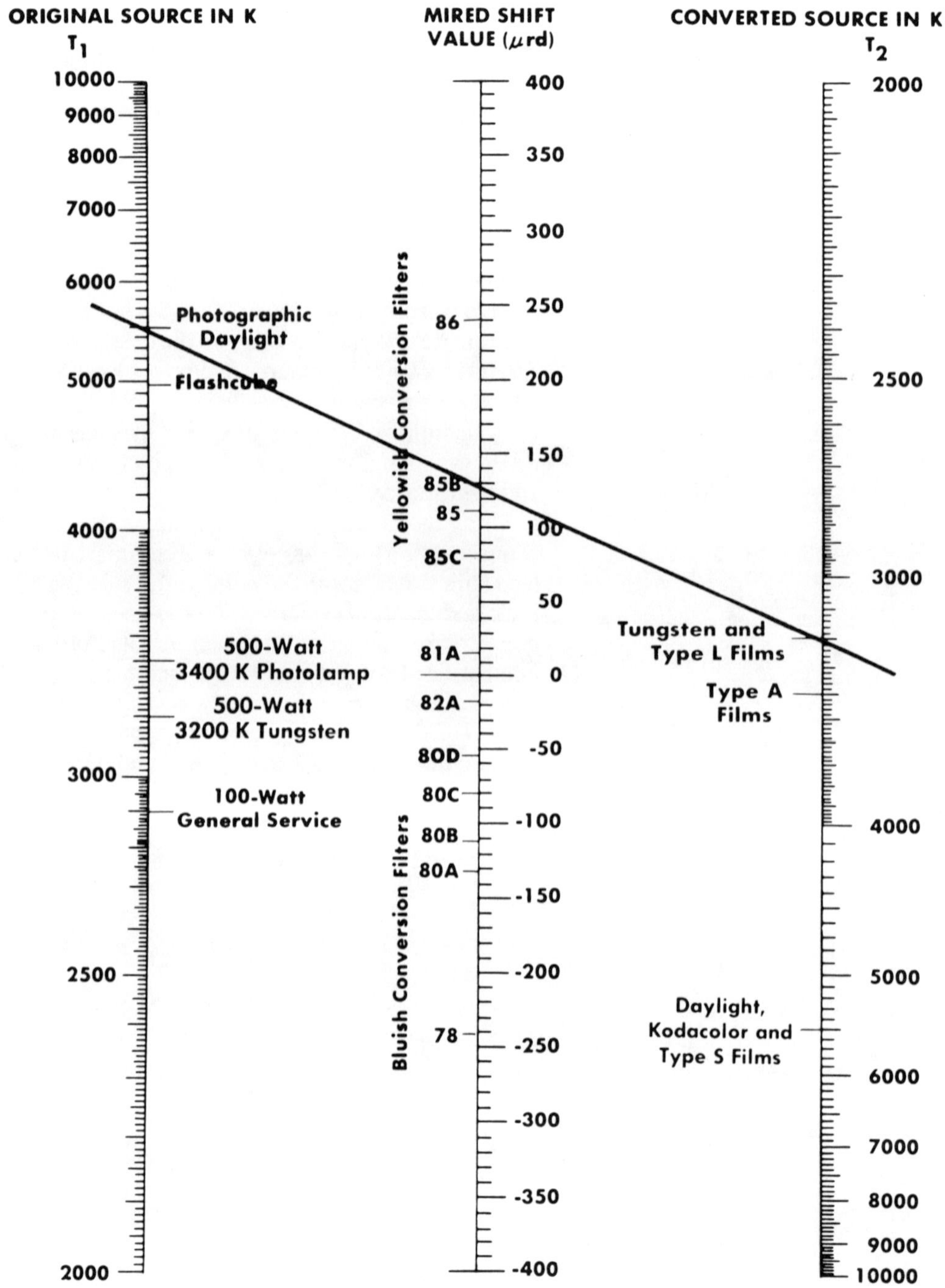
ORIGINAL SOURCE IN K
T_1
MIRED SHIFT VALUE (μrd)
CONVERTED SOURCE IN K
T_2
Photographic Daylight
Flashcube
500-Watt 3400 K Photolamp
500-Watt 3200 K Tungsten
100-Watt General Service
Yellowish Conversion Filters
Bluish Conversion Filters
86
85B
85
85C
81A
82A
80D
80C
80B
80A
78
Tungsten and Type L Films
Type A Films
Daylight, Kodacolor and Type S Films
10000
9000
8000
7000
6000
5000
4000
3000
2500
2000
400
350
300
250
200
150
100
50
0
-50
-100
-150
-200
-250
-300
-350
-400

Appendix J

Film Manufacturers' Addresses

Agfa Corporation
Consumer and Professional Division
100 Challanger Road
Ridgefield Park, NJ 07660
201/440-2500

Forte Film
GMI
1776 New Highway
Farmingdale, NY 11735
516/752-0066

Fuji Film Company
555 Taxter Street
Elmsford, NY 10523
800/843-5976

Ilford Photo Corporation
West 70 Century Road
Paramus, NJ 07653
201/265-6000

Eastman Kodak Company
Information Center
343 State Street
Rochester, NY 14650
800/242-2424

Konica U.S.A., Inc.
440 Sylvan Avenue
Englewood Cliffs, NJ 07632

Polaroid Corporation
575 Technology Square
Cambridge, MA 02139
617/577-2000

Scotch
3M Photo Color Systems Division
Building 223-2S-05, 3M Center
St. Paul, MN 55144
612/733-4608

Index